Microbial Consortia in Plant Science and Sustainable Agriculture

Microbial Consortia in Plant Science and Sustainable Agriculture

Edited by

Ajay Tomar

Amity Institute of Organic Agriculture
Amity University Uttar Pradesh
Noida, India

Ajay Kumar

Amity Institute of Biotechnology
Amity University
Noida, India

Ramsés Ramón González Estrada

Tecnológico Nacional de México/Instituto Tecnológico de Tepic
Coordinación Investigación
Tepic, Nayarit, Mexico

Kshitij Parmar

Amity Institute of Organic Agriculture
Amity University Uttar Pradesh
Noida, India

WILEY

This edition first published 2026

Registered Offices
John Wiley & Sons, Inc., 111 River Street, Hoboken, NJ 07030, USA
John Wiley & Sons Ltd, New Era House, 8 Oldlands Way, Bognor Regis, West Sussex, PO22 9NQ, UK

For details of our global editorial offices, customer services, and more information about Wiley products visit us at www.wiley.com.

The manufacturer's authorized representative according to the EU General Product Safety Regulation is Wiley-VCH GmbH, Boschstr. 12, 69469 Weinheim, Germany, e-mail: Product_Safety@wiley.com.

Wiley also publishes its books in a variety of electronic formats and by print-on-demand. Some content that appears in standard print versions of this book may not be available in other formats.

Library of Congress Cataloging-in-Publication Data

Names: Tomar, Ajay editor | Kumar, Ajay
editor | González Estrada, Ramsés Ramón editor | Parmar, Kshitij editor
Title: Microbial consortia in plant science and sustainable agriculture /
edited by Ajay Tomar, Ajay Kumar, Ramses Ramon Gonzalez Estrada, Kshitij Parmar.
Description: First edition. | Hoboken, NJ : John Wiley & Sons Ltd, 2026. |
Includes bibliographical references and index.
Identifiers: LCCN 2025032544 | ISBN 9781394324972 hardback | ISBN
9781394324996 adobe pdf | ISBN 9781394324989 epub | ISBN 9781394325009
Subjects: LCSH: Plant-microbe relationships | Sustainable agriculture |
Crop improvement | Plant diseases
Classification: LCC QR351.M417 2025
LC record available at https://lccn.loc.gov/2025032544

Cover Design: Wiley
Cover Image: © Gabrielle/stock.adobe.com

Typesetting: Set in 9.5/12.5pt STIXTwoText by Lumina Datamatics

Contents

List of Contributors

Selene Aguilera-Aguirre
Agrobiotechnology and Electrochemistry Laboratory-Integral Food Research Laboratory
National Technological Institute of Mexico/Technological Institute of Tepic
Tepic, Nayarit, México

and

Laboratorio Integral de Investigación en Alimentos/Departamento de Ingeniería Química y Bioquímica
Tecnológico Nacional de México-Instituto Tecnológico de Tepic
Tepic, Nayarit, México

Ashwini Ananda
AICRP-Vegetable Crops, School of Agricultural Sciences, Nagaland University
Medziphema, Nagaland, India

Aparna
Department of Agriculture
Jagan Nath University
Chaksu, Jaipur, India

Aastha
Department of Biotechnology
School of Applied and Life Sciences, Uttaranchal University
Dehradun, Uttarakhand, India

Miriam del Carmen Bañuelos-González
Laboratorio Integral de Investigación en Alimentos/Departamento de Ingeniería Química y Bioquímica
Tecnológico Nacional de México-Instituto Tecnológico de Tepic
Tepic, Nayarit, México

Nidhi Bhatt
Surajmal University Kichha
Udham Singh Nagar, Uttarakhand, India

Gema Cárdenas-Flores
Agrobiotechnology and Electrochemistry Laboratory-Integral Food Research Laboratory
National Technological Institute of Mexico/Technological Institute of Tepic
Tepic, Nayarit, México

and

Laboratorio Integral de Investigación en Alimentos/Departamento de Ingeniería Química y Bioquímica
Tecnológico Nacional de México-Instituto Tecnológico de Tepic
Tepic, Nayarit, México

Mercedes Susana Carranza-Patiño
Facultad de Ciencias Agrarias y Forestales
Universidad Técnica Estatal de Quevedo
Quevedo, Ecuador

Martina Alejandra Chacón-López
Agrobiotechnology and Electrochemistry Laboratory-Integral Food Research Laboratory
National Technological Institute of Mexico/Technological Institute of Tepic
Tepic, Nayarit, México

and

Laboratorio Integral de Investigación en Alimentos/Departamento de Ingeniería Química y Bioquímica
Tecnológico Nacional de México-Instituto Tecnológico de Tepic
Tepic, Nayarit, México

Juan Edgardo Cortés-Aguilera
Agrobiotechnology and Electrochemistry Laboratory-Integral Food Research Laboratory
National Technological Institute of Mexico/Technological Institute of Tepic
Tepic, Nayarit, México

Devashish
Ramakrishna Mission Vivekananda Educational and Research Institute
Kolkata, India

Ana Yareli Flores-Ramírez
Agrobiotechnology and Electrochemistry Laboratory-Integral Food Research Laboratory
National Technological Institute of Mexico/Technological Institute of Tepic
Tepic, Nayarit, México

Ramsés Ramón González Estrada
Tecnológico Nacional de México/I. T. Tepic
Tepic, Nayarit, México

Lucky Gupta
Faculty of Agricultural Sciences
GLA University
Mathura, Uttar Pradesh, India

Benjamín Hernandez-Vazquez
Resilient and Sustainable Agriculture Research Group (ASORE)
SECIHTI-Chapingo Autonomous University
Texcoco, Estado de México, México

Robinson J. Herrera-Feijoo
Facultad de Ciencias Agrarias y Forestales
Universidad Técnica Estatal de Quevedo
Quevedo, Ecuador

Raj Kishor Kapardar
The Energy and Resources Institute
New Delhi, India

Aditya Kumar
Amity Institute of Organic Agriculture, Amity University Noida
Noida, Uttar Pradesh, India

Rocky Kumar
Department of Biotechnology
School of Applied and Life Sciences, Uttaranchal University
Dehradun, Uttarakhand, India

Vikash Kumar
School of Agriculture Science and Technology, Kargil Campus
University of Ladakh
Ladakh, India

Dalpat Lal
College of Agriculture, Agriculture University
Mandor, Jodhpur, India

Ulises Miguel López-García
Agrobiotechnology and Electrochemistry Laboratory-Integral Food Research Laboratory
National Technological Institute of Mexico/Technological Institute of Tepic
Tepic, Nayarit, México

and

Laboratorio Integral de Investigación en Alimentos/Departamento de Ingeniería Química y Bioquímica
Tecnológico Nacional de México-Instituto Tecnológico de Tepic
Tepic, Nayarit, México

Guneshori Maisnam
Department of Rural Development and Planning
Nagaland University
Medziphema Campus, Nagaland, India

Manoj Kumar Maurya
Department of Vegetable Science
Acharya Narendra Dev University of Agriculture and Technology
Kumarganj, Ayodhya, India

Bernardo Murillo-Amador
Agriculture in Dry Zones Program (PAZA)
Centro de Investigaciones Biológicas del Noroeste
La Paz, Baja California Sur, México

María Fernanda Olmedo-Díaz
Laboratorio Integral de Investigación en Alimentos/Departamento de Ingeniería Química y Bioquímica
Tecnológico Nacional de México-Instituto Tecnológico de Tepic
Tepic, Nayarit, México

Rosa Isela Ortiz-Basurto
Agrobiotechnology and Electrochemistry Laboratory-Integral Food Research Laboratory
National Technological Institute of Mexico/Technological Institute of Tepic
Tepic, Nayarit, México

and

Laboratorio Integral de Investigación en Alimentos/Departamento de Ingeniería Química y Bioquímica
Tecnológico Nacional de México-Instituto Tecnológico de Tepic
Tepic, Nayarit, México

Shivani Pal
Department of Biotechnology
School of Applied and Life Sciences,
Uttaranchal University
Dehradun, Uttarakhand, India

Durga Prasad
College of Agriculture, Agriculture University
Mandor, Jodhpur, India

Pablo Preciado-Rangel
Departamento de Horticultura
Universidad Autónoma Agraria Antonio Narro
Periférico Raúl López Sánchez and Carretera Santa Fe S/N
Torreón, México

Riya Raghav
Amity Institute of Biotechnology, Amity University Uttar Pradesh
Noida, Uttar Pradesh, India

Nikhil Raghuvanshi
Department of Agronomy
DDU Gorakhpur University
Gorakhpur, Uttar Pradesh, India

Rakesh
Department of Biotechnology
School of Applied and Life Sciences,
Uttaranchal University
Dehradun, Uttarakhand, India

Agrim Rattan
Atmospheric Sciences
IIT Delhi
Delhi, India

Tomás Rivas-García
Resilient and Sustainable Agriculture
Research Group (ASORE)
SECIHTI-Chapingo Autonomous
University
Texcoco
Estado de México, México

Justine Sanasam
(Agriculture) Agronomy
Amity Institute of Organic Agriculture
Noida, Uttar Pradesh, India

Rita Schwentesius-Rindermann
Resilient and Sustainable Agriculture
Research Group (ASORE)
Chapingo Autonomous University
Texcoco, Estado de México, México

R.S. Sengar
Department of Plant Biotechnology
Sardar Vallabhbhai Patel University of
Agriculture and Technology
Meerut, Uttar Pradesh, India

Abha Sharma
Amity Institute of Organic Agriculture,
Amity University Noida
Noida, Uttar Pradesh, India

Abhay Sharma
Regional Horticultural Research &
Training Station
Kullu, Himachal Pradesh, India

Deepika Sharma
Amity Institute of Organic Agriculture,
Amity University
Noida, Uttar Pradesh, India

Meenakshi Sharma
Department of Botany
Daulat Ram College, University of Delhi
Delhi, India

Saurabh Sharma
Department of Agricultural Extension
Education
Nagaland University
Dimapur, Nagaland, India

Aashita Shree
Amity Institute of Organic Agriculture
Amity University Noida
Noida, Uttar Pradesh, India

Karuna Singh
Department of Biotechnology
School of Applied and Life Sciences,
Uttaranchal University
Dehradun, Uttarakhand, India

Monika Singh
Department of Biotechnology
School of Applied and Life Sciences,
Uttaranchal University
Dehradun, Uttarakhand, India

Sakshi Singh
Young Professional-II, ICAR-Central
Potato Research Institute
Shimla, Himachal Pradesh, India

Saurabh Singh
Faculty of Agricultural Sciences
GLA University
Mathura, Uttar Pradesh, India

Vivek Singh
Department of Plant Pathology
Acharya Narendra Dev University of Agriculture and Technology
Kumarganj, Ayodhya, India

Lokesh Sisodiya
Faculty of Agricultural Sciences
GLA University
Mathura, Uttar Pradesh, India

Mario Alberto Soltero-Rios
Departamento de Genética y Biología Molecular
Centro de Investigación y de Estudios Avanzados del IPN
Ciudad de México, México

Rajpal Srivastav
Amity Institute of Biotechnology, Amity University Uttar Pradesh
Noida, Uttar Pradesh, India

Umalaxmi Thingujam
Department of Soil Science and Agricultural Chemistry, PSB
Visva Bharati
West Bengal, India

Victor Thingujam
ICAR Research Complex for Eastern Region-Farming System Research Centre for Hill and Plateau Region
Ranchi, Jharkhand, India

Sunidhi Tiwari
Department of Agriculture
Jagan Nath University
Chaksu, Jaipur, India

Ajay Tomar
Amity Institute of Organic Agriculture
Amity University
Noida, Uttar Pradesh, India

Juan Antonio Torres-Rodríguez
Faculty of Agricultural and Forestry Sciences
Universidad Técnica Estatal de Quevedo
Quevedo, Ecuador

Mayra Carolina Velez Ruiz
Facultad de Ciencias Agrarias y Forestales
Universidad Técnica Estatal de Quevedo
Quevedo, Ecuador

Krishnakumari Yendrembam
Lovely Professional University
Jalandhar, Punjab, India

1

Microbial Consortia Used as Biofertilizers and Biocontrol Agents in Agriculture

Tomás Rivas-García[1], Pablo Preciado-Rangel[2], Bernardo Murillo-Amador[3], Rita Schwentesius-Rindermann[4], Ramsés Ramón González Estrada[5], Benjamín Hernandez-Vazquez[1], and Juan Antonio Torres-Rodríguez[6]*

[1] *Resilient and Sustainable Agriculture Research Group (ASORE), SECIHTI-Chapingo Autonomous University, Texcoco, Estado de México, México*
[2] *Departamento de Horticultura, Universidad Autónoma Agraria Antonio Narro, Periférico Raúl López Sánchez and Carretera Santa Fe S/N, Torreón, México*
[3] *Agriculture in Dry Zones Program (PAZA), Centro de Investigaciones Biológicas del Noroeste, La Paz, Baja California Sur, México*
[4] *Resilient and Sustainable Agriculture Research Group (ASORE), Chapingo Autonomous University, Texcoco, Estado de México, México*
[5] *Tecnológico Nacional de México/I. T. Tepic, Tepic, Nayarit, México*
[6] *Faculty of Agricultural and Forestry Sciences, Universidad Técnica Estatal de Quevedo, Quevedo, Ecuador*

* Corresponding author: tomas.rivas@secihti.mx

1.1 Introduction

The growing global concerns (e.g. rising food demand, chemical fertilizer shortages, and climate change) have pushed a shift toward more sustainable approaches (Pandey & Mishra, 2024). Microbial consortia are promising sustainable alternatives to pesticide inputs because they improve crop output and nutrient usage efficiency, plant stress tolerance, disease resistance, and adaptation to climate change (Li et al., 2024). Furthermore, some microbial consortia have proven to be beneficial in bioremediation processes (Zhao et al., 2023). In nature, microbial organisms have evolved by interactions with other microbes associated with their host and rarely interact as single species (Lyu et al., 2021). These consortia interact with plants, often forming positive or negative relationships (Espinosa-Palomeque et al., 2025). Their co-adaptation may play significant agroecological roles. Combining multiple species with complementary qualities, termed a consortium, has been gaining interest in the previous decade (Negi et al., 2024).

Microbial Consortia in Plant Science and Sustainable Agriculture, First Edition. Edited by Ajay Tomar, et al.

Microbial consortia can overcome the biological limits of single organisms. For example, microbial consortia are more viable for establishment, adaptation, and development in a wide range of biological and environmental conditions (Rivas-García et al., 2022). Combining microorganisms is advantageous because they usually have different growth characteristics, nutritional requirements, and modes of action, which usually makes synergistic interactions occur (Chaudhary et al., 2023). Additionally, microbial consortia can minimize cost and promote plant and soil biodiversity (Aguilar-Paredes et al., 2020). Nonetheless, implementing microbial consortia presents obstacles such as niche competition, incompatibility between selected strains, and habitat disturbance (Lahiri et al., 2022). As a result, microbial consortia with diverse plant growth properties are superior to single microbial strain applications in sustainable agriculture.

1.2 Microbial Consortia

Traditionally, axenic culture has been utilized to create and synthesize microbial strains for agricultural concerns (Negi et al., 2024). The global array of microbial-based fertilizers and biopesticides, which includes single strains, reflects this strategy (Murugan et al., 2022). Nonetheless, axenic cultures usually cannot compete for space and nutrients with microbial communities and are more susceptible to environmental conditions. Furthermore, single strain–based bioformulations usually have limited mechanisms of action, which decrease their efficacy under field real conditions (Rivas-Garcia et al., 2019). Microbial consortia–based formulations, on the other hand, have many mechanisms of action (e.g. nitrogen fixation, phosphate solubilization, nutrient competition, mycoparasitism, antibiosis, induction of plant resistance, and protection against abiotic stresses), as well as establishment, adaptability, and development to a wide range of niches and environmental conditions (Emami et al., 2023).

Microbial consortia with bacteria, fungi, oomycetes, and viruses are naturally found in diverse environments such as humans and animals health (Peixoto et al., 2021; Aggarwal et al., 2023), aquatic ecosystems (Sehnal et al., 2021), and soil and plant systems (Hartmann & Six, 2022; Compant et al., 2024). Ancient civilizations used natural microbial consortia to ferment foods and beverages. Moreover, it was demonstrated that growing plants inoculated with single strains had a slower development than those with natural microbial consortia on them (Carvalhais et al., 2013).

Microbial consortia interactions can be roughly characterized as cooperative or noncooperative. Specifically, microbial relationships are classified as positive (commensalism, symbiosis, and mutualism) or negative (amensalism, predation, and parasitism) (Khan, 2022). Microbial consortia can also be classified as natural, artificial, synthetic, or semisynthetic (Santoyo et al., 2021). Artificial ones are composed of multiple strains that do not normally coexist but are chosen for specific characteristics and positive interactions (Mewalal et al., 2017). While synthetic ones are made up of two or more microorganisms that have been genetically modified, semisynthetic ones include both microorganisms from natural environments and genetically modified sources (Negi et al., 2024) (Figure 1.1).

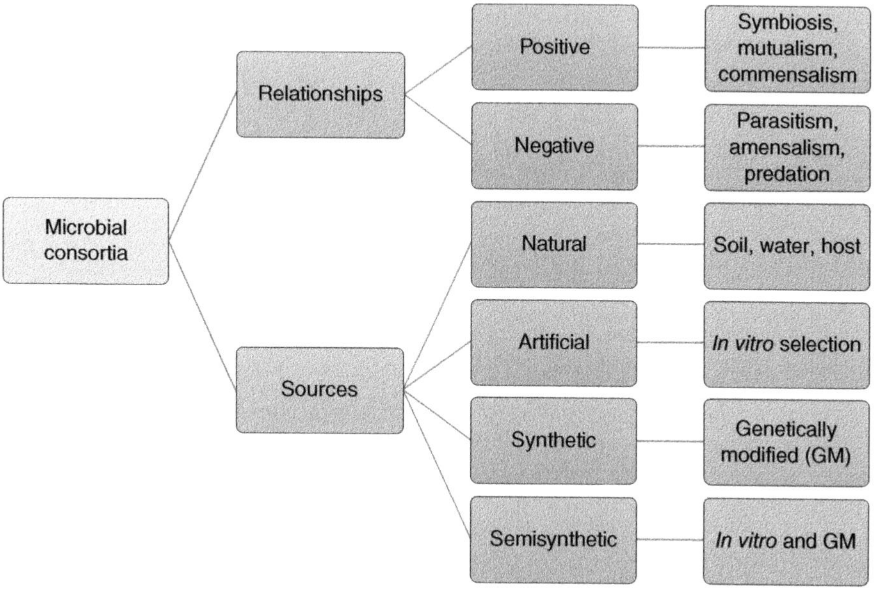

Figure 1.1 Microbial consortia classification according to their relationships and source.

The development and administration of microbial consortia that may be strategically applied in agricultural applications are made easier by this classification. Nonetheless, developing viable microbial consortia requires both *in vitro* experimentation and field validation based on real conditions of crop production.

1.3 Microbial Consortia for Plant Nutrition

Fertilization is crucial in plant nutrition, and the agricultural sector is primarily concerned with its proper application. The commonly called green revolution not only promoted the generalized use of agrochemicals to enhance crop productivity but also posed environmental dangers and limited water and soil resources (Rivas-Garcia et al., 2022a).

As a result, continuing to use traditional practices endangers food security as well as soil and water resource integrity. Researchers are looking for strategies to boost crop productivity while preserving the quality and safety of agricultural products and environmental resources (e.g. biodiversity, soil health, water).

Biofertilizers are one possible strategy for mitigating the unfavorable effects of the increasing use of agrochemical products. They are based on microorganisms' formulation that increased bioavailability and absorption of nutrients (e.g. nitrogen fixing) and enhancing plant–root systems (e.g. hormone synthesis), thus boosting crop yield and productivity (Table 1.1) (Daniel et al., 2022).

In recent years, the impact of numerous microbial consortia as biofertilizers has been well explored with promising results for future formulation and commercialization (Rivas-Garcia et al., 2022b). Table 1.2 provides an overview of the microbial consortia that have been tested as biofertilizers.

Table 1.1 Mechanisms of action of microbial-based biofertilizers.

Biofertilizer	Mechanism
Nitrogen fixing	Increase N_2 levels in the soil through the transformation of atmospheric nitrogen into a form usable by plants.
Phosphorus mobilization	With many applications, move phosphorus from the soil substrate to the root cortex.
Phosphorus solubilization	Release organic acids to reduce soil pH to transform insoluble phosphorus to soluble forms.
Potassium solubilization	Promote metal remotion and potassium solubilization ions.
Potassium mobilization	Move potassium by transforming it into plant-usable forms.
Micronutrient	Some micronutrients like zinc can be solubilized by redox processes, chelation, and proton addition.
Plant growth promotion	Release hormones that increase crop yields and plant protection.

Source: Daniel et al. (2022)/MDPI/CC BY 4.0.

Table 1.2 Microbial consortia tested as biofertilizers on different crops.

Crop	Microbial consortia	Results	References
Ashwagandha	*Azospirillum* sp., *Azotobacter chroococcum, Bacillus megaterium, Pseudomonas fluorescens*	Improved growth (root/plant length) and alkaloid content	Rajasekar and Elango (2011)
Chilli	*Erwinia persicina, Halomonas aquamarina, Pseudomonas extremorientalis*	Improved growth (root/plant length, biomass, number of leaves, fruit per plant), physiological parameters (chlorophylls, soluble sugar), and bioactive compounds (phenolics, and flavonoids)	Devi et al. (2022b)
Faba bean	*Siccibacter colletis, Enterobacter huaxiensis, Pantoea* sp.	Improved growth (root/plant biomasses), P uptake, and chlorophyll content	Chamkhi et al. (2023)
Foxtail millet	*Stenotrophomonas rhizophila, Pseudomonas marginalis, Bacillus* sp.	Improved growth and physiological parameters	Kaur et al. (2023)
Lupines	*Pseudomonas* sp., *Ochrobactrum cytisi, Bradyrhizobium* sp.	Improved plant yield and N absorption and decreased plant metal accumulation	Dary et al. (2010)

Crop	Microbial consortia	Results	References
Pea	*Pseudomonas aeruginosa, Trichoderma harzianum, Bacillus subtilis*	Improved growth (plant length, biomass, number of leaves), physiological parameters (chlorophyll and carotenoids), and nodulation promotion	Jain et al. (2015)
Sunflower	*Bacillus* sp., *Pseudomonas aeruginosa, Bacillus licheniformis, Streptomyces fradiae*	Improved plant growth and yield	Srinivasan and Mathivanan (2011)
Tomato	*Achromobacter* sp., *Bacillus* sp. *Enterobacter* sp., *Delftia* sp.	Improved plant biomass, root/plant weight, and chlorophyll content	Kapadia et al. (2021)
Wheat	*Bacillus megaterium, Arthrobacter chlorophenolicus, Enterobacter* sp., *Pseudomonas aeruginosa*	Improved growth (plant/root length and biomass), physiological parameters (chlorophyll content, carotenoids, total soluble sugar, and flavonoids)	Kumar et al. (2021)

1.4 Microbial Consortia for Plant Protection

In addition to the indiscriminate use of agrochemicals, land urbanization and climate change might produce unfavorable conditions for plant–microbial interactions in crop production systems (Vimal et al., 2017). Specifically, those stressors could have a negative impact on plant behavior (e.g. morpho-physiology, gene regulation and expression, biochemistry), generating the loss of crop yield and plant productivity (Chaudhary et al., 2022). Abiotic stress is brought on by nonliving things like salinity, drought, high temperatures, and pollutants, while biotic stress is brought on by living things like viruses, bacteria, fungi, nematodes, insects, arachnids, and weeds (Kumar & Nautiyal, 2022).

Biological control agents (BCAs) can protect plants against diseases by direct and indirect mechanisms. Direct protection occurs when the BCA acts on the disease-causing agent (phytopathogen). This can be accomplished through parasitism, predation, antibiosis, or the creation of lytic enzymes, and it can inhibit pathogens both before and during invasion. In contrast, indirect action modifies the environment to reduce the presence of pathogens and the likelihood of disease development. This can be accomplished with a variety of strategies, including developing plant resistance or competing with the BCA and pathogens (Figure 1.2; Köhl et al., 2019).

Recently, the impact of numerous microbial consortia as BCAs has been well explored with promising results for future formulation and commercialization. Table 1.3 provides an overview of the microbial consortia that have been tested as BCAs.

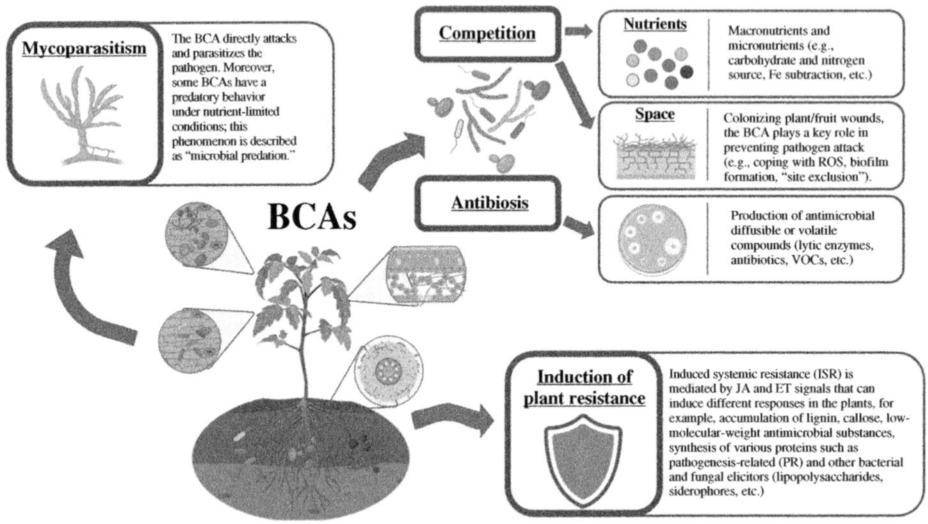

Figure 1.2 Mechanisms of action exerted by microbial consortia used as biological control agents. *Source:* Palmieri et al. (2022)/MDPI/CC BY 4.0.

Table 1.3 Microbial consortia tested as biological control agents against phytopathogens on different crops.

Host plant	Microbial consortia	Pathogen	References
Rice	*Bacillus firmus, Bacillus cereus, Serratia marcescens, Pseudomonas aeruginosa*	*Pyricularia oryzae*	Suryadi et al. (2013)
Potato	*P. aeruginosa, S. marcescens, Alcaligenes faecalis*	*Fusarium oxysporum*	Devi et al. (2018)
Lettuce	*Trichoderma viride, P. fluorescens, Bacillus thuringiensis*	*Ralstonia solanacearum*	Khan et al. (2018)
Tomato	*Funneliformis mosseae, Glomus fasciculatum, Bacillus* sp.	*F. oxysporum*	Devi et al. (2022a)
Pepper	*Serratia plymuthica, Chromobacterium* sp., *Lysobacter enzymogenes*	*Phytophthora capsici*	Kim et al. (2008)
Wheat	*Trichoderma gamsii, Trichoderma harzianum*	*Fusarium pseudograminearum*	Stummer et al. (2022)
Muskmelon	*Debaryomyces hansenii, Stenotrophomonas rhizophila*	*Fusarium proliferatum*	Rivas-Garcia et al. (2022c)
Beet	*Chitinophaga* sp., *Flavobacterium* sp.	*Rhizoctonia solani*	Carrión et al. (2019)

Host plant	Microbial consortia	Pathogen	References
Arabidopsis	*Xanthomonas* spp., *Microbacterium* spp., *Stenotrophomonas* spp.	*Hyaloperonospora arabidopsidis*	Berendsen et al. (2018)
Pea	*P. aeruginosa, B. subtilis, Trichoderma* sp.	*Sclerotinia sclerotiorum*	Jain et al. (2015)
Chickpea	*Bacillus amyloliquefaciens, S. marcescens, Rahnella aquatilis, P. fluorescens*	*F. oxysporum*	Palmieri et al. (2017)
Cape gooseberry	*Trichoderma virens, Trichoderma* spp., *Bacillus velezensis*	*F. oxysporum*	Izquierdo-García et al. (2021)
Kiwifruit	*Lysobacter enzymogenes, Bacillus safensis*	*Pseudomonas syringae*	Lin et al. (2023)

1.5 Microbial Consortia Development and Production

The development and production of microorganisms involves numerous phases, including isolation and identification of desirable strains; selection and optimization of culture media and industrial production; formulation and stabilization; and quality control and field testing (Díaz-Rodríguez et al., 2025). Selecting microorganisms for development and commercialization is an important first step in the bioformulation starting process (Khan et al., 2023). Isolation of possible candidates takes place from a variety of microbiological sources, including soil, water, and plant tissues. Recent studies have demonstrated that using native microorganisms can boost plant growth considerably, implying that these microorganisms are adapted to the niche environmental requirements and conditions. This increases the possibility of microbial establishment and may also improve plant growth even under previous plant disturbance scenarios (Li et al., 2023).

Preliminary, *in vitro* studies are performed to preselect prospective strains with desirable promoting growth capabilities and diverse mechanisms of action (Santoyo et al., 2019). However, *in vitro* tests, because they are carried out with nutrient-rich culture media, do not guarantee their effectiveness or the same mechanisms of action under real *in vivo* conditions where available nutrients are usually limited and highly variable (Chakraborty, 2020). Therefore, these *in vitro* studies are complemented by field tests under conditions very similar to natural ones, where the microbial consortia will be used (Li et al., 2023). Before fermentation, it is crucial to select an optimal culture medium and the conditions for fermentation for obtaining the best results. The aforementioned conditions influence not only the efficacy of the microbial inocula but also the cost of production and losses in later processes such as preservation (Villarreal-Delgado et al., 2024).

Then, to improve their stability and shelf life, microbial consortia are formulated with some liquid and solid carriers. The carrier material promotes a protected habitat and acts as an energy source for microbial growth and development, ensuring the effectiveness of microbial cells upon inoculation (Gopalakrishnan et al., 2016). To achieve maximum efficacy, seed treatment carriers should have a high adhesion capacity for the seeds (Bashan et al., 2014). The selected carriers for microbial formulations should have some characteristics such as chemical stability, safety, cost-effectiveness, and microbial cell viability (at least two to three months) for storage and commercialization (Bashan et al., 2014). The formulation used is determined by the individual microbe, its intended application, and the desired qualities of the final product.

Methods to ensure quality during microbial production are essential for ensuring uniformity and quality in microbial compositions (Khan et al., 2023). To assure quality, standards for metrics such as effectiveness, viable cell counts, and purity should be defined and followed throughout the production process, from isolation to commercialization (Mishra & Barolia, 2020). The efficacy of bioformulations must also be evaluated to estimate their economic use; thus, several variables are investigated, including application timings, intervals, and concentrations (Elena & Köhl, 2020). Finally, after field validation, the bioformulation must be registered as a patent before commercialization (Nunes et al., 2024).

1.6 Microbial Consortia's Challenges and Limitations

Recently, there has been a growing interest in using microbial consortia to produce microbial consortia-based formulations as sustainable alternatives for increasing plant yield and protecting them against pathogens. This alternative has seen great success, but it is not without its obstacles and restrictions. The complexity of duplicating their favorable impacts on plants under ever-changing environmental circumstances in the field represents a significant challenge. There is a need to promote awareness among farming communities about the scientific procedures used to apply microbial bioformulations in the field, as well as the ecological significance of these formulations. It can be difficult to select the appropriate microbial strains for various agricultural contexts. Due to the complexity of the natural and agricultural plant-associated systems and many elements impacting the end outcome, there is no general procedure for constructing potential microbial consortia (Díaz-Rodríguez et al., 2025). The efficiency of these strains varies depending on the soil type, temperature, pH, and moisture levels. Another drawback is that microbial compositions have a limited shelf life. Over time, the viability of microorganisms in these formulations can deteriorate, limiting their usefulness in the field. Most microbial bioformulations are also selective in their actions, which limits their compatibility with specific chemical pesticides or fertilizers. This can have an impact on integrated pest control or nutrient management programs. To address these obstacles and limits, scientists, agricultural practitioners, and policymakers must conduct ongoing research, development, and collaboration.

1.7 Conclusions and Future Perspectives

The adoption and utilization of microbial consortia in agriculture is continuously growing, as indicated by the increasing patents and commercialization of biological products, including many species/strains. Proper microbial mixture selection can enhance application efficiency and consistency. Creating an effective bioformulation demands several essential steps: identification of beneficial strains; culture media, optimization of conditions, and mass cultivation methods; formulation and stabilization; and quality control and field testing. While there are more study articles on this topic, there is still a lack of in-depth field studies. For the future, scientists, agricultural practitioners, and policymakers must conduct ongoing research, development, and collaboration to face the aforementioned limitations and challenges.

References

Aggarwal, N., Kitano, S., Puah, G.R.Y., Kittelmann, S., Hwang, I.Y., & Chang, M.W. 2023. Microbiome and human health: current understanding, engineering, and enabling technologies. *Chem. Rev.* 123(1), 31–72.

Aguilar-Paredes, A., Valdés, G., & Nuti, M. 2020. Ecosystem functions of microbial consortia in sustainable agriculture. *Agronomy* 10(12), 1902.

Bashan, Y., de-Bashan, L.E., Prabhu, S.R., & Hernandez, J.P. 2014. Advances in plant growth-promoting bacterial inoculant technology: formulations and practical perspectives (1998–2013). *Plant. Soil.* 378(1–2), 1–33.

Berendsen, R.L., Vismans, G., Yu, K., Song, Y., De Jonge, R., Burgman, W.P., Burmølle, M., Herschend, J., Bakker, P.A.H.M., & Pieterse, C.M.J. 2018. Disease-induced assemblage of a plant-beneficial bacterial consortium. *ISME J.* 12(6), 1496–1507.

Carrión, V.J., Perez-Jaramillo, J., Cordovez, V., Tracanna, V., De Hollander, M., Ruiz-Buck, D., Mendes, L.W., van Ijcken, W.F.J., Gomez-Exposito, R., Elsayed, S.S., Mohanraju, P., Arifah, A., van der Oost, J., Paulson, J.N., Mendes, R., van Wezel, G.P., Medema, M.H., & Raaijmakers, J.M. 2019. Pathogen-induced activation of disease-suppressive functions in the endophytic root microbiome. *Science* 366(6465), 606–612.

Carvalhais, L.C., Muzzi, F., Tan, C.H., Hsien-Choo, J., & Schenk, P.M. 2013. Plant growth in Arabidopsis is assisted by compost soil-derived microbial communities. *Front. Plant. Sci.* 4, 49996.

Chakraborty, A.P. 2020. Carrier based bioformulations of PGPR-characteristics, shelf life and application in improving health status of crop plants-a mini review. *Int. J. Robot. Res.* 7, 4.

Chamkhi, I., Zwanzig, J., Ibnyasser, A., Cheto, S., Geistlinger, J., Saidi, R., Zeroual, Y., Kouisni, L., Bargaz, A., & Ghoulam, C. 2023. Siccibacter colletis as a member of the plant growth-promoting rhizobacteria consortium to improve faba-bean growth and alleviate phosphorus deficiency stress. *Front. Sustain. Food. Sys.* 7, 1134809.

Chaudhary, P., Singh, S., Chaudhary, A., Sharma, A., & Kumar, G. 2022. Overview of biofertilizers in crop production and stress management for sustainable agriculture. *Fronti. Plant. Sci.* 13, 930340.

Chaudhary, P., Xu, M., Ahamad, L., Chaudhary, A., Kumar, G., Adeleke, B.S., Verma, K.K., Hu, D.M., Širić, I., Kumar, P., Popescu, S.M., & Abou Fayssal, S. 2023. Application of synthetic consortia for improvement of soil fertility, pollution remediation, and agricultural productivity: a review. *Agronomy* 13(3), 643.

Compant, S., Cassan, F., Kostić, T., Johnson, L., Brader, G., Trognitz, F., & Sessitsch, A. 2024. Harnessing the plant microbiome for sustainable crop production. *Nat. Rev. Microbiol.* 23(1), 9–23.

Daniel, A.I., Fadaka, A.O., Gokul, A., Bakare, O.O., Aina, O., Fisher, S., Burt, A.F., Mavumengwana, V., Keyster, M., & Klein, A. 2022. Biofertilizer: the future of food security and food safety. *Microorganisms* 10(6), 1220.

Dary, M., Chamber-Pérez, M.A., Palomares, A.J., & Pajuelo, E. 2010. "*In situ*" phytostabilisation of heavy metal polluted soils using *Lupinus luteus* inoculated with metal resistant plant-growth promoting rhizobacteria. *J. Hazard. Mater.* 177(1–3), 323–330.

Devi, A.R., Sharma, G.D., Majumdar, P.B., & Pandey, P. 2018. A multispecies consortium of bacteria having plant growth promotion and antifungal activities, for the management of Fusarium wilt complex disease in potato (*Solanum tuberosum* L.). *Biocatal. Agric. Biotechnol.* 16, 614–624.

Devi, N.O., Tombisana Devi, R.K., Debbarma, M., Hajong, M., & Thokchom, S. 2022a. Effect of endophytic *Bacillus* and arbuscular mycorrhiza fungi (AMF) against Fusarium wilt of tomato caused by *Fusarium oxysporum* f. sp. lycopersici. *Egypt. J. Biol. Pest Control.* 32(1), 1–14.

Devi, R., Kaur, T., Kour, D., Yadav, A.N., & Suman, A. 2022b. Potential applications of mineral solubilizing rhizospheric and nitrogen fixing endophytic bacteria as microbial consortium for the growth promotion of chilli (*Capsicum annum* L.). *Biologia* 77(10), 2933–2943.

Díaz-Rodríguez, A.M., Parra Cota, F.I., Cira Chávez, L.A., García Ortega, L.F., Estrada Alvarado, M.I., Santoyo, G., & de los Santos-Villalobos, S. 2025. Microbial inoculants in sustainable agriculture: advancements, challenges, and future directions. *Plants* 14(2), 191.

Elena, G., & Köhl, J. 2020. Screening strategies for selection of new microbial antagonists of plant pathogens. In: A. De Cal, P. Melgarejo, & N. Magan (Eds.), *Progress in Biological Control* (pp. 165–181). Cham: Springer.

Emami, S., Alikhani, H.A., & Kardgar, V. 2023. Sustainable improvement of productivity and quality of agricultural crops using a microbial consortium. *Microorganisms for Sustainability* (Vol. 43, pp. 129–149). Springer.

Espinosa-Palomeque, B., Jiménez-Pérez, O., Ramírez-Gottfried, R.I., Preciado-Rangel, P., Buendía-García, A., Sifuentes, G.Z., Sariñana-Navarrete, M.A., & Rivas-García, T. 2025. Biocontrol of phytopathogens using plant growth promoting rhizobacteria: bibliometric analysis and systematic review. *Horticulturae* 11(3), 271.

Gopalakrishnan, S., Sathya, A., Vijayabharathi, R., & Srinivas, V. 2016. Formulations of plant growth-promoting microbes for field applications. In: D. Singh, H. Singh, & R. Prabha (Eds.), *Microbial Inoculants in Sustainable Agricultural Productivity. Functional Applications* (pp. 239–252).

Hartmann, M., & Six, J. 2022. Soil structure and microbiome functions in agroecosystems. *Nat. Rev. Earth. Environ.* 4(1), 4–18.

Izquierdo-García, L.F., Cotes, A.M., & Moreno-Velandia, C.A. 2021. Screening for effective microbial consortia against Fusarium wilt of cape gooseberry (*Physalis peruviana*). *Bio. Control.* 66(5), 713–725.

Jain, A., Singh, A., Singh, S., & Singh, H.B. 2015. Biological management of Sclerotinia sclerotiorum in pea using plant growth promoting microbial consortium. *J. Basic. Microbiol.* 55(8), 961–972.

Kapadia, C., Sayyed, R.Z., Enshasy, H.A. El, Vaidya, H., Sharma, D., Patel, N., Malek, R.A., Syed, A., Elgorban, A.M., Ahmad, K., & Zuan, A.T.K. 2021. Halotolerant microbial consortia for sustainable mitigation of salinity stress, growth promotion, and mineral uptake in tomato plants and soil nutrient enrichment. *Sustainability* 13(15), 8369.

Kaur, T., Devi, R., Kumar, S., Kour, D., & Yadav, A.N. 2023. Synergistic effect of endophytic and rhizospheric microbes for plant growth promotion of foxtail millet (*Setaria italica* L.). *Nat. Acad. Sci. Lett.* 46(1), 27–30.

Khan, A., Singh, A.V., Gautam, S.S., Agarwal, A., Punetha, A., Upadhayay, V.K., Kukreti, B., Bundela, V., Jugran, A.K., & Goel, R. 2023. Microbial bioformulation: a microbial assisted biostimulating fertilization technique for sustainable agriculture. *Front. Plant. Sci.* 14, 1270039.

Khan, P., Bora, L.C., Borah, P.K., Bora, P., & Talukdar, K. 2018. Efficacy of microbial consortia against bacterial wilt caused by Ralstonia solanacearum in hydroponically grown lettuce plant. *IJCMAS* 7(06), 3046–3055.

Khan, S.T. 2022. Consortia-based microbial inoculants for sustaining agricultural activities. *Appl. Soil. Ecol.* 176, 104503.

Kim, Y.C., Jung, H., Kim, K.Y., & Park, S.K. 2008. An effective biocontrol bioformulation against Phytophthora blight of pepper using growth mixtures of combined chitinolytic bacteria under different field conditions. *Eur. J. Plant. Pathol.* 120(4), 373–382.

Köhl, J., Kolnaar, R., & Ravensberg, W.J. 2019. Mode of action of microbial biological control agents against plant diseases: Relevance beyond efficacy. *Front. Plant. Sci.* 10, 454982.

Kumar, A., Maurya, B.R., & Raghuwanshi, R. 2021. The microbial consortium of indigenous rhizobacteria improving plant health, yield and nutrient content in wheat (*Triticum aestivum*). *J. Plant. Nutr.* 44(13), 1942–1956.

Kumar, V., & Nautiyal, C.S. 2022. Plant abiotic and biotic stress alleviation: from an endophytic microbial perspective. *Curr. Microbiol.* 79(10), 1–14.

Lahiri, D., Nag, M., Ghosh, S., Dey, A., & Ray, R.R. 2022. Microbial consortium and crop improvement: Advantages and limitations. In: R. Soni, & A.N. Yadav (Eds.), *Trends of Applied Microbiology for Sustainable Economy* (pp. 109–123). Academic Press Inc.

Li, C., Chen, X., Jia, Z., Zhai, L., Zhang, B., Grüters, U., Ma, S., Qian, J., Liu, X., Zhang, J., & Müller, C. 2024. Meta-analysis reveals the effects of microbial inoculants on the biomass and diversity of soil microbial communities. *Nat. Ecol. Evol.* 8(7), 1270–1284.

Li, J., Wang, J., Liu, H., Macdonald, C.A., & Singh, B.K. 2023. Microbial inoculants with higher capacity to colonize soils improved wheat drought tolerance. *Microb. Biotechnol.* 16(11), 2131–2144.

Lin, L., Li, L., Tao, M., Wu, Q., Zhou, L., Wang, B., Wang, L., Shao, X., Zhong, C., & Qian, G. 2023. Assembly of an active microbial consortium by engineering compatible combinations containing foreign and native biocontrol bacteria of kiwifruit. *Comput. Struct. Biotechnol. J.* 21, 3672–3679.

Lyu, D., Msimbira, L.A., Nazari, M., Antar, M., Pagé, A., Shah, A., Monjezi, N., Zajonc, J., Tanney, C.A.S., Backer, R., & Smith, D.L. 2021. The coevolution of plants and microbes underpins sustainable agriculture. *Microorganisms* 9(5), 1036.

Mewalal, R., Rai, D.K., Kainer, D., Chen, F., Külheim, C., Peter, G.F., & Tuskan, G.A. 2017. Plant-derived terpenes: a feedstock for specialty biofuels. *Trends. Biotechnol.* 35(3), 227–240.

Mishra, B.K., & Barolia, S.K. 2020. Quality assessment of microbial inoculants as biofertilizer. *IJCMAS* 9(10), 3715–3729.

Murugan, A.C., Perumal, V., & Kannan, S. 2022. Production, cost analysis, and marketing of agricultural effective microorganisms. In: N. Amaresan, D. Dharumadurai, & D.R. Cundell (Eds.), *Industrial Microbiology Based Entrepreneurship. Microorganisms for Sustainability* (Vol. 42, pp. 199–223). Springer.

Negi, R., Sharma, B., Jan, T., Kaur, T., Chowdhury, S., Kapoor, M., Singh, S., Kumar, A., Rai, A.K., Rustagi, S., Shreaz, S., Kour, D., Ahmed, N., Kumar, K., & Yadav, A.N. 2024. Microbial consortia: promising tool as plant bioinoculants for agricultural sustainability. *Curr. Microbiol.* 81(8), 1–14.

Nunes, P.S.O., Lacerda-Junior, G. V., Mascarin, G.M., Guimarães, R.A., Medeiros, F.H.V., Arthurs, S., & Bettiol, W. 2024. Microbial consortia of biological products: do they have a future? *Biol. Control.* 188, 105439.

Palmieri, D., Ianiri, G., Del Grosso, C., Barone, G., De Curtis, F., Castoria, R., & Lima, G. 2022. Advances and perspectives in the use of biocontrol agents against fungal plant diseases. *Horticulturae* 8(7), 577.

Palmieri, D., Vitullo, D., De Curtis, F., & Lima, G. 2017. A microbial consortium in the rhizosphere as a new biocontrol approach against *Fusarium* decline of chickpea. *Plant. Soil.* 412(1–2), 425–439.

Pandey, D.K., & Mishra, R. 2024. Towards sustainable agriculture: harnessing AI for global food security. *Artif. Intell. Agric.* 12, 72–84.

Peixoto, R.S., Harkins, D.M., & Nelson, K.E. 2021. Advances in microbiome research for animal health. *Annu. Rev. Anim. Biosci.* 9(9), 289–311.

Rajasekar, S., & Elango, R. 2011. Effect of microbial consortium on plant growth and improvement of alkaloid content in *Withania somnifera* (Ashwagandha). *Curr. Bot.* 2(8), 27–30.

Rivas-Garcia, T., Espinosa-Calderón, A., Hernández-Vázquez, B., & Schwentesius-Rindermann, R. 2022a. Overview of environmental and health effects related to glyphosate usage. *Sustainability* 14(11), 6868.

Rivas-García, T., Hernández-Montiel, L.G., González-Estrada, R.R., Murillo-Amador, B., & Ledea-Rodríguez, J.L. 2022b. Enhancement of biocontrol agents activity by compatible treatments against postharvest disease of fruits. In: A. Kumar (Ed.) *Microbial Biocontrol: Food Security and Post Harvest Management.* (Vol. 2, pp. 245–264). Cham: Springer.

Rivas-Garcia, T., Murillo-Amador, B., Nieto-Garibay, A., Rincon-Enriquez, G., Chiquito-Contreras, R.G., & Hernandez-Montiel, L.G. 2019. Enhanced biocontrol of fruit rot on muskmelon by combination treatment with marine *Debaryomyces hansenii* and *Stenotrophomonas rhizophila* and their potential modes of action. *Postharvest Biol. Technol.* 151, 61–67.

Rivas-Garcia, T., Murillo-Amador, B., Preciado-Rangel, P., Ávila-Quezada, G.D., Hernandez-Montiel, L.G., Reyes-Pérez, J.J., Chiquito-Contreras, R.G., & Lara-Capistran, L. 2022c. *Debaryomyces hansenii, Stenotrophomonas rhizophila*, and Ulvan as Biocontrol Agents of Fruit Rot Disease in Muskmelon (*Cucumis melo* L.). *Plants.* 11(2), 184.

Santoyo, G., Guzmán-Guzmán, P., Parra-Cota, F.I., de los Santos-Villalobos, S., Orozco-Mosqueda, M.D.C., & Glick, B.R. 2021. Plant growth stimulation by microbial consortia. *Agronomy* 11(2), 219.

Santoyo, G., Sánchez-Yáñez, J.M., & de los Santos-Villalobos, S. 2019. Methods for detecting biocontrol and plant growth-promoting traits in rhizobacteria. In: D. Reinhardt, & A. Sharma (Eds.) *Methods in Rhizosphere Biology Research. Rhizosphere Biology.* (pp. 133–149). Singapore: Springer.

Sehnal, L., Brammer-Robbins, E., Wormington, A.M., Blaha, L., Bisesi, J., Larkin, I., Martyniuk, C.J., Simonin, M., & Adamovsky, O. 2021. Microbiome composition and function in aquatic vertebrates: small organisms making big impacts on aquatic animal health. *Front. Microbiol.* 12, 567408.

Srinivasan, K., & Mathivanan, N. 2011. Plant growth promoting microbial consortia mediated classical biocontrol of sunflower necrosis virus disease. *J. Biopest.* 4(1), 65–72.

Stummer, B.E., Zhang, X., Yang, H., & Harvey, P.R. 2022. Co-inoculation of Trichoderma gamsii A5MH and Trichoderma harzianum Tr906 in wheat suppresses in planta abundance of the crown rot pathogen Fusarium pseudograminearum and impacts the rhizosphere soil fungal microbiome. *Biol. Control.* 165, 104809.

Suryadi, Y., Susilowati, D.N., Riana, E., & Mubarik, N.R. 2013. Management of rice blast disease (Pyricularia oryzae) using formulated bacterial consortium. *Emir. J. Food. Agr.* 25(5), 349–357.

Villarreal-Delgado, M.F., Díaz-Rodríguez, A.M., & De los Santos-Villalobos, S. 2024. Microbial inoculants production systems. In: S. De los Santos-Villalobos (Ed.), *New Insights, Trends, and Challenges in the Development and Applications of Microbial Inoculants in Agriculture.* (pp. 99–104). Academic Press.

Vimal, S.R., Singh, J.S., Arora, N.K., & Singh, S. 2017. Soil-plant-microbe interactions in stressed agriculture management: a review. *Pedosphere* 27(2), 177–192.

Zhao, S., Wang, H., & Wang, J. 2023. Synthesis and application of a compound microbial inoculant for effective soil remediation. *Enviro. Sci. Pollut. Res. Int.* 30(57), 120915–120929.

2

Microbial Consortia to Increase Crop Resilience to Drought Stress

Juan Antonio Torres-Rodriguez[1*]*, Mercedes Susana Carranza-Patiño*[1]*, Mayra Carolina Vélez Ruiz*[1]*, Robinson J. Herrera-Feijoo*[1]*, and Tomás Rivas-García*[2]

[1] *Facultad de Ciencias Agrarias y Forestales, Universidad Técnica Estatal de Quevedo, Quevedo, Ecuador*
[2] *SECIHTI-Universidad Autonoma Chapingo, Carretera México-Texcoco km 38.5, Chapingo, Texcoco, Estado de Mexico, Mexico*

* Corresponding author: jatorres@uteq.edu.ec

2.1 Introduction

Global climate change has become an increasingly important threat to food security (Malhi et al., 2021). Among its effects, drought is one of the main factors limiting the productivity of agricultural crops. Drought negatively affects agriculture by altering soil structure, reducing soil productivity, and limiting plants' ability to uptake water and nutrients. It affects plant growth by causing loss of turgor, damage to cell membranes, metabolic disturbance, and oxidative stress. Moreover, drought creates imbalances in infiltration and oxido-reduction processes, leading to plant death (Kalefetoğlu & Ekmekçi, 2010; Farrell et al., 2017).

In recent years, microorganisms have been used to improve plant productivity under drought conditions (Li et al., 2024; Singh et al., 2024). Among these, microbial consortia stand out as a promising alternative to mitigate the adverse effects of drought stress in agricultural systems.

Microbial consortia consist of communities of microorganisms that coexist and collaborate in a shared environment, be they either natural or artificially engineered. Interactions between these microbial communities enable the execution of complex processes that individual organisms cannot achieve independently (Ram et al., 2022). Their composition may include bacteria, fungi, and archaea.

Microbial consortia are classified according to their origin (natural, artificial, semisynthetic, or synthetic), their modes of interaction (synergistic, commensal, or mutualistic), and their specific functions (environmental maintenance, nutrient exchange, substrate facilitation, or signal exchange) (Bhatia et al., 2018; Massot et al., 2022).

Microbial Consortia in Plant Science and Sustainable Agriculture, First Edition. Edited by Ajay Tomar, et al.

In agriculture, they have been studied as biocontrol agents, in bioremediation processes, in promoting plant growth, and in improving crop resilience to stress conditions. With advances in genetic engineering, metabolic engineering, and synthetic biology, researchers can improve microbial consortia to achieve higher productivity (Kaya et al., 2024; Solanki et al., 2024).

Microbial consortia play a crucial role in mitigating the adverse effects of drought on plants by promoting resilience through various mechanisms. These communities, capable of forming biofilms and participating in quorum sensing, enhance plant uptake of water and nutrients by stabilizing soil structure and facilitating metabolic processes. Their coordinated interactions, such as metabolite exchange and participation in nitrogen and carbon cycling, help reduce oxidative stress, restore turgor, and maintain metabolic balance in plants suffering from drought, thus improving agricultural productivity (Antoniewicz, 2020; Penesyan et al., 2021).

Therefore, alleviating abiotic stress in plants through the application of microbial consortia represents a promising avenue for improving agricultural productivity. This chapter summarizes the main action mechanisms, interactions between microbial communities and inert powders, biotechnological advances, and the strategic design of microbial consortia, highlighting their role in improving plant productivity and agricultural sustainability in drought conditions.

2.2 Selection and Design of Microbial Consortia to Increase Crop Resistance to Drought

The use of microbial consortia in water stress conditions provides benefits superior to those of an individual microbial inoculum due to the synergy of the microorganisms that comprise them (Aguilar-Paredes et al., 2020). The selection of microbial consortia is a key process in the development of sustainable strategies to improve agricultural productivity, especially for plants under abiotic stress such as drought. To select microbial consortia, it is fundamental to take into account the compatibility between microbial species, their individual and collective functionality, and the conditions of the environment in which they will be implemented (Behera et al., 2021; Nunes et al., 2024).

The modes of interaction between microorganisms within a microbial consortium reflect their functional compatibility. Interactions between microorganisms are synergistic, commensal, and mutualistic. In synergistic interactions, microorganisms divide their efforts to use different substrates in a complementary way, achieving a greater effect than they could achieve separately. In commensal interactions, one microorganism benefits without affecting the other. Mutualistic interactions benefit all species involved by sharing nutrients, enzymes, and/or metabolites (Bernstein & Carlson, 2012; Behera et al., 2021; Wang et al., 2023b).

Microorganisms belonging to the same species or taxonomic group tend to be more compatible due to similar multicellular behavior patterns (Cano, 2011). However, species from different groups may evolve strategies to share resources and maintain a balanced coexistence.

In natural consortia, compatibility arises intrinsically, resulting from the co-evolution of species in habitats such as the rhizosphere, where nitrogen-fixing bacteria and mycorrhizal fungi collaborate to promote plant growth (Barea et al., 2005; Yang et al., 2024). On the other hand, artificial consortia require experimental testing to assess the ability of selected microorganisms to interact without competing to ensure metabolic synergy (Efremenko et al., 2024). Semisynthetic consortia combine natural and genetically modified microorganisms, a process that requires careful design to maintain stability and functionality. Synthetic consortia, designed entirely by synthetic biology, enable the precise control of metabolic interactions through *in silico* simulations, ensuring functional compatibility even under controlled conditions (Che & Men, 2019).

The selection of microbial consortia based on their functionality focuses on identifying microorganisms whose metabolic capabilities enable them to achieve specific targets in a given context. This approach ensures that the combined metabolic activities of consortium members are adapted to perform complex tasks more efficiently than individual species (Li et al., 2021b). For example, in agriculture, a functional consortium may include microorganisms capable of improving nutrient availability, such as breaking down complex organic matter or solubilizing essential minerals, while others increase crop tolerance to abiotic stresses by producing osmoprotectants or antioxidant enzymes.

The design of artificial microbial consortia must be based on specific criteria to ensure their effectiveness and stability. It is crucial that selected microorganisms do not compete for the same carbon source, as this can affect both their growth and productivity. In addition, it is essential to establish appropriate inoculum ratios to prevent the depletion of energy sources and to prevent one species from dominating the consortium. Likewise, growth parameters such as pH and temperature must be kept within appropriate physiological ranges, and it is essential to understand the nutritional needs of the microorganisms to design a balanced medium that favors their growth without disproportionately benefiting one species (Bhatia et al., 2018). *In silico* tools are helpful, and these include constraint-based reconstruction and analysis, flux balance analysis, dynamic Bayesian networks, and Lotka–Volterra-based models. The tools enable the analysis of metabolic interactions and temporal dynamics within microbial communities, optimizing consortium design by identifying efficient metabolic pathways, minimizing competition for resources, and ensuring functional stability in diverse environments (Srinivasan et al., 2024).

Technologies such as metagenomics, transcriptomics, proteomics, and metabolomics, together with qualitative experiments, can now provide information even on non-culturable microorganisms, which represent the majority of those microorganisms currently known (Brenner et al., 2008; Duncker et al., 2021; Srinivasan et al., 2024).

Most studies focus primarily on the effects of individual microbial species or strains, without addressing possible intraspecific interactions (Shang & Liu, 2021). While it is important to assess the efficacy of individual strains, relatively few studies compare the efficacy of microbial consortia with that of single-strain applications (Xu et al., 2011). As a result, the synergistic benefits of using microbial consortia, including interactions with native microbes and the surrounding environment, remain difficult to quantify.

2.3 Action Mechanisms of Microbial Consortia Against Drought Stress

Microbial consortia play a crucial role in plant adaptation to drought conditions by modifying physiological, structural, and biochemical processes (Vurukonda et al., 2016). They contribute to optimizing plant water and nutrient uptake. Their activity includes the production of secondary metabolites such as exopolysaccharides and phytohormones, which regulate plant growth and improve resistance to water stress (de Vries et al., 2020). They favor the formation of biofilms and mycelial networks, which increases moisture retention in the rhizosphere and provides a more stable microenvironment for roots, even under drought conditions (Shaffique et al., 2022).

In agricultural ecosystems, natural microbial consortia are a fundamental tool for maintaining productivity in arid soils. However, recent advances have made it possible to design artificial and semisynthetic consortia that combine selected microorganisms with natural communities, thus maximizing their beneficial effects (Kaushal & Wani, 2016). These engineered consortia enhance critical functions such as nitrogen fixation, the production of osmoprotectants, and the regulation of gene expression related to water stress. For example, the combination of *Bacillus* bacteria and mycorrhizal fungi improves soil water retention and enhances antioxidant production, which protects plants from the negative effects of drought (Mamun et al., 2024).

Microbial consortia have been shown to enhance tolerance to water stress through the activation of biochemical mechanisms, such as an increased activity of antioxidant enzymes. These enzymes, such as superoxide dismutase (SOD), catalase (CAT), and peroxidase (POD), play a crucial role in the detoxification of reactive oxygen species (ROS), which are produced in excess during droughts and cause oxidative damage to plant cells (Saleem et al., 2021; Tyagi et al., 2023; Nivedita et al., 2024).

In this context, Nivedita et al. (2024) showed that *Streptomyces* spp. consortia promoted stress tolerance in rice by increasing SOD, CAT, and POD activity and improving growth and photosynthesis. Similarly, Pérez-Moncada et al. (2024) reported that a consortium of *Claroideoglomus claroideum, Naganishia albida*, and *Burkholderia caledonica* in strawberry plants reduced oxidative damage, with a 32% decrease in malondialdehyde levels and an increase in antioxidant activity, resulting in higher chlorophyll and nutrient content. Krishna et al. (2022) reported that a hexa-PGPM consortium composed of *Bacillus megaterium, Pseudomonas* spp., and *Trichoderma harzianum* in tomato plants not only increased CAT and SOD activity but also promoted the expression of stress response-related genes, such as DREB1 and P5CS, which contributed to cell protection and improved soil fertility.

During drought, microbial consortia, consisting of plant growth-promoting rhizobacteria (PGPR) and arbuscular mycorrhizal fungi (AMF), collaborate to increase water use efficiency in plants. These consortia induce physiological changes in the root, such as the development of root hairs, elongation of tap roots, and the formation of biofilms and mycelial networks. These changes allow plants to explore a larger volume of soil in search of water, improving their ability to withstand prolonged drought conditions. For example, in corn, inoculation with a consortium of *Serendipita indica, Rhizophagus intraradices*, and *Azotobacter chroococcum* increased root and stem growth and improved water content and membrane stability,

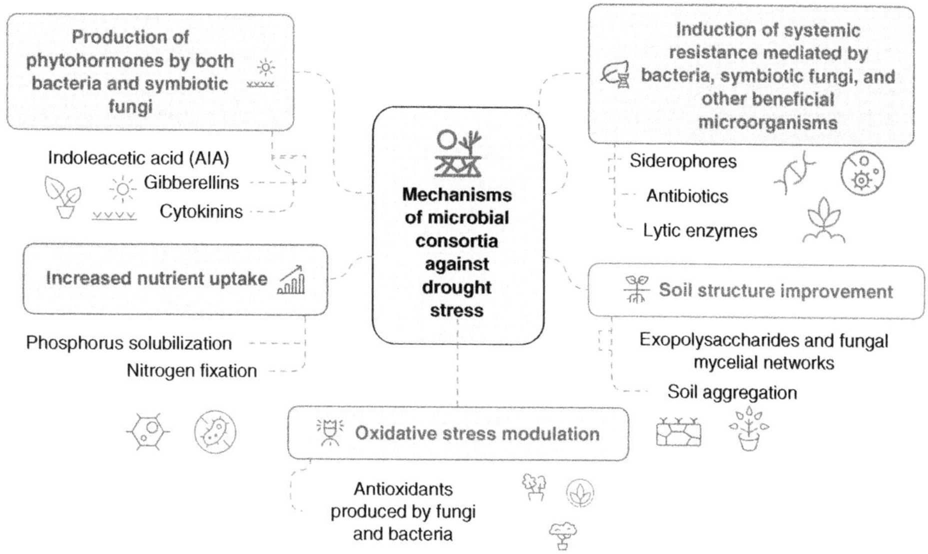

Figure 2.1 Mechanisms of microbial consortia against drought stress.

which reduced drought damage (Tyagi et al., 2023). In peanut crops in drought conditions, a bacterial consortium consisting of *Aneurinibacillus deltaense, Rhizobium* spp., *Caballeronia zhejiangensis, Burkholderia dolosa*, and *Bacillus safensis* increased germination by 91% and improved nutrient uptake and yield (Prasanna Kumar et al., 2022).

What is more, the exopolysaccharides produced by the microbial consortia improve soil structure by increasing aggregation and water retention, creating a more stable and humid environment in the rhizosphere. Figure 2.1 shows the main mechanisms by which microbial consortia improve plant tolerance to drought stress.

2.4 Plant–Microbe Communication

The plant–microorganism interaction is based on a chemical and molecular dialogue, facilitated by the root exudates that plants release and the response of microorganisms to these signals. These exudates include compounds such as sugars, amino acids, organic acids, phenols, and flavonoids, which attract beneficial bacteria and fungi in the rhizosphere (Bais et al., 2006). Under drought conditions, the composition of root exudates changes significantly. Plants increase the release of certain compounds, such as flavonoids and organic acids, which are known to attract PGPR and AMF (Jones et al., 2009). These compounds not only act as chemical signals but also provide a carbon source for the microorganisms, encouraging their colonization of the roots. Cooperation between bacteria and fungi in microbial consortia allows for more efficient colonization, promoting metabolic synergies that optimize the use of available resources and enhance tolerance to water stress.

For example, indole-3-acetic acid (IAA), a phytohormone produced by both plants and bacteria such as *Azospirillum brasilense*, promotes secondary root growth and enhances the plant's ability to explore the soil for water (Spaepen & Vanderleyden, 2011). Along with responding to root exudates, microorganisms use coordinated mechanisms such as quorum sensing to maximize their effectiveness.

A key mechanism in the communication of microbial consortia is quorum sensing, a process by which bacteria detect and respond to their population density through the release of signaling molecules. These molecules, such as acyl-homoserine lactones, enable the coordination of the collective behavior of bacteria, facilitating biofilm formation and root colonization (Venturi & Keel, 2016). In drought conditions, quorum sensing plays an important role in the formation of protective biofilms around roots, which helps to retain moisture in the rhizosphere and reduce soil desiccation.

Phytohormones, such as abscisic acid (ABA), IAA, and jasmonic acid, are vital mediators in the communication between plants and microorganisms. In response to drought, plants increase ABA production, which induces stomata closure to reduce water loss. However, ABA also acts as a signal that modulates interactions with beneficial microorganisms in the rhizosphere (Sah et al., 2016). For example, certain rhizosphere bacteria can produce ABA or influence plant ABA signaling, helping to regulate water stress responses.

Indole-3-acetic acid (IAA), produced by both plants and rhizobacteria such as *Pseudomonas fluorescens*, stimulates root development, increasing the plant's ability to absorb water and nutrients (Glick, 2012). This hormone can also influence root morphology, which promotes the formation of root hairs that enhance soil exploration. The production of IAA by bacteria is induced by specific compounds present in root exudates, reflecting a complex molecular dialogue that benefits both parties.

During drought, the plant can also release allelopathic compounds that inhibit the growth of pathogenic microorganisms in the rhizosphere, allowing the proliferation of beneficial microorganisms. This defensive strategy is part of a broader adaptive mechanism involving the active modulation of the microbial community in response to changes in the environment (Hu et al., 2016). Plants' ability to influence their root microbiome highlights the importance of plant–microbe communication as a tool for enhancing resilience to environmental stress. In addition to its role as a source of nutrients, the rhizosphere facilitates symbiotic interactions that optimize plants' tolerance to water stress.

2.5 Microbial Consortia Applied in Drought Stress Mitigation

Several studies have evaluated the efficacy of microbial consortia in promoting plant growth, improving nutrient uptake, and increasing tolerance to water stress.

Cirillo et al. (2023) highlighted that the consortium of *A. chroococcum* and *T. afroharzianum* improved water stress tolerance in tomato plants, increasing yield by 48.5% and stem fresh weight by 15%. Also, Bradáčová et al. (2019) showed that microbial consortia, compared to individual strains, performed better under adverse conditions, such as saline and low-fertility soils. Higher phosphorus uptake and superior vegetative growth were highlighted.

Endophytic microbial consortia significantly improved the survival and growth of *Pseudotsuga menziesii* and *Thuja plicata* under water stress and degraded soils. Inoculation with Salicaceae-derived consortia doubled the survival rate of *T. plicata* seedlings under drought conditions (Aghai et al., 2019). In onion plants, the application of a microbial consortium, in combination with reduced NPK fertilizer doses, resulted in 50% fertilizer and 35% water savings, without compromising yield (36 $t{\cdot}ha^{-1}$). In addition, water use efficiency was increased by 47% (Blanco et al., 2021).

Furthermore, in *Brachypodium distachyon*, root exudates stimulated colonization by endophytic bacteria and the formation of biofilms, increasing alginate production by 160% in consortia subjected to water stress, which is evidence of its ability to adapt to unfavorable environments (Saleh et al., 2020).

Table 2.1 presents a summary of 23 studies that investigated the benefits of using microbial consortia on 15 crop species, including cereals (*Triticum aestivum* L., *Zea mays* L., *Oryza sativa* L., and *Avena sativa* L.), legumes (*Cicer arietinum* L., *Vigna radiata* L., and *Phaseolus vulgaris* L.), vegetables (*Solanum lycopersicum* L. and *Solanum tuberosum* L.), oilseeds (*Arachis hypogaea* L.), and fruit species (*Fragaria* spp.). The consortia used combine PGPB, AMF, and other symbiotic microorganisms, highlighting strains such as *Bacillus subtilis, A. brasilense, P. fluorescens*, and *Glomus intraradices*, selected for their high tolerance to abiotic stress conditions. Among the main findings, it was found that the microbial consortia increased aerial and root biomass, chlorophyll and proline content, and antioxidant activity (SOD, CAT, and POD), improving tolerance to water stress in all crops studied.

Table 2.1 Main microbial consortia used in different types of crops to improve drought stress resistance.

Species	Microbial consortium	Main results	References
Arachis hypogaea L.	*Aneurinibacillus deltaense* AMT1 + *Rhizobium* sp. + *Caballeronia zhejiangensis* BPT9 + *Burkholderia dolosa* BPT8 + *Bacillus safensis* BPT6	This improved soil aggregate formation, relative water content, chlorophyll content, and grain yield in water deficit conditions.	Prasanna Kumar et al. (2022)
Avena sativa L.	*Azospirillum brasilense* + *Herbaspirillum* sp. + *Rhizobium leguminosarum*	This increased tolerance to water deficit, maintained chlorophyll b and carotenoid levels, and enabled rapid recovery after rehydration.	Villagran et al. (2024)
Cicer arietinum L.	*Mesorhizobium ciceri* + *Bacillus subtilis* + *Bacillus mojavensis* + *Enterobacter cloacae* + *Providencia vermicola*	This improved yields through IAA, ACC-deaminase, exopolysaccharides, and antioxidant enzymes, highlighting its efficacy in extreme drought.	Muniyappan et al. (2024)
Eclipta prostrata L.	*Funneliformis mosseae* + *Trichoderma harzianum* + *Bacillus megaterium*	This increased wedelolactone levels in drought conditions, showing the highest potential for secondary metabolites.	Aulakh et al. (2023)

(Continued)

Table 2.1 (Continued)

Species	Microbial consortium	Main results	References
Fragaria spp.	*F. mosseae* + *C. lamellosum* + *C. claroideum* + *P. frederiksbergensis* + *B. tequilensis* + *B. caledonica* + *Rhodotorula mucilaginosa* + *Candida guilliermondii* + *N. albida*	This increased tolerance to water stress and improved the plants' physiological, biochemical, and nutritional capacity.	Sinha & Raghuwanshi (2020)
Oryza sativa L.	*Achromobacter spanius* + *B. pumilus* + *B. cereus* + *B. altitudinis* + *Bacillus* sp.	This reduced ethylene; increased chlorophyll, proline, and soluble sugar content; and significantly increased yield components.	Pérez-Moncada et al. (2024)
	S. chilikensis + *S. griseoincarnatus* + *S. barkulensis* + *S. chitinovorans*	This mitigated salinity and drought effects and increased shoot and root length, chlorophyll and proline content, and the activity of antioxidant enzymes.	Saleem et al. (2021)
	Glomus intraradices + *A. brasilense* AZ-39	This improved water use efficiency, relative water content, and tolerance to water stress.	Rafedzi et al. (2024)
Quercus suber L.	Russula/Burkholderaceae + Lactarius/Bacillaceae	This improved resilience and adaptation in arid environments under water stress.	Reis et al. (2021)
Solanum lycopersicum L.	*Azotobacter chroococcum* + *Trichoderma afroharzianum*	This increased yields (+48.5%) and improved water potential (+9.5%).	Cirillo et al. (2023)
	B. megaterium + *Pseudomonas fluorescens* + *Pseudomonas aeruginosa* + *Pseudomonas putida* + *Paenibacillus polymyxa* + *Trichoderma harzianum*	In drought conditions, this improved growth, yield, and soil properties while reducing cell damage.	Damodaran et al. (2023)
	Azotobacter vinelandii + *Clostridium* sp. + *Lactobacillus* sp. + *Bacillus velezensis* + *P. fluorescens* + *T. harzianum*	This reversed the reduction in bacterial diversity and increased tolerance to salinity and water stress.	Krishna et al. (2022)
	Lysinibacillus spp. + *Bacillus* spp.	This improved vegetative parameters, yield, and lycopene content and mitigated salt stress by excluding Na^+ and increasing K^+.	Mamun et al. (2024)

Species	Microbial consortium	Main results	References
Solanum tuberosum L.	*Rhizophagus irregularis* + *Bacillus amyloliquefaciens* + *Pseudomonas brassicacearum*	This increased tuber biomass by up to 951% and improved nutritional status, ROS detoxification, and osmotic adjustment through glycine betaine.	Inostroza et al. (2017)
Triticum aestivum L.	*Bacillus* sp. AD-54A + *Serratia* sp. AD-7.6 + *Bacillus* sp. AC-225	This improved wheat growth in water deficit (30% maximum water holding capacity [MWHC]), highlighting its usefulness in arid and marginal lands.	Zoppellari et al. (2014)
Vigna radiata L.	*B. subtilis* + *Bacillus* sp. + *Bacillus licheniformis*	This promoted plant growth and improved physiological attributes in drought conditions.	Aghai et al. (2019)
	Rhizobium sp. + *Candida tropicalis* + *Paenibacillus staichungensis* + AMF	This improved resistance to water stress and increased the activity of antioxidant enzymes such as catalase, peroxidase, and superoxide dismutase.	Verma et al. (2018)
Zea mays L.	*B. subtilis* + *Bacillus* sp. + *Glomus* sp. + *Acaulospora* sp.	This increased aerial and root biomass and grain yield, reduced hydrogen peroxide and malondialdehyde, and improved root colonization and osmolyte accumulation.	Kalaji et al. (2018)
	Serendipita indica + *Rhizophagus intraradices* + *A. chroococcum*	This increased root and stem growth and antioxidant enzymes and improved physiological parameters under water stress.	Molina-Romero et al. (2017)
	Bacillus sp. MN-54 + *Enterobacter* sp. FD-17 + *P. fluorescens* 1	This increased photosynthetic activity by 47–61% and antioxidant systems by 42–62%, improving tolerance to water stress.	Tyagi et al. (2023)
	A. brasilense + *P. putida* + *Acinetobacter* sp. + *Sphingomonas* sp.	This improved growth and mitigated the effects of water stress after desiccation prior to germination.	Justin et al. (2020)
	A. brasilense sp. 7 + *P. putida* KT2440 + *Acinetobacter* sp. EMM02 + *Sphingomonas* sp. OF178	This improved growth under water deficit and promoted physiological attributes.	Thijs et al. (2014)
	AMF + *Halomonas* sp. PS30 + *B. subtilis* BA30	This improved water stress tolerance and biomass yield, suggesting effectiveness in drought conditions.	Rizzo et al. (2023)

2.6 Genomic and Biotechnological Advances in Microbial Consortia

Recent advances in genomics and biotechnology allow a deeper understanding of the molecular mechanisms underlying microbial–plant consortia interactions, especially in water stress conditions (Compant et al., 2019; de Vries et al., 2020). These tools facilitate the identification and manipulation of key genes responsible for the production of metabolites such as exopolysaccharides, phytohormones, and osmoprotectants, which are essential for increasing crop resilience (Vurukonda et al., 2016; Mikiciuk et al., 2024). These discoveries have laid the groundwork for optimizing the use of microorganisms in sustainable agriculture, enabling environmental challenges such as water scarcity and climate change to be addressed (Jalal et al., 2023).

Mass genome sequencing is a crucial tool in identifying genes associated with water stress tolerance. For example, the EPS gene in *B. subtilis* is involved in the production of exopolysaccharides, compounds that enhance the formation of biofilms in the rhizosphere. These biofilms act as moist microenvironments, providing a more favorable environment for roots in dry soils (Kaushal & Wani, 2016). Likewise, in bacteria such as *A. brasilense*, genes such as ipdC are related to the biosynthesis of IAA, a phytohormone that stimulates root growth, which improves water and nutrient uptake in drought conditions (Spaepen & Vanderleyden, 2011).

What is more, gene editing tools, such as CRISPR-Cas9, have revolutionized our ability to engineer microorganisms to optimize their functionality in the rhizosphere. These technologies allow microbial genes to be precisely modified to enhance the synthesis of bioactive compounds. For example, modified strains of *P. fluorescens* showed an increased ability to produce siderophores and antioxidant enzymes, which help reduce oxidative stress in plants during adverse conditions (Borrelli et al., 2018). These innovations not only increase the efficiency of microbial consortia but also improve their adaptability in various agricultural environments.

Synthetic biology has enabled the assembly of metabolic pathways designed to optimize key processes in microbial consortia. Applications such as enhanced nitrogen fixation and phosphorus solubilization are examples of the impact of these technologies in agriculture. These strategies, based on the design of microbial communities, are promising tools for mitigating the effects of water stress and ensuring agricultural sustainability (García-Fraile et al., 2015). Together, these advances offer innovative solutions that combine science and technology to address the challenges of food security and sustainability within a changing world.

2.7 Interactions Between Arthropods and Microbial Consortia in Drought Conditions

Drought stress is one of the most significant abiotic challenges affecting plant growth and substantially reduces crop yields across various regions globally. Approximately one-third of soils worldwide are estimated to experience drought conditions, making it critical to ensure adequate measures are in place to support normal plant development (Calvo-Polanco et al., 2016; Wu & Zou, 2017).

Soil organisms, encompassing bacteria, fungi, archaea, insects, annelids, and other invertebrates, are integral to sustaining soil health and agricultural productivity. They facilitate nutrient cycling, organic matter decomposition, and disease suppression, thereby enhancing crop growth and resilience (Culliney, 2013; Neemisha, 2020; Wang et al., 2024). The interactions between these diverse soil organisms create a dynamic ecosystem that supports sustainable agricultural production. Maintaining a healthy and diverse soil biota is essential for optimizing crop yields and ensuring long-term soil fertility (Neemisha, 2020).

The sustainability of agricultural systems in arid environments relies on the synergistic interaction between soil organisms such as arthropods and microbial consortia (Flórián et al., 2019; Bogati & Walczak, 2022; Frąc et al., 2022) (Figure 2.2).

Microbial consortia can be classified in terms of their construction, modes of interaction, and functions (Bhatia et al., 2018). They consist of AMF, PGPR, and actinobacteria, which work together to support sustainable agricultural practices and enhance the quality and health of crop production (Aguilar-Paredes et al., 2020). Some examples of microorganisms that comprise microbial consortia in the soil, including nitrogen-fixing bacteria (e.g. *Rhizobium* spp.), siderophore-producing bacteria (e.g. *P. fluorescens, P. putida, B. subtilis, A. vinelandii,* and *Streptomyces* spp.), antagonistic fungi (e.g. *Trichoderma* spp.), and AMF, which are symbiotic associations formed between fungi of the Glomeromycota group and the roots of most terrestrial plants (Stürmer et al., 2018).

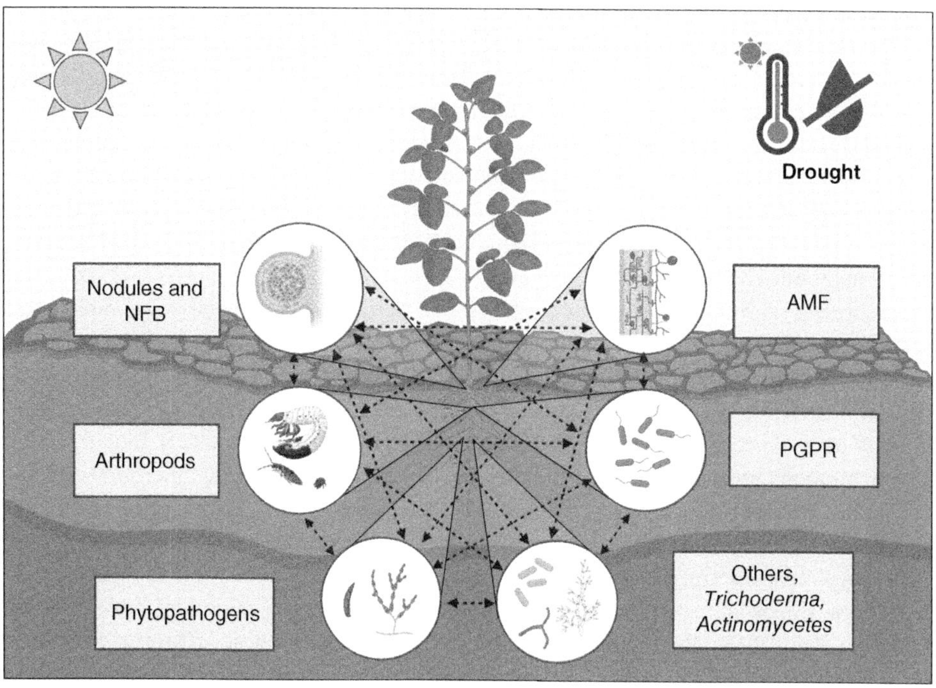

Figure 2.2 Schematic representation of rhizosphere interactions between microorganisms, arthropods, and plants under drought stress. NFB: nitrogen-fixing bacteria; PGPR: plant growth-promoting rhizobacteria; AMF: arbuscular mycorrhizal fungi.

The rhizosphere microbial consortia play a significant role in enhancing crop resilience in drought conditions. Many microorganisms contribute to improving plant growth and stress tolerance through various mechanisms, such as nitrogen fixation, osmolyte accumulation, and biotic priming, which collectively help mitigate the adverse effects of drought stress (Igiehon & Babalola, 2021; Nishu et al., 2022; Shah & Khan, 2024).

Specific examples highlight the effectiveness of these consortia under drought stress conditions in different crops. In soybean plants, *Rhizobium* spp. and AMF species improved yield by enhancing nutrient uptake and stress tolerance (Igiehon & Babalola, 2021). In rice, the soil application of *Trichoderma* spp., particularly *T. harzianum* and *T. viride*, reduced the negative impacts of water stress and promoted shoot and root growth as well as biomass production (Shah & Khan, 2024). Biotic priming with *P. fluorescens* induced drought tolerance in *Abelmoschus esculentus* (Pravisya et al., 2019), while *P. putida* enhanced drought resilience in *Cicer arietinum* (Tiwari et al., 2016). Similarly, *B. subtilis* improved growth in *Phleum pratense* by modulating osmolyte accumulation in roots and shoots (Gagné-Bourque et al., 2016). PGPR, such as *Azotobacter vinelandii*, increased drought tolerance in *Festuca ovina* by improving germination indices, dry weight, stem length, and root length (Rigi et al., 2023). Finally, the application of *Streptomyces* spp. enhanced drought resistance in *T. aestivum* through mechanisms that likely involve soil nutrient cycling and plant hormone modulation (Li et al., 2020).

AMF and PGPR are probably the most studied organisms in the context of resilience under drought conditions. AMF are fundamental in dry soils, as their hyphae enable the absorption of water and nutrients beyond the root depletion zone (Cheng et al., 2021). Furthermore, they secrete glomalin, a key protein that improves soil aggregation and its water retention capacity, mitigating the effects of water stress on plants (Latef et al., 2016; Thomopoulos et al., 2023). These fungi also induce defense mechanisms in plant roots, strengthening their resilience to adverse conditions, such as high temperatures and nutrient deficiencies (Ortiz et al., 2015; Santoyo et al., 2021). PGPR play a crucial role in inducing systemic resistance, synthesizing plant hormones, and improving plant resilience to abiotic stresses such as drought. Additionally, PGPR facilitate enhanced mycorrhizal colonization (Santoyo et al., 2021; Singh et al., 2022).

Moreover, soil arthropods, including springtails, mites, and beetles, play a complementary role in soil ecosystems by decomposing organic matter, releasing essential nutrients, and contributing to the dissemination of viable spores of ectomycorrhizal, AMF, and beneficial bacteria. They are also potential vectors of entomopathogenic and saprophytic fungi (Bonkowski et al., 2009; Gehring & Bennett, 2009; Lakshmi et al., 2020). Detritivorous microarthropods break down detritus into smaller particles, enhancing microbial colonization and substrate moisture, which stimulates microbial activity. However, feces make the surface area subject to colonization by microorganisms (Neher & Barbercheck, 2019). Arthropods' mobility through soil pores enhances aeration and redistributes microorganisms in the rhizosphere, promoting a functional balance among microbial communities (Philippot et al., 2013; Dhooria, 2016). Additionally, in dry soils, these arthropods regulate microbial populations by consuming fungi and bacteria, promoting functional diversity and reducing the dominance of specific species.

Research suggests that some arthropods feed directly on the mycelium or spores of mycorrhizal fungi, reducing fungal biomass and affecting the ability of the fungus to colonize new roots. As an example, the larvae of *Phyllophaga vetula* were found to reduce plant growth, nutrition, and AMF root colonization (Zitlalpopoca-Hernandez et al., 2017), and the vine weevil *Otiorhynchus sulcatus* reduced AMF colonization by 61% in *Plantago lanceolata* (Johnson & Rasmann, 2015). Nevertheless, the literature does not entirely support the notion that insects or other root arthropods always disrupt symbiosis (Gange, 2000; Johnson & Rasmann, 2015).

The interactions between arthropods and microbial consortia are complex and can be direct or indirect, influencing soil health, nutrient recycling, pest control, and agricultural productivity (Bengtsson et al., 1996). Microbial communities interact with arthropods often in a positive feedback relationship that reinforces tolerance to abiotic stress and enhances nutrient cycling (Naylor & Coleman-Derr, 2018; Wang et al., 2023a). The synergy between arthropods and microbial consortia creates a cascading effect on the soil ecosystem. Improvements in soil structure and functionality resulting from these interactions not only increase soil water retention capacity but also contribute to the biological control of phytopathogens and pests (Gehring & Bennett, 2009; Kumar et al., 2024). Additionally, predatory arthropods that feed on microarthropods contribute to balancing trophic chains, creating a more resilient ecosystem against climatic fluctuations (Koltz et al., 2024). However, the available information on interactions between beneficial rhizosphere microorganisms from different functional groups, as well as interactions between soil fauna and rhizosphere microorganisms, remains relatively unexplored (Larsen et al., 2015).

Understanding the interactions between arthropods and microbial consortia offers valuable insights for developing sustainable agricultural strategies in arid environments or under plant drought stress (Peng et al., 2024). These interactions play a fundamental role in soil recovery following water stress by improving soil structure, enhancing its capacity to retain water and nutrients, and restoring beneficial microbiota. The integration of microbial consortia application and arthropod biodiversity management has the potential to become a key practice for maintaining agricultural productivity in the face of climate change (Flórián et al., 2019; Allison, 2023). Future research should prioritize exploring the long-term effects of these interactions on soil health and their role in enhancing soil resilience to extreme environmental conditions (Kumar et al., 2024).

2.8 Inert Dusts and Their Synergy with Microbial Consortia in Water Retention

Water scarcity is one of the primary challenges faced by modern agriculture, particularly in arid and semiarid regions (Tubeileh & Thomas, 2023). Among the strategies to mitigate this issue, the use of inert dusts (IDs) has stood out due to their specific properties that can contribute to water retention in the soil (Petoumenou, 2023). Each type of IDs exhibits unique physical and chemical characteristics, such as adsorption capacity and porosity, which influence their effectiveness in conserving soil moisture and reducing evaporation (Subramanyam & Roesli, 2000; Kadhim Al-Musawi, 2024).

IDs have various industrial and agricultural uses, one of the most significant being as insecticide dust diluents and carriers (Subramanyam & Roesli, 2000). Additionally, the use of IDs has gained relevance as a sustainable strategy to control pests, improve the physical and chemical properties of soil, particularly its water retention capacity, and provide a favorable environment for plant growth by increasing moisture content and nutrient availability (Kadhim Al-Musawi, 2024).

In agriculture, IDs like bentonite, kaolin, zeolites, silicates, basalt dust, and agricultural gypsum have been widely studied because they not only act as soil conditioners but also interact synergistically with microbial consortia to enhance soil functionality (Datta et al., 2020; Li et al., 2022). The choice of these dusts depends on factors such as soil texture, climate, and the crops involved, enabling a tailored management approach for agricultural systems.

The clay mineral bentonite has received considerable attention for its potential to reduce the harmful effects of drought on soil health and biological activity. It also has a high cation-exchange capacity, which allows it to retain water and nutrients in its layered structure (Murray, 2006; Al-Taey et al., 2023). Kaolin, on the other hand, forms a protective barrier against direct evaporation. Kaolin particles catalyze the degradation of ozone and increase the surface area of the particle film, providing a habitat for microbial populations (Brito et al., 2019). Zeolites, with their microporous crystalline network, improve soil porosity, facilitating water infiltration and storage (Eroglu et al., 2017). Moreover, they increase soil microbial biomass and the incorporation of added ^{14}C into microbial biomass (Cataldo et al., 2021). Silicates and basalt dust supply essential minerals, stabilize soil aggregates, and promote microbial development (Rogers & Bennett, 2004; Li et al., 2021a; Brindavathy, 2022). Finally, agricultural gypsum acts as an amendment to correct sodic soils, increasing permeability and retaining water in critical root zones (Green et al., 2023; El-Sharkawy et al., 2024).

Meanwhile, microbial consortia, composed of bacteria, fungi, and actinobacteria, play a crucial role in soil water dynamics (Gautam et al., 2024). These microorganisms not only degrade organic matter to release nutrients but also produce polysaccharides that improve soil aggregation (Agnihotri et al., 2022). Soil microorganisms, such as PGPR and AMF, produce exudates that can interact with inert powder particles. IDs can act as adsorbents, retaining nutrients or organic compounds, and can also nucleate and precipitate on cell walls, leading to biomineralization processes in which living organisms produce or induce minerals (Cygan & Tazaki, 2014).

The incorporation of IDs can amplify these microbial activities by providing porous surfaces that host diverse indigenous microorganisms, often spore-forming (Bartak et al., 2023). For example, zeolites and basalt dust enrich the soil with micronutrients that stimulate microbial metabolism, while kaolin and bentonite stabilize moisture conditions that favor microbial activity (Cygan & Tazaki, 2014; Eroglu et al., 2017; Li et al., 2019; Li et al., 2021a). These materials also reduce water and nutrient leaching by forming microenvironments that promote moisture accumulation in soil pores (Soca & Daza-Torres, 2016). Additionally, compounds released by microorganisms, such as organic acids, interact with IDs to increase water retention through physicochemical bonds (Nakhli et al., 2017; Fomina & Skorochod, 2020).

Moreover, zeolite plays a significant role in promoting the activity and diversity of plant growth-promoting microorganisms (Enamorado-Horrutiner et al., 2016; Ibrahim et al., 2023; Khan et al., 2024). Zeolites have been shown to increase microbial diversity in the soil by serving as a "home" for microorganisms, similar to the role clays play in natural ecosystems (Budiyanto, 2020). Under conditions of water stress, microorganisms can take refuge within the porous structure of zeolite particles, as they do with clays, ensuring their survival until environmental conditions improve. In fact, under conditions of water stress, microorganisms take refuge inside the clay particles until the environmental conditions are suitable to colonize the soil once more (Domenico, 2020).

In vineyard soils, for instance, zeolite application has been shown to enhance the cation-exchange capacity, thereby improving nutrient availability. This also induces changes in soil organic matter and stimulates microbial metabolic activity, as evidenced by increased dehydrogenase activity (Doni et al., 2021; Cataldo et al., 2021; Khan et al., 2024). Additionally, Bikkinina et al. (2020) evaluated the effects of nanostructured zeolites on soil productivity, confirming that their use increases soil respiratory activity, likely due to the growth of microbial biomass. Similarly, Wolny-Koładka et al. (2023) studied the effects of zeolite-carbon composites, leonardite, and lignite on soil microorganisms. They concluded that the combination of NPK fertilization and mineral-organic mixtures containing lignite, leonardite, and zeolite-carbon composites significantly influenced microbial populations in comparison to unfertilized soil in crops such as spring rape and spring wheat. The abundance of various microbial groups, including bacteria, mold fungi, actinomycetes, *Azotobacter* spp., and ammonifiers, varied depending on the type and amount of fertilizer applied as well as the cultivated plant species.

Zeolites also play a role in stimulating AMF root colonization (Baghaie et al., 2019). However, the interaction between AMF symbiosis and zeolite remains poorly understood. Sarmiento-López et al. (2025) demonstrated that combining nitrogen-enriched zeolite with AMF symbiosis enhances maize growth. This effect is attributed to AMF activating high-affinity mycorrhiza-specific ammonium transporters during symbiosis, improving nitrogen uptake via the mycorrhizal pathway. This growth promotion is directly linked to the high ion-exchange capacity of zeolite, which facilitates nutrient availability.

Integrating IDs with microbial consortia represents a promising strategy to address agricultural challenges associated with climate change and water scarcity. Future research should focus on optimizing the proportions and combinations of these materials and microorganisms, as well as developing formulations tailored to different soil types and crops. It is crucial to adopt sustainable water and nutrient management practices to ensure the long-term viability of agricultural systems, supporting food security and the livelihoods of farming communities (Sabir et al., 2024).

2.9 Leading Countries in Microbial Consortia Research

Authors who have contributed to research on microbial consortia aimed at increasing crop resilience to drought have affiliations in a total of 31 countries across five continents (Figure 2.3). The geographical distribution of these publications reflects a notable concentration of research efforts in Asia, the Americas, and Europe, which is evidence of the global interest in developing advanced biotechnological strategies to improve agricultural resilience to water stress.

The concentration of research is related to significant agricultural challenges in countries such as India, China, Chile, Australia, and Germany, whose inhabitants face conditions of extreme climate variability, aridity, and high agricultural demand. These factors directly influence the interest in developing adaptive biotechnology solutions.

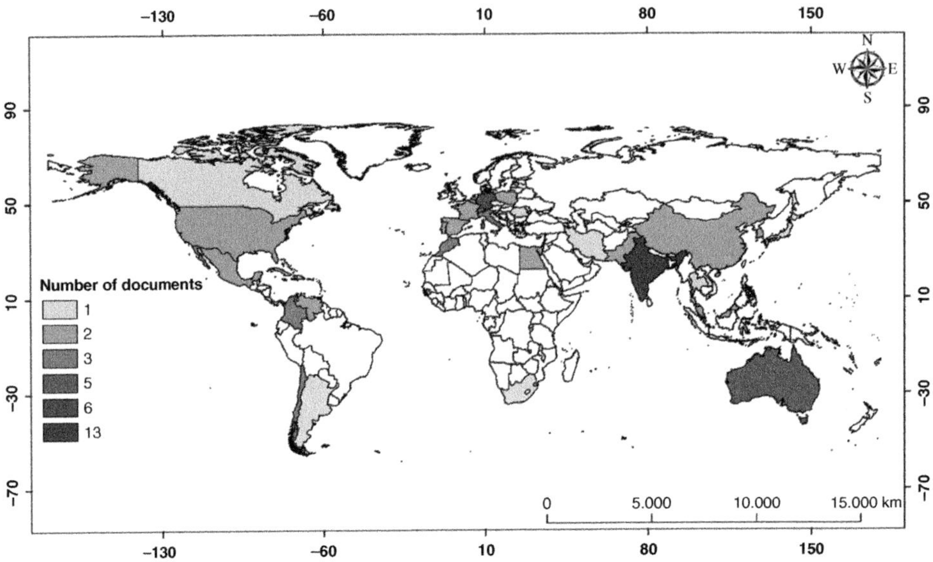

Figure 2.3 Scientific production according to authors' countries of affiliation.

In Asia, India and China's leadership responds to the need to maintain agricultural productivity in environments with limited access to water. Similarly, in the Americas, countries such as Chile and Mexico have focused on improving crop resilience due to their high dependence on species that are vulnerable to climate change, especially in arid regions. In Oceania, Australia stands out for its research aimed at mitigating the effects of extreme drought in low-fertility environments. Finally, Germany leads the way in Europe, showing sustained interest in the development of biological technologies that can be applied to sustainable agriculture. Together, these countries have promoted research aimed at ensuring productive stability and the adaptation of agricultural systems to severe water stress conditions.

2.10 Temporal Dynamics of Keywords in Microbial Consortia Studies

Figure 2.4 presents the evolution of the main research topics related to microbial consortia for water stress, showing the permanence of the term "climate change" as a central theme from the early years to the present day. This term reflects the continuing interest of the scientific community in researching biotechnological solutions to the effects of climate change on agriculture. From 2014 to 2017, studies focused on exploring microbial biodiversity and its potential use in mitigating the effects of climate change, with prominent terms including "bacterial diversity," "microbial diversity," and "*Bacillus*."

From 2018 onward, a shift toward applied research was identified. This was reflected in terms such as biofertilizers, biostimulants, and plant growth promotion, indicating a growing interest in the development of microbial technologies that promote plant growth and improve crop resistance to adverse conditions. In the most recent period (2020–2024),

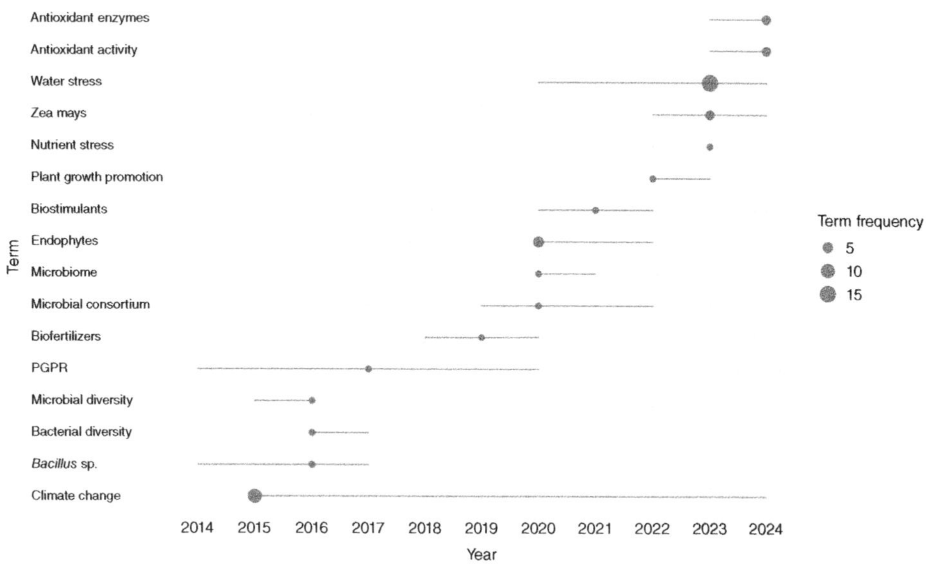

Figure 2.4 Temporal evolution of keywords.

predominant terms such as antioxidant enzymes, antioxidant activity, endophytes, and microbiome suggest a focus on the biochemical and physiological mechanisms by which microbial consortia contribute to crop resistance and adaptation to hostile environments.

2.11 Conclusions or Future Prospective

Microbial consortia represent a promising biotechnological strategy for improving agricultural resilience to drought stress. These communities of microorganisms interact synergistically to optimize water and nutrient uptake, mitigate oxidative stress, and stabilize soil structure, thereby enhancing crop productivity even in arid environments. Their ability to produce secondary metabolites, such as exopolysaccharides and phytohormones, plays a crucial role in protecting plants against abiotic stress factors. Furthermore, the integration of microbial consortia with IDs offers an innovative strategy to improve soil water retention and enhance microbial activity, maximizing its benefits on crop growth and resilience.

Despite these advances, the mechanisms underlying the interactions between microbial consortia, plants, and soils are still not fully understood. It is essential to investigate the functional and metabolic compatibility of consortia members, as well as the impact of environmental factors on their performance. Emerging biotechnological tools offer great potential for optimizing the design of synthetic and semisynthetic consortia tailored to specific agricultural needs.

Future research should focus on the genetic, biochemical, and ecological factors that influence these interactions. This integrated approach will allow us to understand the full potential of microbial consortia, contributing to sustainable agricultural practices and ensuring global food security in the face of climate change.

References

Aghai, M.M., Khan, Z., Joseph, M.R., Stoda, A.M., Sher, A.W., Ettl, G.J., & Doty, S.L. 2019. The effect of microbial endophyte consortia on *Pseudotsuga menziesii* and *Thuja plicata* survival, growth, and physiology across edaphic gradients. *Front. Microbiol.* 10, 1353. https://doi.org/10.3389/fmicb.2019.01353

Agnihotri, R., Sharma, M.P., Prakash, A., Ramesh, A., Bhattacharjya, S., Patra, A.K., Manna, M.C., Kurganova, I., & Kuzyakov, Y. 2022. Glycoproteins of arbuscular mycorrhiza for soil carbon sequestration: review of mechanisms and controls. *Sci. Total. Environ.* 806, 150571. https://doi.org/10.1016/j.scitotenv.2021.150571

Aguilar-Paredes, A., Valdés, G., & Nuti, M. 2020. Ecosystem functions of microbial consortia in sustainable agriculture. *Agronomy* 10, 1902. https://doi.org/10.3390/agronomy10121902

Allison, S.D. 2023. Microbial drought resistance may destabilize soil carbon. *Trends. Microbiol.* 31, 780–787. https://doi.org/10.1016/j.tim.2023.03.002

Al-Taey, D.K.A., Hussain, A.J., & Kadhum, H.J. 2023. Bentonite impact on soil properties and biological activity in the face of drought : a review. *IOP Conf. Ser. Earth. Environ. Sci.* 1262, 042058. https://doi.org/10.1088/1755-1315/1262/4/042058

Antoniewicz, M.R. 2020. A guide to deciphering microbial interactions and metabolic fluxes in microbiome communities. *Curr. Opin. Biotechnol.*, Analytical Biotechnology 64, 230–237. https://doi.org/10.1016/j.copbio.2020.07.001

Aulakh, A.M., Qadir, G., Hassan, F.U., Hayat, R., Sultan, T., Bibi, R., Ahmed, S., Yunas, M., Itikhar, M., & Malik, S.N. 2023. Role of resilient soil microbial consortia on growth and yield attributes of chickpea under water stress. *Egypt. J. Agron.* 45, 99–110. https://doi.org/10.21608/agro.2023.159807.1336

Baghaie, A.H., Aghili, F., & Jafarinia, R. 2019. Soil-indigenous arbuscular mycorrhizal fungi and zeolite addition to soil synergistically increase grain yield and reduce cadmium uptake of bread wheat (through improved nitrogen and phosphorus nutrition and immobilization of Cd in roots). *Environ. Sci. Pollut. Res.* 26, 30794–30807. https://doi.org/10.1007/s11356-019-06237-0

Bais, H.P., Weir, T.L., Perry, L.G., Gilroy, S., & Vivanco, J.M. 2006. The role of root exudates in rhizosphere interactions with plants and other organisms. *Annu. Rev. Plant Biol.* 57, 233–266. https://doi.org/10.1146/annurev.arplant.57.032905.105159

Barea, J.M., Azcón, R., & Azcón-Aguilar, C. 2005. Interactions between mycorrhizal fungi and bacteria to improve plant nutrient cycling and soil structure. In: A. Varma, & F. Buscot (Eds.), *Microorganisms in Soils: Roles in Genesis and Functions* (pp. 195–212). Berlin, Heidelberg: Springer. https://doi.org/10.1007/3-540-26609-7_10

Bartak, D., Bedrníková, E., Kašpar, V., Říha, J., Hlaváčková, V., Večerník, P., Šachlová, Š., & Černá, K. 2023. Survivability and proliferation of microorganisms in bentonite with implication to radioactive waste geological disposal: strong effect of temperature and negligible effect of pressure. *World J. Microbiol. Biotechnol.* 40, 41. https://doi.org/10.1007/s11274-023-03849-0

Behera, B., Das, T.K., Raj, R., Ghosh, S., Raza, M.B., & Sen, S. 2021. Microbial consortia for sustaining productivity of non-legume crops: prospects and challenges. *Agric. Res.* 10, 1–14. https://doi.org/10.1007/s40003-020-00482-3

Bengtsson, J., Setälä, H., & Zheng, D.W. 1996. Food webs and nutrient cycling in soils: interactions and positive feedbacks. In: G.A. Polis, & K.O. Winemiller (Eds.), *Food Webs: Integration of Patterns & Dynamics* (pp. 30–38). US, Boston, MA: Springer. https://doi.org/10.1007/978-1-4615-7007-3_3

Bernstein, H.C., & Carlson, R.P. 2012. Microbial consortia engineering for cellular factories: *in vitro* to *in silico* systems. *Comput. Struct. Biotechnol. J.* 3, e201210017. https://doi.org/10.5936/csbj.201210017

Bhatia, S.K., Bhatia, R.K., Choi, Y.-K., Kan, E., Kim, Y.-G., & Yang, Y.-H. 2018. Biotechnological potential of microbial consortia and future perspectives. *Crit. Rev. Biotechnol.* 38, 1209–1229. https://doi.org/10.1080/07388551.2018.1471445

Bikkinina, L.M.-H., Ezhkov, V.O., Faizrakhmanov, R.N., Gazizov, R.R., & Ezhkova, A.M. 2020. Effect of zeolites on soil modification and productivity. *BIO Web Conf.* 17, 00117. https://doi.org/10.1051/bioconf/20201700117

Blanco, E.L., Rada, F., Paolini, J., & Guerrero, J.A. 2021. Effects of induced water deficit and biofertilization on growth dynamics and bulb yield of onion (*Allium cepa* L.) in a neotropical semi-arid environment. *Can. J. Soil Sci.* 101, 494–506. https://doi.org/10.1139/cjss-2021-0011

Bogati, K., & Walczak, M. 2022. The impact of drought stress on soil microbial community, enzyme activities and plants. *Agronomy* 12, 189. https://doi.org/10.3390/agronomy12010189

Bonkowski, M., Villenave, C., & Griffiths, B. 2009. Rhizosphere fauna: the functional and structural diversity of intimate interactions of soil fauna with plant roots. *Plant. Soil.* 321, 213–233. https://doi.org/10.1007/s11104-009-0013-2

Borrelli, V.M.G., Brambilla, V., Rogowsky, P., Marocco, A., & Lanubile, A. 2018. The enhancement of plant disease resistance using CRISPR/Cas9 technology. *Front. Plant Sci.* 9, 1245. https://doi.org/10.3389/fpls.2018.01245

Bradáčová, K., Florea, A.S., Bar-Tal, A., Minz, D., Yermiyahu, U., Shawahna, R., Kraut-Cohen, J., Zolti, A., Erel, R., Dietel, K., Weinmann, M., Zimmermann, B., Berger, N., Ludewig, U., Neumann, G., & Poşta, G. 2019. Microbial consortia versus single-strain inoculants: an advantage in PGPM-assisted tomato production? *Agronomy* 9, 105. https://doi.org/10.3390/agronomy9020105

Brenner, K., You, L., & Arnold, F.H. 2008. Engineering microbial consortia: a new frontier in synthetic biology. *Trends Biotechnol.* 26, 483–489. https://doi.org/10.1016/j.tibtech.2008.05.004

Brindavathy, R. 2022. Silicate minerals induced by microorganisms. In: A. Berenjian, & M. Seifan (Eds.), *Mineral Formation by Microorganisms: Concepts and Applications* (pp. 125–159). Cham: Springer International Publishing. https://doi.org/10.1007/978-3-030-80807-5_5

Brito, C., Dinis, L.-T., Moutinho-Pereira, J., & Correia, C. 2019. Kaolin, an emerging tool to alleviate the effects of abiotic stresses on crop performance. *Sci. Hortic.* 250, 310–316. https://doi.org/10.1016/j.scienta.2019.02.070

Budiyanto, G. 2020. The application of zeolite to increase nitrogen use efficiency in corn vegetative growth in coastal sandy soils. *PLANTA Trop.* 8, 1–6. https://doi.org/10.18196/pt.2020.107.1-6

Calvo-Polanco, M., Sánchez-Romera, B., Aroca, R., Asins, M.J., Declerck, S., Dodd, I.C., Martínez-Andújar, C., Albacete, A., & Ruiz-Lozano, J.M. 2016. Exploring the use of recombinant inbred lines in combination with beneficial microbial inoculants (AM fungus and PGPR) to improve drought stress tolerance in tomato. *Environ. Exp. Bot.* 131, 47–57. https://doi.org/10.1016/j.envexpbot.2016.06.015

Cano, M.A. 2011. Interacción de microorganismos benéficos en plantas: micorrizas, *Trichoderma* spp. y *Pseudomonas* spp. UNA revisión. *Rev. UDCA Actual. Divulg. Científica*. 14, 15–31.

Cataldo, E., Salvi, L., Paoli, F., Fucile, M., Masciandaro, G., Manzi, D., Masini, C.M., & Mattii, G.B. 2021. Application of zeolites in agriculture and other potential uses: a review. *Agronomy* 11, 1547. https://doi.org/10.3390/agronomy11081547

Che, S., & Men, Y. 2019. Synthetic microbial consortia for biosynthesis and biodegradation: promises and challenges. *J. Ind. Microbiol. Biotechnol.* 46, 1343–1358. https://doi.org/10.1007/s10295-019-02211-4

Cheng, S., Zou, Y.-N., Kuča, K., Hashem, A., Abd_Allah, E.F., & Wu, Q.-S. 2021. Elucidating the mechanisms underlying enhanced drought tolerance in plants mediated by arbuscular mycorrhizal fungi. *Front. Microbiol.* 12, 809473. https://doi.org/10.3389/fmicb.2021.809473

Cirillo, V., Romano, I., Woo, S.L., Di Stasio, E., Lombardi, N., Comite, E., Pepe, O., Ventorino, V., & Maggio, A. 2023. Inoculation with a microbial consortium increases soil microbial diversity and improves agronomic traits of tomato under water and nitrogen deficiency. *Front. Plant. Sci.* 14, 1304627. https://doi.org/10.3389/fpls.2023.1304627

Compant, S., Samad, A., Faist, H., & Sessitsch, A. 2019. A review on the plant microbiome: Ecology, functions, and emerging trends in microbial application. *J. Adv. Res.* 19, 29–37. https://doi.org/10.1016/j.jare.2019.03.004

Culliney, T.W. 2013. Role of arthropods in maintaining soil fertility. *Agriculture* 3, 629–659. https://doi.org/10.3390/agriculture3040629

Cygan, R.T., & Tazaki, K. 2014. Interactions of kaolin minerals in the environment. *Elements* 10, 195–200. https://doi.org/10.2113/gselements.10.3.195

Damodaran, T., Jha, S.K., Kumari, S., Gupta, G., Mishra, V.K., Sharma, P.C., Gopal, R., Singh, A., & Jat, H.S. 2023. Development of halotolerant microbial consortia for salt stress mitigation and sustainable tomato production in sodic soils: an enzyme mechanism approach. *Sustainability* 15, 5186. https://doi.org/10.3390/su15065186

Datta, R., Holatko, J., Latal, O., Hammerschmiedt, T., Elbl, J., Pecina, V., Kintl, A., Balakova, L., Radziemska, M., Baltazar, T., Skarpa, P., Danish, S., Zafar-ul-Hye, M., Vyhnanek, T., & Brtnicky, M. 2020. Bentonite-Based organic amendment enriches microbial activity in agricultural soils. *Land* 9, 258. https://doi.org/10.3390/land9080258

de Vries, F.T., Griffiths, R.I., Knight, C.G., Nicolitch, O., & Williams, A. 2020. Harnessing rhizosphere microbiomes for drought-resilient crop production. *Science* 368, 270–274. https://doi.org/10.1126/science.aaz5192

Dhooria, M.S. 2016. Soil mites. In: M.S. Dhooria (Ed.), *Fundamentals of Applied Acarology* (pp. 197–206). Singapore: Springer. https://doi.org/10.1007/978-981-10-1594-6_10

Domenico, P. 2020. Optimised fertilisation with zeolitites containing Plant Growth Promoting Rhizobacteria (PGPR) in *Ranunculus asiaticus*. *GSC Biol. Pharm. Sci.* 10, 096–102. https://doi.org/10.30574/gscbps.2020.10.1.0011

Doni, S., Gispert, M., Peruzzi, E., Macci, C., Mattii, G.B., Manzi, D., Masini, C.M., & Grazia, M. 2021. Impact of natural zeolite on chemical and biochemical properties of vineyard soils. *Soil. Use. Manag.* 37, 832–842. https://doi.org/10.1111/sum.12665

Duncker, K.E., & Holmes, Z.A. You, L., 2021. Engineered microbial consortia: strategies and applications. *Microb. Cell. Factories.* 20, 211. https://doi.org/10.1186/s12934-021-01699-9

Efremenko, E., Stepanov, N., Senko, O., Aslanli, A., Maslova, O., & Lyagin, I. 2024. Using fungi in artificial microbial consortia to solve bioremediation problems. *Microorganisms* 12, 470. https://doi.org/10.3390/microorganisms12030470

El-Sharkawy, M., Alotaibi, M.O., Li, J., Mahmoud, E., Ghoneim, A.M., Ramadan, M.S., & Shabana, M. 2024. Effect of nano-zinc oxide, rice straw compost, and gypsum on wheat (*Triticum aestivum* L.) yield and soil quality in saline–sodic soil. *Nanomaterials* 14, 1450. https://doi.org/10.3390/nano14171450

Enamorado-Horrutiner, Y., Villanueva-Tagle, M.E., Behar, M., Rodríguez-Fuentes, G., Ferraz Dias, J., & Pomares-Alfonso, M.S. 2016. Cuban zeolite for lead sorption: application for water decontamination and metal quantification in water using nondestructive techniques. *Int. J. Environ. Sci. Technol.* 13, 1245–1256. https://doi.org/10.1007/s13762-016-0956-9

Eroglu, N., Emekci, M., & Athanassiou, C.G. 2017. Applications of natural zeolites on agriculture and food production. *J. Sci. Food. Agric.* 97, 3487–3499. https://doi.org/10.1002/jsfa.8312

Farrell, C., Szota, C., & Arndt, S.K. 2017. Does the turgor loss point characterize drought response in dryland plants? *Plant. Cell. Environ.* 40, 1500–1511. https://doi.org/10.1111/pce.12948

Flórián, N., Ladányi, M., Ittzés, A., Kröel-Dulay, G., Ónodi, G., Mucsi, M., Szili-Kovács, T., Gergócs, V., Dányi, L., & Dombos, M. 2019. Effects of single and repeated drought on soil microarthropods in a semi-arid ecosystem depend more on timing and duration than drought severity. *PLoS One* 14, e0219975. https://doi.org/10.1371/journal.pone.0219975

Fomina, M., & Skorochod, I. 2020. Microbial interaction with clay minerals and its environmental and biotechnological implications. *Minerals.* 10, 861. https://doi.org/10.3390/min10100861

Frąc, M., Hannula, E.S., Bełka, M., Salles, J.F., & Jedryczka, M. 2022. Soil mycobiome in sustainable agriculture. *Front. Microbiol.* 13. https://doi.org/10.3389/fmicb.2022.1033824

Gagné-Bourque, F., Bertrand, A., Claessens, A., Aliferis, K.A., & Jabaji, S. 2016. Alleviation of drought stress and metabolic changes in timothy (*Phleum pratense* L.) colonized with *Bacillus subtilis* B26. *Front. Plant Sci.* 7, 584. https://doi.org/10.3389/fpls.2016.00584

Gange, A. 2000. Arbuscular mycorrhizal fungi, Collembola and plant growth. *Trends. Ecol. Evol.* 15, 369–372. https://doi.org/10.1016/S0169-5347(00)01940-6

García-Fraile, P., Menéndez, E., Rivas, R., García-Fraile, P., Menéndez, E., & Rivas, R. 2015. Role of bacterial biofertilizers in agriculture and forestry. *AIMS Bioeng.* 2, 183–205. https://doi.org/10.3934/bioeng.2015.3.183

Gautam, S., Rathikannu, S., Bhagat, P.S., Singh, V.V., & Lakra, A.R. 2024. Dynamics of microbial ecology, particulate matter, and bacterial aerosols in agriculture: implications for health and sustainability. *Water. Air. Soil Pollut.* 235, 763. https://doi.org/10.1007/s11270-024-07569-5

Gehring, C., & Bennett, A. 2009. Mycorrhizal fungal–plant–insect interactions: the importance of a community approach*. *Environ. Entomol.* 38, 93–102. https://doi.org/10.1603/022.038.0111

Glick, B.R. 2012. Plant growth-promoting bacteria: mechanisms and applications. *Sci.* 2012, 963401. https://doi.org/10.6064/2012/963401

Green, H., Larsen, P., Koci, J., Edwards, W., & Nelson, P.N. 2023. Long-term effects of gypsum on the chemistry of sodic soils. *Soil. Tillage. Res.* 233, 105780. https://doi.org/10.1016/j.still.2023.105780

Hu, J., Wei, Z., Friman, V.-P., Gu, S., Wang, X., Eisenhauer, N., Yang, T., Ma, J., Shen, Q., Xu, Y., & Jousset, A. 2016. Probiotic diversity enhances rhizosphere microbiome function and plant disease suppression. *mBio* 7, e01790-16. https://doi.org/10.1128/mbio.01790-16

Ibrahim, E.A., El-Sherbini, M.A.A., & Selim, E.-M.M. 2023. Effects of biochar, zeolite and mycorrhiza inoculation on soil properties, heavy metal availability and cowpea growth in a multi-contaminated soil. *Sci. Rep.* 13, 6621. https://doi.org/10.1038/s41598-023-33712-z

Igiehon, O.N., & Babalola, O.O. 2021. Rhizobium and mycorrhizal fungal species improved soybean yield under drought stress conditions. *Curr. Microbiol.* 78, 1615–1627. https://doi.org/10.1007/s00284-021-02432-w

Inostroza, N.G., Barra, P.J., Wick, L.Y., Mora, M.L., & Jorquera, M.A. 2017. Effect of rhizobacterial consortia from undisturbed arid- and agro-ecosystems on wheat growth under different conditions. *Lett. Appl. Microbiol.* 64, 158–163. https://doi.org/10.1111/lam.12697

Jalal, A., Oliveira, C.E. da S., Rosa, P.A.L., Galindo, F.S., & Teixeira Filho, M.C.M. 2023. Beneficial microorganisms improve agricultural sustainability under climatic extremes. *Life.* 13, 1102. https://doi.org/10.3390/life13051102

Johnson, S.N., & Rasmann, S. 2015. Root-feeding insects and their interactions with organisms in the rhizosphere. *Annu. Rev. Entomol.* 60, 517–535. https://doi.org/10.1146/annurev-ento-010814-020608

Jones, D.L., Nguyen, C., & Finlay, R.D. 2009. Carbon flow in the rhizosphere: carbon trading at the soil–root interface. *Plant. Soil.* 321, 5–33. https://doi.org/10.1007/s11104-009-9925-0

Justin, S., Lihan, S., Elvis-Sulang, M., & Chiew, T. 2020. Formulated microbial consortium as inoculant for agarwood induction. *J. Trop. For. Sci.* 32, 161–169.

Kadhim Al-Musawi, A.A. 2024. Evaluating effectiveness of some inert dusts and chemical pesticides in *Tribolium castaneum* coleoptera: tenebrindae under laboratory conditions. *IOP Conf. Ser. Earth. Environ. Sci.* 1371, 032025. https://doi.org/10.1088/1755-1315/1371/3/032025

Kalaji, H.M., Elshery, N., & Mojski, J. 2018. Effect of microbial consortia on photosynthetic efficiency of Arabidopsis thaliana under drought stress. *Chiang. Mai. J. Sci.* 45, 1–10.

Kalefetoğlu, T., & Ekmekçi, Y. 2010. The effects of drought on plants and tolerance mechanisms. *Gazi. Univ. J. Sci.* 18, 723–740.

Kaushal, M., & Wani, S.P. 2016. Plant-growth-promoting rhizobacteria: drought stress alleviators to ameliorate crop production in drylands. *Ann. Microbiol.* 66, 35–42. https://doi.org/10.1007/s13213-015-1112-3

Kaya, C., Uğurlar, F., Ashraf, M., Hou, D., Kirkham, M.B., & Bolan, N. 2024. Microbial consortia-mediated arsenic bioremediation in agricultural soils: current status, challenges, and solutions. *Sci. Total. Environ.* 917, 170297. https://doi.org/10.1016/j.scitotenv.2024.170297

Khan, M.R., Tarafder, M.A., & Priti, M.S. 2024. Potential use of zeolites in agriculture: a review. *SAARC J. Agric.* 22, 17–30. https://doi.org/10.3329/sja.v22i1.72433

Koltz, A.M., Koyama, A., & Wallenstein, M. 2024. Warming alters cascading effects of a dominant arthropod predator on fungal community composition in the Arctic. *mBio* 15, e00590-24. https://doi.org/10.1128/mbio.00590-24

Krishna, R., Jaiswal, D.K., Ansari, W.A., Singh, S., Soumia, P.S., Singh, A.K., Kumari, B., Singh, M., & Verma, J.P. 2022. Potential microbial consortium mitigates drought stress in tomato (*Solanum lycopersicum* L.) plant by up-regulating stress-responsive genes and improving fruit yield and soil properties. *J. Soil Sci. Plant Nutr.* 22, 4598–4615. https://doi.org/10.1007/s42729-022-00929-2

Kumar, D., Kour, S., Ali, M., Sharma, R., Parkirti, Vikram, Changotra, H., Manhas, R.K., & Ohri, P. 2024. Interactions between arbuscular mycorrhizal fungi and other microorganisms in the rhizosphere and hyphosphere. In: G.J. Ahammed, & R. Hajiboland (Eds.), *Arbuscular Mycorrhizal Fungi and Higher Plants: Fundamentals and Applications* (pp. 37–66). Singapore: Springer Nature. https://doi.org/10.1007/978-981-99-8220-2_3

Lakshmi, G., Okafor, B.N., & Visconti, D. 2020. Soil microarthropods and nutrient cycling. In: S. Fahad, M. Hasanuzzaman, M. Alam, H. Ullah, M. Saeed, I. Ali Khan, & M. Adnan (Eds.), *Environment, Climate, Plant and Vegetation Growth* (pp. 453–472). Cham: Springer International Publishing. https://doi.org/10.1007/978-3-030-49732-3_18

Larsen, J., Jaramillo-López, P., Nájera-Rincon, M., & González-Esquivel, C.E. 2015. Biotic interactions in the rhizosphere in relation to plant and soil nutrient dynamics. *J. Soil. Sci. Plant. Nutr.* 15, 449–463. https://doi.org/10.4067/S0718-95162015005000039

Latef, A.A.H.A., Hashem, A., Rasool, S., Abd_Allah, E.F., Alqarawi, A.A., Egamberdieva, D., Jan, S., Anjum, N.A., & Ahmad, P. 2016. Arbuscular mycorrhizal symbiosis and abiotic stress in plants: a review. *J. Plant. Biol.* 59, 407–426. https://doi.org/10.1007/s12374-016-0237-7

Li, G.L., Zhou, C.H., Fiore, S., & Yu, W.H. 2019. Interactions between microorganisms and clay minerals: new insights and broader applications. *Appl. Clay. Sci.* 177, 91–113. https://doi.org/10.1016/j.clay.2019.04.025

Li, H., Guo, Q., Jing, Y., Liu, Z., Zheng, Z., Sun, Y., Xue, Q., & Lai, H. 2020. Application of Streptomyces pactum Act12 enhances drought resistance in wheat. *J. Plant. Growth. Regul.* 39, 122–132. https://doi.org/10.1007/s00344-019-09968-z

Li, J., Mavrodi, D.V., & Dong, Y. 2021a. Effect of rock dust-amended compost on the soil properties, soil microbial activity, and fruit production in an apple orchard from the Jiangsu province of China. *Arch. Agron. Soil. Sci.* 67, 1313–1326. https://doi.org/10.1080/03650340.2020.1795136

Li, S., Li, G., Peng, K., Yang, L., Huang, X., Lu, L., & Liu, J. 2022. The combined effect of *Diversispora versiformis* and sodium bentonite contributes to the colonization of *Phragmites* in cadmium-contaminated soil. *Chemosphere.* 293, 133613. https://doi.org/10.1016/j.chemosphere.2022.133613

Li, W., Wang, H., Lv, G., Wang, J., & Li, J. 2024. Regulation of drought stress on nutrient cycle and metabolism of rhizosphere microorganisms in desert riparian forest. *Sci. Total. Environ.* 954, 176148. https://doi.org/10.1016/j.scitotenv.2024.176148

Li, X., Wu, S., Dong, Y., Fan, H., Bai, Z., & Zhuang, X. 2021b. Engineering microbial consortia towards bioremediation. *Water* 13, 2928. https://doi.org/10.3390/w13202928

Malhi, G.S., Kaur, M., & Kaushik, P. 2021. Impact of climate change on agriculture and its mitigation strategies: a review. *Sustainability* 13, 1318. https://doi.org/10.3390/su13031318

Mamun, A.A., Neumann, G., Moradtalab, N., Ahmed, A., Dupuis, B., Darbon, G., Nawaz, F., Declerck, S., Mai, K., Vogt, W., Ludewig, U., & Weinmann, M. 2024. Microbial consortia versus single-strain inoculants as drought stress protectants in potato affected by the form of N supply. *Horticulturae* 10, 102. https://doi.org/10.3390/horticulturae10010102

Massot, F., Bernard, N., Alvarez, L.M.M., Martorell, M.M., Mac Cormack, W.P., & Ruberto, L.A.M. 2022. Microbial associations for bioremediation. What does "microbial consortia" mean? *Appl. Microbiol. Biotechnol.* 106, 2283–2297. https://doi.org/10.1007/s00253-022-11864-8

Mikiciuk, G., Miller, T., Kisiel, A., Cembrowska-Lech, D., Mikiciuk, M., Łobodzińska, A., & Bokszczanin, K. 2024. Harnessing beneficial microbes for drought tolerance: a review of ecological and agricultural innovations. *Agriculture* 14, 2228. https://doi.org/10.3390/agriculture14122228

Molina-Romero, D., Baez, A., Quintero-Hernández, V., Castañeda-Lucio, M., Fuentes-Ramírez, L.E., Bustillos-Cristales, M. del R., Rodríguez-Andrade, O., Morales-García, Y.E., Munive, A., & Muñoz-Rojas, J. 2017. Compatible bacterial mixture, tolerant to desiccation, improves maize plant growth. *PLoS One* 12, e0187913. https://doi.org/10.1371/journal.pone.0187913

Muniyappan, V.K., Sundaralingam, K., Sivakumar, U., Geetha, V.V., Jerlin, R., Shobana, N., Prasad, V.B.R., & Murugeshwari, T. 2024. Enhancing plant resilience and drought stress in green gram through seed priming with nodule-associated plant probiotics. *Plant. Sci. Today* 11(4), 1406–1414. 11. https://doi.org/10.14719/pst.4603

Murray, H.H. 2006. Chapter 6 bentonite applications. In: H.H. Murray (Ed.), *Developments in Clay Science, Applied Clay Mineralogy* (pp. 111–130). Elsevier. https://doi.org/10.1016/S1572-4352(06)02006-X

Nakhli, S.A.A., Delkash, M., Bakhshayesh, B.E., & Kazemian, H. 2017. Application of zeolites for sustainable agriculture: a review on water and nutrient retention. *Water. Air. Soil. Pollut.* 228, 464. https://doi.org/10.1007/s11270-017-3649-1

Naylor, D., & Coleman-Derr, D. 2018. Drought stress and root-associated bacterial communities. *Front. Plant. Sci.* 8, 2223. https://doi.org/10.3389/fpls.2017.02223

Neemisha, 2020. Role of soil organisms in maintaining soil health, ecosystem functioning, and sustaining agricultural production. In: B. Giri, & A. Varma (Eds.), *Soil Health* (pp. 313–335). Cham: Springer International Publishing. https://doi.org/10.1007/978-3-030-44364-1_17

Neher, D.A. & Barbercheck, M.E. 2019. Soil microarthropods and soil health: intersection of decomposition and pest suppression in agroecosystems. *Insects* 10, 414. https://doi.org/10.3390/insects10120414

Nishu, S.D., No, J.H., & Lee, T.K. 2022. Transcriptional response and plant growth promoting activity of *Pseudomonas fluorescens* DR397 under drought stress conditions. *Microbiol. Spectr.* 10, e00979-22. https://doi.org/10.1128/spectrum.00979-22

Nivedita, S., Behera, S.S., Behera, P.K., Parwez, Z., Giri, S., Behera, H.T., & Ray, L. 2024. Salt-resistant *Streptomyces* consortia promote growth of rice (*Oryza sativa* var. Swarna) alleviating salinity and drought stress tolerance by enhancing photosynthesis, antioxidant function, and proline content. *The Microbe* 4, 100124. https://doi.org/10.1016/j.microb.2024.100124

Nunes, P.S.O., Lacerda-Junior, G.V., Mascarin, G.M., Guimarães, R.A., Medeiros, F.H.V., Arthurs, S., & Bettiol, W. 2024. Microbial consortia of biological products: do they have a future? *Biol. Control.* 188, 105439. https://doi.org/10.1016/j.biocontrol.2024.105439

Ortiz, N., Armada, E., Duque, E., Roldán, A., & Azcón, R. 2015. Contribution of arbuscular mycorrhizal fungi and/or bacteria to enhancing plant drought tolerance under natural soil conditions: effectiveness of autochthonous or allochthonous strains. *J. Plant. Physiol.* 174, 87–96. https://doi.org/10.1016/j.jplph.2014.08.019

Penesyan, A., Paulsen, I.T., Kjelleberg, S., & Gillings, M.R. 2021. Three faces of biofilms: a microbial lifestyle, a nascent multicellular organism, and an incubator for diversity. *Npj Biofilms. Microbiomes.* 7, 1–9. https://doi.org/10.1038/s41522-021-00251-2

Peng, Y., Xu, H., Shi, J., Wang, Z., Lv, J., Li, L., & Wang, X. 2024. Soil microbial composition, diversity, and network stability in intercropping versus monoculture responded differently to drought. *Agric. Ecosyst. Environ.* 365, 108915. https://doi.org/10.1016/j.agee.2024.108915

Pérez-Moncada, U.A., Santander, C., Ruiz, A., Vidal, C., Santos, C., & Cornejo, P. 2024. Design of microbial consortia based on arbuscular mycorrhizal fungi, yeasts, and bacteria to improve the biochemical, nutritional, and physiological status of strawberry plants growing under water deficits. *Plants* 13, 1556. https://doi.org/10.3390/plants13111556

Petoumenou, D.G. 2023. Enhancing yield and physiological performance by foliar applications of chemically inert mineral particles in a rainfed vineyard under mediterranean conditions. *Plants* 12, 1444. https://doi.org/10.3390/plants12071444

Philippot, L., Raaijmakers, J.M., Lemanceau, P., & van der Putten, W.H. 2013. Going back to the roots: the microbial ecology of the rhizosphere. *Nat. Rev. Microbiol.* 11, 789–799. https://doi.org/10.1038/nrmicro3109

Prasanna Kumar, B., Trimurtulu, N., Vijaya Gopal, A., & Nagaraju, Y. 2022. Impact of culturable endophytic bacteria on soil aggregate formation and peanut (*Arachis hypogaea* L.) growth and yield under drought conditions. *Curr. Microbiol.* 79, 308. https://doi.org/10.1007/s00284-022-03000-6

Pravisya, P., Jayaram, K.M., & Yusuf, A. 2019. Biotic priming with *Pseudomonas fluorescens* induce drought stress tolerance in *Abelmoschus esculentus* (L.) Moench (Okra). *Physiol. Mol. Biol. Plants.* 25, 101–112. https://doi.org/10.1007/s12298-018-0621-5

Rafedzi, E.A.K., Bharudin, I., Kamaruddin, S., Bakar, F.D.A., & Murad, A.M.A. 2024. Effectiveness of the Drought-tolerant Plant Growth Promoting Rhizobacteria (PGPR) that Supports Paddy Growth in Drought Condition. | EBSCOhost [WWW Document]. https://doi.org/10.22207/JPAM.18.1.13

Ram, R.M., Debnath, A., Negi, S., & Singh, H.B. 2022. Chapter 22 - use of microbial consortia for broad spectrum protection of plant pathogens: regulatory hurdles, present status and future prospects. In: A. Rakshit, V.S. Meena, P.C. Abhilash, B.K. Sarma, H.B. Singh, L. Fraceto, M. Parihar, & A.K. Singh (Eds.), *Biopesticides, Advances in Bio-Inoculant Science.* (pp. 319–335). Woodhead Publishing. https://doi.org/10.1016/B978-0-12-823355-9.00017-1

Reis, F., Magalhães, A.P., Tavares, R.M., Baptista, P., & Lino-Neto, T. 2021. Bacteria could help ectomycorrhizae establishment under climate variations. *Mycorrhiza* 31, 395–401. https://doi.org/10.1007/s00572-021-01027-4

Rigi, F., Saberi, M., & Ebrahimi, M. 2023. Improved drought tolerance in *Festuca ovina* L. using plant growth promoting bacteria. *J. Arid. Land.* 15, 740–755. https://doi.org/10.1007/s40333-023-0015-6

Rizzo, G.F., Al Achkar, N., Treccarichi, S., Malgioglio, G., Infurna, M.G., Nigro, S., Tribulato, A., & Branca, F. 2023. Use of bioinoculants affects variation in snap bean yield grown under deficit irrigation. *Agriculture* 13, 865. https://doi.org/10.3390/agriculture13040865

Rogers, J.R., & Bennett, P.C. 2004. Mineral stimulation of subsurface microorganisms: release of limiting nutrients from silicates. *Chem. Geol.* 203, 91–108. https://doi.org/10.1016/j.chemgeo.2003.09.001

Sabir, R.M., Sarwar, A., Shoaib, M., Saleem, A., Alhousain, M.H., Wajid, S.A., Rasul, F., Adnan Shahid, M., Anjum, L., Safdar, M., Muhammad, N.E., Waqas, R.M., Zafar, U., & Raza, A. 2024. Managing water resources for sustainable agricultural production. In: S. Kanga, S.K. Singh, K. Shevkani, V. Pathak, & B. Sajan (Eds.), *Transforming Agricultural Management for a Sustainable Future: Climate Change and Machine Learning Perspectives* (pp. 47–74). Switzerland, Cham: Springer Nature. https://doi.org/10.1007/978-3-031-63430-7_3

Sah, S.K., Reddy, K.R., & Li, J. 2016. Abscisic acid and abiotic stress tolerance in crop plants. *Front. Plant Sci.* 7, 571. https://doi.org/10.3389/fpls.2016.00571

Saleem, M., Nawaz, F., Hussain, M.B., & Ikram, R.M. 2021. Comparative effects of individual and consortia plant growth promoting bacteria on physiological and enzymatic mechanisms to confer drought tolerance in maize (*Zea mays* L.). *J. Soil. Sci. Plant. Nutr.* 21, 3461–3476. https://doi.org/10.1007/s42729-021-00620-y

Saleh, D., Sharma, M., Seguin, P., & Jabaji, S. 2020. Organic acids and root exudates of *Brachypodium distachyon*: effects on chemotaxis and biofilm formation of endophytic bacteria. *Can. J. Microbiol.* 66, 562–575. https://doi.org/10.1139/cjm-2020-0041

Santoyo, G., Guzmán-Guzmán, P., Parra-Cota, F.I., Santos-Villalobos, S. de los, Orozco-Mosqueda, M. del C., & Glick, B.R. 2021. Plant growth stimulation by microbial consortia. *Agronomy* 11, 219. https://doi.org/10.3390/agronomy11020219

Sarmiento-López, L.G., Matos-Alegria, A., Cesario-Solis, M.E., Tapia-Maruri, D., Goodwin, P.H., Quinto, C., Santana, O., & Cardenas, L. 2025. Combination of nitrogen-enriched zeolite and arbuscular mycorrhizal symbiosis to improve growth of maize (*Zea mays* L.). *Agronomy* 15, 156. https://doi.org/10.3390/agronomy15010156

Shaffique, S., Khan, M.A., Imran, M., Kang, S.-M., Park, Y.-S., Wani, S.H., & Lee, I.-J. 2022. Research progress in the field of microbial mitigation of drought stress in plants. *Front. Plant Sci.* 13, 870626. https://doi.org/10.3389/fpls.2022.870626

Shah, M.H., & Khan, M.R. 2024. Effectiveness of local isolates of *Trichoderma* spp. in imparting drought tolerance in rice, *Oryza. sativa*. *Sci. Rep.* 14, 17672. https://doi.org/10.1038/s41598-024-67991-x

Shang, J., & Liu, B. 2021. Application of a microbial consortium improves the growth of Camellia sinensis and influences the indigenous rhizosphere bacterial communities. *J. Appl. Microbiol.* 130, 2029–2040. https://doi.org/10.1111/jam.14927

Singh, A.K., Zhu, X., Chen, C., Wu, J., Yang, B., Zakari, S., Jiang, X.J., Singh, N., & Liu, W. 2022. The role of glomalin in mitigation of multiple soil degradation problems. *Crit. Rev. Environ. Sci. Technol.* 52, 1604–1638. https://doi.org/10.1080/10643389.2020.1862561

Singh, D., Thapa, S., Singh, J.P., Mahawar, H., Saxena, A.K., Singh, S.K., Mahla, H.R., Choudhary, M., Parihar, M., Choudhary, K.B., & Chakdar, H. 2024. Prospecting the potential of plant growth-promoting microorganisms for mitigating drought stress in crop plants. *Curr. Microbiol.* 81, 84. https://doi.org/10.1007/s00284-023-03606-4

Sinha, S., & Raghuwanshi, R. 2020. Estimation of wedelolactone in *Eclipta prostrata* (l.) l by high performance liquid chromatography after microbial. *Plant. Cell. Biotechnol. Mol. Biol.* 21, 89–96.

Soca, M., & Daza-Torres, M.C. 2016. Evaluación de fracciones granulométricas y dosis de zeolita para la agricultura. *Agrociencia* 50, 965–976.

Solanki, M.K., Joshi, N.C., Singh, P.K., Singh, S.K., Santoyo, G., de Azevedo, L.C.B. & Kumar, A. 2024. From concept to reality: transforming agriculture through innovative rhizosphere engineering for plant health and productivity. *Microbiol. Res.* 279, 127553.

Spaepen, S., & Vanderleyden, J. 2011. Auxin and plant-microbe interactions. *Cold. Spring. Harb. Perspect. Biol.* 3, a001438. https://doi.org/10.1101/cshperspect.a001438

Srinivasan, S., Jnana, A., & Murali, T.S. 2024. Modeling microbial community networks: methods and tools for studying microbial interactions. *Microb. Ecol.* 87, 56. https://doi.org/10.1007/s00248-024-02370-7

Stürmer, S.L., Bever, J.D., & Morton, J.B. 2018. Biogeography of arbuscular mycorrhizal fungi (*Glomeromycota*): a phylogenetic perspective on species distribution patterns. *Mycorrhiza* 28, 587–603. https://doi.org/10.1007/s00572-018-0864-6

Subramanyam, B., & Roesli, R. 2000. Inert dusts. In: B. Subramanyam, & D.W. Hagstrum (Eds.), *Alternatives to Pesticides in Stored-Product IPM* (pp. 321–380). US, Boston, MA: Springer. https://doi.org/10.1007/978-1-4615-4353-4_12

Thijs, S., Weyens, N., Sillen, W., Gkorezis, P., Carleer, R., & Vangronsveld, J. 2014. Potential for plant growth promotion by a consortium of stress-tolerant 2,4-dinitrotoluene-degrading bacteria: isolation and characterization of a military soil. *Microb. Biotechnol.* 7, 294–306. https://doi.org/10.1111/1751-7915.12111

Thomopoulos, S., Elsgaard, L., Munkholm, L.J., & Ravnskov, S. 2023. Evaluation of the relation between soil biomass of arbuscular mycorrhizal fungi and glomalin-related soil protein in conservation agriculture. *Soil. Biol. Biochem.* 187, 109222. https://doi.org/10.1016/j.soilbio.2023.109222

Tiwari, S., Lata, C., Chauhan, P.S., & Nautiyal, C.S. 2016. *Pseudomonas putida* attunes morphophysiological, biochemical and molecular responses in *Cicer arietinum* L. during drought stress and recovery. *Plant. Physiol. Biochem.* 99, 108–117. https://doi.org/10.1016/j.plaphy.2015.11.001

Tubeileh, A.M., & Thomas, R.J. 2023. The management of soil systems in arid and semi-arid climates. In: *Biological Approaches to Regenerative Soil Systems.* CRC Press.

Tyagi, J., Mishra, A., Kumari, S., Singh, S., Agarwal, H., Pudake, R.N., Varma, A., & Joshi, N.C. 2023. Deploying a microbial consortium of *Serendipita indica, Rhizophagus intraradices*, and *Azotobacter chroococcum* to boost drought tolerance in maize. *Environ. Exp. Bot.* 206, 105142. https://doi.org/10.1016/j.envexpbot.2022.105142

Venturi, V., & Keel, C. 2016. Signaling in the rhizosphere. *Trends. Plant. Sci.* 21, 187–198. https://doi.org/10.1016/j.tplants.2016.01.005

Verma, J.P., Jaiswal, D.K., Krishna, R., Prakash, S., Yadav, J., & Singh, V. 2018. Characterization and screening of thermophilic Bacillus strains for developing plant growth promoting consortium from hot spring of Leh and Ladakh region of India. *Front. Microbiol.* 9, 1293. https://doi.org/10.3389/fmicb.2018.01293

Villagran, E., Toro-Tobón, G., Velázquez, F.A., & Estrada-Bonilla, G.A. 2024. Integration of IoT technologies and high-performance phenotyping for climate control in greenhouses and mitigation of water deficit: a study of high-andean oat. *Agric. Eng.* 6, 4011–4040. https://doi.org/10.3390/agriengineering6040227

Vurukonda, S.S.K.P., Vardharajula, S., Shrivastava, M., & SkZ, A. 2016. Enhancement of drought stress tolerance in crops by plant growth promoting rhizobacteria. *Microbiol. Res.* 184, 13–24. https://doi.org/10.1016/j.micres.2015.12.003

Wang, M., Wang, Z., Guo, M., Qu, L., & Biere, A. 2023a. Effects of arbuscular mycorrhizal fungi on plant growth and herbivore infestation depend on availability of soil water and nutrients. *Front. Plant. Sci.* 14, 1101932. https://doi.org/10.3389/fpls.2023.1101932

Wang, X., Chi, Y., & Song, S. 2024. Important soil microbiota's effects on plants and soils: a comprehensive 30-year systematic literature review. *Front. Microbiol.* 15, 1347745. https://doi.org/10.3389/fmicb.2024.1347745

Wang, Z., Solanki, M.K., Kumar, A., Solanki, A.C., Pang, F., Ba, Z.X., Niu, J.Q. & Ren, Z.X. 2023b. Promoting plant resilience against stress by engineering root microenvironment with Streptomyces inoculants. *Microbiol. Res.* 277, 127509.

Wolny-Koładka, K., Marcińska-Mazur, L., Jarosz, R., Juda, M., Lošák, T., & Mierzwa-Hersztek, M. 2023. Effect of soil application of zeolite-carbon composite, leonardite and lignite on the microorganisms. *Ecol. Chem. Eng. S.* 29, 553–563. https://doi.org/10.2478/eces-2022-0040

Wu, Q.-S., & Zou, Y.-N. 2017. Arbuscular mycorrhizal fungi and tolerance of drought stress in plants. In: Q.-S. Wu (Ed.), *Arbuscular Mycorrhizas and Stress Tolerance of Plants* (pp. 25–41). Singapore: Springer. https://doi.org/10.1007/978-981-10-4115-0_2

Xu, X.-M., Jeffries, P., Pautasso, M., & Jeger, M.J. 2011. Combined use of biocontrol agents to manage plant diseases in theory and practice. *Phytopathology®*. 101, 1024–1031. https://doi.org/10.1094/PHYTO-08-10-0216

Yang, M., Song, Y., Ma, H., Li, Z., Ding, J., Yin, T., Niu, K., Sun, S., Qi, J., Lu, G., Fazal, A., Yang, Y., & Wen, Z. 2024. Unveiling the hidden world: how arbuscular mycorrhizal fungi and its regulated core fungi modify the composition and metabolism of soybean rhizosphere microbiome. *Environ. Microbiome.* 19, 78. https://doi.org/10.1186/s40793-024-00624-y

Zitlalpopoca-Hernandez, G., Najera-Rincon, M.B., del-Val, E., Alarcon, A., Jackson, T., & Larsen, J. 2017. Multitrophic interactions between maize mycorrhizas, the root feeding insect *Phyllophaga vetula* and the entomopathogenic fungus *Beauveria bassiana*. *Appl. Soil. Ecol.* 115, 38–43. https://doi.org/10.1016/j.apsoil.2017.03.014

Zoppellari, F., Malusà, E., Chitarra, W., Lovisolo, C., Spanna, F., & Bardi, L. 2014. Improvement of drought tolerance in maize (*Zea mays* L.) by selected rhizospheric microorganisms. *Ital. J. Agrometeorol.* 18, 5–18.

3

Impediments and Limitations in the Production and Use of Biofertilizers

*Gema Cárdenas-Flores, Ana Yareli Flores-Ramírez, Juan Edgardo Cortés-Aguilera, Martina Alejandra Chacón-López, Selene Aguilera-Aguirre, Rosa Isela Ortiz-Basurto, and Ulises Miguel López-García**

Agrobiotechnology and Electrochemistry Laboratory-Integral Food Research Laboratory, National Technological Institute of Mexico/Technological Institute of Tepic, Tepic, Nayarit, México

* Corresponding author: ulopez@tepic.tecnm.mx

3.1 Introduction

A continuously growing global population has placed increasing pressure on agriculture and ecosystems to meet food production demands (Vyas & Bansal, 2023). As a result, modern agriculture relies heavily on agrochemicals. However, concerns about their use primarily arise from improper applications, often by inexperienced or uninformed farmers. For instance, fertilizers should be applied only after assessing soil mineral deficiencies to ensure plants receive the necessary nutrients for growth (Kuila & Ghosh, 2022). However, in many cases, application rates far exceed recommended levels, even though plants can absorb only a fraction of these inputs. Excess agrochemicals leach into the environment, contaminating soil and water, disrupting the soil microbiome, and posing risks to animal and human health (Hu et al., 2024).

Increasing awareness of the environmental impact of chemical inputs has driven interest in using beneficial microorganisms to support plant health. These microorganisms can function as biocontrol agents, enhance plant disease resistance, or improve nutrition through biofertilizers, which increase nutrient availability via natural biological processes (Li et al., 2023; Wang et al., 2023; Pourhosseini et al., 2024). Biofertilizers optimize the absorption of essential nutrients such as nitrogen (N), phosphorus (P), and potassium (K) (Mejía-Guerra et al., 2025). Notable examples include (i) plant growth-promoting bacteria, which fix nitrogen and produce phytohormones (Mateos-Naranjo et al., 2025; Vafa et al., 2024); (ii) arbuscular mycorrhizal fungi, which enhance water and nutrient uptake (González-Mancilla et al., 2024); and (iii) algae, which act as biostimulants and improve resistance to environmental stress (Chatterjee et al., 2017). Despite the promising

Microbial Consortia in Plant Science and Sustainable Agriculture, First Edition. Edited by Ajay Tomar, et al.

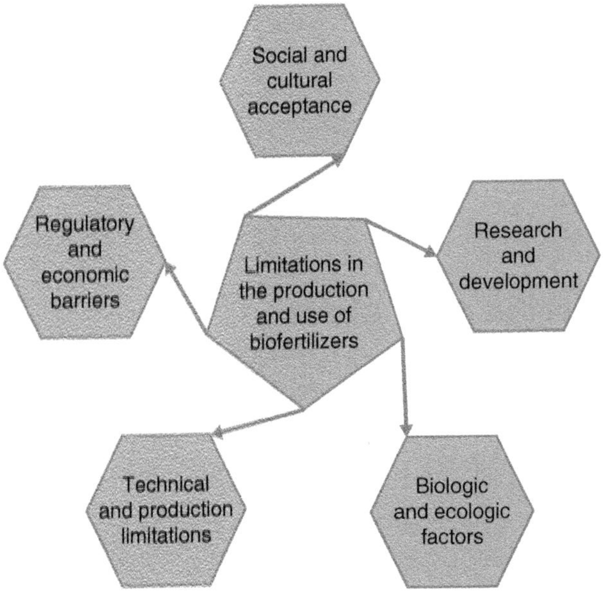

Figure 3.1 Biological, technical, economic, and regulatory barriers limit the production and application of biofertilizers in agriculture.

benefits of biofertilizers, their widespread adoption faces multiple challenges that must be addressed to ensure their effectiveness and large-scale implementation. These challenges include biological limitations related to microbial survival and efficacy under diverse soil conditions, technical constraints in formulation and storage, regulatory hurdles affecting approval and commercialization, economic barriers due to high production costs and scalability issues, and social acceptance influenced by farmer awareness and market demand (Figure 3.1).

3.2 Biological and Ecological Factors

Biological and ecological factors significantly influence the formulation and application of biofertilizers, directly affecting their effectiveness and safety (Figure 3.2). To ensure optimal performance, biofertilizers must be effective and compatible with target crops, soil conditions, and climate. Additionally, their potential effects on the native soil microbiota, rhizosphere dynamics, and human or animal health must be thoroughly evaluated.

3.2.1 Microbial Interactions in the Rhizosphere

The rhizosphere is a competitive environment in which native microorganisms often outcompete introduced biofertilizer strains, limiting their colonization and activity. Conversely, non-native strains may also disrupt the native microbiota, leading to either antagonistic or synergistic interactions (Mawarda et al., 2020; Solanki et al., 2024).

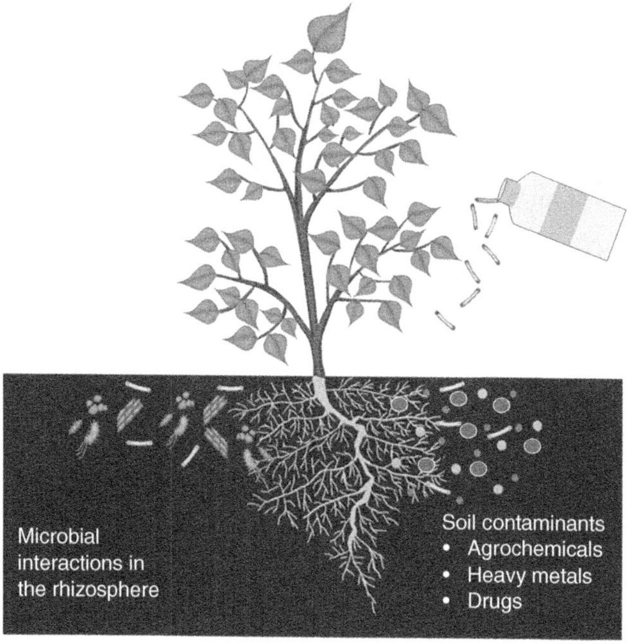

Figure 3.2 Factors affecting the efficacy of biofertilizers in field applications, including microbial interactions and soil contaminants (agrochemicals, heavy metals, and pharmaceuticals).

For example, *Bacillus* spp. have been shown to restructure the soil microbiome and improve resistance to *Fusarium*-induced diseases (Xiong et al., 2017). However, other studies have reported minimal changes in the native microbial communities (Piromyou et al., 2013; Kröber et al., 2014). In some cases, introducing non-native strains, such as *Rhizobium* spp., can increase bacterial richness but reduce overall diversity (Zhang et al., 2010; Trabelsi et al., 2011). Interestingly, substances such as antibiotics play a dual role in microbial interactions. For instance, pyocyanin, an antibiotic produced by *Pseudomonas aeruginosa*, promotes biofilm formation in nodule-forming rhizobia at subinhibitory concentrations, enhancing competitiveness within ecological niches (Khare & Arora, 2011).

Effective colonization often depends on the diversity of native microbiota. Bioinoculants tend to perform better in degraded soils with low microbial diversity (Tardy et al., 2014; da Costa et al., 2020). However, in more diverse soils, their effects may diminish within 35–60 days without reapplication (Zhang et al., 2019; Wang et al., 2021).

One promising strategy to enhance biofertilizer efficacy is the use of native microbial strains in combination with introduced inoculants. Tailored formulations that complement native microbiota can improve compatibility, soil adaptation, and functional diversity. For instance, combining *Pseudomonas* spp. as a biocontrol agent with nutrient-enhancing microorganisms such as *Bacillus* spp. or arbuscular mycorrhizal fungi has shown synergistic effects. However, these highly specific systems face significant technical and economic barriers to large-scale implementation (Allison & Martiny, 2008; Wang et al., 2024).

While it is necessary to consider microbial–microbial interactions in the rhizosphere as a potential limitation to the use of microbial inoculants, other molecules continuously accumulating in the soil, such as agrochemicals, could also significantly limit the performance of commercial bioinoculant strains in the field.

3.2.2 Effect of Soil Contaminants on Biofertilizer Efficacy

Soil contamination is a major challenge affecting agricultural sustainability, as various pollutants can disrupt soil microbiota and reduce the effectiveness of biofertilizers. Among these contaminants, agrochemicals, pharmaceuticals, and heavy metals pose significant risks by altering microbial activity, impacting nutrient cycling, and hindering plant–microbe interactions.

3.2.2.1 Agrochemicals

One of the most significant soil contaminants affecting biofertilizer efficacy is agrochemicals. In recent years, the excessive use of agrochemicals has become a pressing environmental concern. Studies show that their overapplication contributes to air, water, and soil contamination, significantly impacting plants, animals, and microorganisms and disrupting the ecological balance critical for ecosystem sustainability (Elumalai et al., 2025).

In response, bioinoculants have emerged as a sustainable alternative to agrochemicals (Tang et al., 2025). These microorganisms enhance plant nutrition and serve as biocontrol agents against pathogens. However, it is crucial to consider that bioinoculants are often applied to fields that have been repeatedly treated with fertilizers and pesticides. These agrochemicals accumulate in the topsoil, coexisting with the plant rhizosphere microbiome. Consequently, residual agrochemicals may have detrimental effects on introduced bioinoculants (He et al., 2013).

Excessive accumulation of mineral fertilizers can significantly impact not only the biomass and diversity of soil microorganisms (Francioli et al., 2016) but also the survival of subsequently applied biofertilizers. This effect is primarily attributed to changes in soil properties, such as pH, since the overuse of fertilizers can lead to soil acidification or alkalinization (Amin et al., 2024). Furthermore, pesticides, including herbicides, insecticides, and fungicides, tend to accumulate in the soil due to their constant use in agriculture, potentially affecting biofertilizer performance. Herbicides, in particular, have been shown to alter microbial communities by disrupting biosynthetic mechanisms, modifying rhizosphere enzymatic activity, and affecting protein biosynthesis and growth regulator transport (Kremer & Means, 2009). The extent of these alterations depends on herbicide properties, such as bioactivity, bioavailability, degradability, and toxicity, as well as soil factors, including moisture, organic matter content, composition, and vegetation (Yadav et al., 2017). Furthermore, some herbicides have been reported to have fewer negative effects on microorganisms than others. For instance, glyphosate has been documented to have fewer detrimental effects than herbicides such as diclosulam and imazapic (Angelini et al., 2013). Additionally, certain microorganisms, particularly nitrogen-fixing cyanobacteria, exhibit heightened sensitivity to herbicides due to their physiological similarities to higher plants, such as their photosynthetic system, which can be inhibited by these chemicals (Vyas & Bansal, 2023).

On the other hand, insecticides such as diamides, organophosphates, and pyrethroids have been reported to directly affect microbial populations, impacting their development and growth. For instance, Rathjen et al. (2020) demonstrated that imidacloprid inhibited the growth of *Rhizobium* and *Mesorhizobium* species at all tested concentrations. Furthermore, these negative effects of insecticides on rhizobia have been previously documented, showing that even recommended doses of pyriproxyfen reduced nodulation in legumes and affected the growth attributes of these species (Ahmad et al., 2024).

Pesticides are typically formulated with herbicides, insecticides, and fungicides, with the latter often present in higher proportions, as they are the primary agents for controlling plant diseases (Vyas & Bansal, 2023). An example is Standak® Top, which combines two fungicides and an insecticide. Research has shown that this formulation significantly impacts specific beneficial rhizosphere microbes, reducing the cell counts of *Bradyrhizobium japonicum, Bradyrhizobium elkanii*, and *Azospirillum brasilense*. Additionally, it has been found to directly affect plant root development (Rodrigues et al., 2020; Santos et al., 2020).

Despite the impact of pesticides on certain microbial species, some microorganisms are compatible with these agrochemicals, allowing for normal growth or even conferring selective advantages. For example, certain species within the Rhizobiaceae family have demonstrated high resistance to common herbicides (Drouin et al., 2010). Similarly, members of this family have been found to tolerate various fungicides. Strains of *Bradyrhizobium* sp. and *Mesorhizobium* sp., for instance, have been reported to withstand relevant concentrations of fungicides such as kitazin, metalaxyl, glyphosate, and imidacloprid (Ahemad & Khan, 2009; Ahemad & Khan, 2011).

Moreover, identifying strains that not only survive in the presence of pesticides but also act synergistically with them is of particular interest. For instance, the biocontrol strains *Bacillus megaterium* and *Burkholderia cepacia* exhibit antifungal activity against *Fusarium oxysporum* in tomato while maintaining tolerance to the fungicide carbendazim. This compatibility suggests that combining these biocontrol agents with fungicides could enhance their overall effectiveness against fungal pathogens (Omar et al., 2006).

While research has identified bioinoculant strains compatible with agrochemicals, a broader challenge remains: most bioinoculants are negatively affected by these chemical compounds. Potential solutions include enhancing strain resistance to agrochemicals, developing genetically modified (GM) plants that reduce pesticide dependence, and investing in novel biopesticides to gradually decrease reliance on synthetic agrochemicals (Vyas & Bansal, 2023).

3.2.2.2 Heavy Metals

Heavy metals are among the most significant environmental pollutants, ranking as the second-most harmful contaminants in agriculture due to their persistence and accumulation in soil (Balkrishna et al., 2024; Khan et al., 2024; Sharma et al., 2024). Their increasing presence is attributed to anthropogenic, industrial, and agricultural activities, as well as natural processes such as volcanic eruptions and rock weathering (Balkrishna et al., 2024; Hussein et al., 2024).

The most commonly found heavy metals in contaminated agricultural soil include aluminum, arsenic, cadmium, cobalt, copper, chromium, lithium, manganese, mercury, nickel, lead, selenium, thallium, tellurium, and zinc (Singh & Kostova, 2024; Hussein et al., 2024). These metals disrupt ecosystems at multiple levels, affecting plants, animals, humans, and soil microbiota (Higueras-Valdivia et al., 2025). In plants, heavy metals hinder photosynthesis, respiration, and nutrient absorption, leading to bioaccumulation in tissues and potential entry into the food chain. Additionally, soil microorganisms are severely impacted, as heavy metal toxicity inhibits cellular functions and enzymatic activity, reducing nutrient cycling and soil fertility (Kurniawan et al., 2022; Mallick et al., 2025).

Bioremediation, the use of biological agents to neutralize or remove pollutants, is a key strategy for remediating heavy metal-contaminated soils (Karnwal et al., 2024). Microorganisms, particularly those isolated from highly contaminated environments, play a crucial role due to their ability to adapt to metal-rich conditions. Bacteria from the Firmicutes, Proteobacteria, and Actinobacteria phyla, along with arbuscular mycorrhizal fungi, have demonstrated resilience in these environments (Mishra et al., 2017).

Recent research focuses not only on microbial strains that facilitate bioremediation but also on those that enhance plant tolerance to metal-induced stress and promote growth in contaminated soils (Haroun et al., 2023). However, commercial biofertilizers may be less effective under high heavy metal concentrations, as most strains are not adapted to such conditions. For instance, certain rhizobia exhibit reduced nitrogenase activity and nodulation in the presence of heavy metals (Mishra et al., 2017). Likewise, Sharma and Kumar (2020) reported that mercury exposure negatively affects the efficacy of *Trichoderma* and *Rhizobium* biofertilizers. Although biofertilizers offer significant benefits in enhancing plant nutrition and can even confer resistance to contaminated soils, certain microbial species may be sensitive to such conditions. As a result, heavy metal pollution in agricultural soils can compromise the effectiveness of some biofertilizers in the field. Ensuring their viability requires selecting and developing strains capable of thriving in metal-contaminated environments while maintaining their plant growth-promoting functions. Further research is essential to optimize biofertilizer formulations and enhance their resilience, making them a reliable tool for sustainable agriculture in polluted soils.

3.2.2.3 Drugs

The increasing demand for pharmaceuticals, both prescription and illicit, has made them significant emerging environmental contaminants due to their low degradability and persistence in ecosystems (Bertin et al., 2020). Soil contamination primarily occurs through untreated wastewater from municipalities and hospitals, which is often discharged into rivers and subsequently used for agricultural irrigation, introducing pharmaceutical residues into the soil. Additionally, the administration of medications and antibiotics to livestock leads to their excretion in manure, which, when used as fertilizer, further contributes to contamination. Over time, these compounds can leach into the soil and reach groundwater, amplifying their environmental impact (Pawłowska & Biczak, 2024).

Various studies have detected a wide range of pharmaceuticals in agricultural soils. Since these compounds are designed to influence biological processes in target organisms, their presence in the environment, particularly through contaminated irrigation water, allows them to accumulate in soil microbiomes or be absorbed by plants. Their impact depends on their sorption capacity and persistence in the soil (Gworek et al., 2021; Menacherry et al., 2022). Pharmaceutical contamination can severely affect plants by acting as endocrine disruptors, interfering with hormonal regulation, and impairing growth, development, and reproduction. These compounds can alter key phytohormones such as auxins, gibberellins, and cytokinins, thereby compromising essential physiological processes (Aliste et al., 2024). Additionally, they negatively affect soil microbiota by promoting antibiotic resistance and reducing the availability of essential nutrients like nitrogen (N), phosphorus (P), and potassium (K). This disruption of biogeochemical cycles and organic matter decomposition leads to genetic mutations, chromosomal rearrangements, and gene transfer among microorganisms, ultimately diminishing soil fertility (Buta et al., 2021; Kovacs et al., 2024).

Pharmaceuticals must be considered a critical factor affecting biofertilizer performance in the field, mainly due to the common agricultural practice of using wastewater for irrigation (Pawłowska & Biczak, 2024). While bacteria are highly adaptable to adverse conditions, including pharmaceutical pollution, not all species respond equally. For instance, *Pseudomonas* species generally exhibit low sensitivity to antibiotics, while certain microbial strains with biofertilizer potential are highly susceptible. *Cyanobacteria*, which have gained attention as promising biofertilizers, are particularly sensitive to flumequine, lomefloxacin, and oxolinic acid (Le Page et al., 2017). Similarly, Revellin et al. (2018) reported that certain *Rhizobium* species, particularly *Bradyrhizobium liaoningense*, show susceptibility to oxytetracycline, sulfamethoxazole, and tylosin despite being generally considered resistant. This sensitivity likely alters their interactions with plants, affecting nitrogen-fixing symbiosis and reducing their effectiveness as biofertilizers.

3.3 Technical and Production Limitations

The effectiveness of bioinoculants in promoting plant growth and disease resistance is influenced not only by biological and environmental factors but also by technical aspects that determine their performance in the field. Key considerations such as formulation, storage, distribution, and application methods significantly impact their efficacy outside controlled laboratory settings.

Bioinoculants are available in various formulations, including solid, liquid, and encapsulated forms. In some cases, biofertilizers also contain microbial derivatives such as cell-free supernatants and bioactive metabolites (Tewari et al., 2020; Khan et al., 2023). Each formulation type presents distinct advantages and challenges that must be carefully assessed to ensure its successful application.

3.3.1 Solid Formulations

Solid biofertilizers utilize various carrier materials, including soil-based substrates (e.g. clay and peat), organic materials (e.g. sawdust, wheat, oat, and soybean residues), composts (e.g. animal manure and vermicompost), and inorganic substances (e.g. vermiculite, perlite, silicates, and biochar) (Arora & Mishra, 2016). These formulations provide a stable, nutrient-rich environment for microbial strains, facilitating their storage and distribution.

However, solid formulations also have limitations. They may not be suitable for certain microbial species with specific survival requirements, are prone to contamination during handling and storage, and can be challenging to distribute uniformly in the field.

Once applied, their effectiveness may be influenced by environmental conditions such as moisture levels and extreme temperatures. Additionally, the slow release of microbial strains from the solid matrix can result in inconsistent field performance (Berninger et al., 2018; Pirttilä et al., 2021).

3.3.2 Liquid Formulations

Liquid biofertilizers typically consist of microbial strains suspended in carrier fluids, along with surfactants and dispersants. The carrier may be water-based, oil-based, or a combination of both. These formulations are classified into concentrated suspensions, ultralow-volume oil suspensions, oil dispersions, and miscible fluid concentrates (Rai et al., 2024).

One major advantage of liquid formulations is their ease of manufacturing, transportation, and storage. They also allow precise dosing and convenient application via spraying or irrigation systems. However, microbial viability in liquid formulations is highly susceptible to external factors such as light exposure, contamination, and temperature fluctuations. Preservatives are often required to enhance stability (Häder et al., 2015; Fadiji et al., 2024). Additionally, these formulations pose a higher risk of spills, which may affect non-target organisms and raise environmental concerns. Improper application can also lead to overdosing, potentially causing plant health issues, or underdosing, necessitating frequent reapplications and reducing overall effectiveness (Pirttilä et al., 2021).

3.3.3 Metabolite-based Formulations

Some biofertilizers are formulated with microbial metabolites or cell-free supernatants. These products contain bioactive molecules that promote plant growth, stimulate defense mechanisms, or exhibit antimicrobial and insecticidal activity against pests (Fadiji et al., 2024).

One key advantage of these formulations is their extended shelf life, as they do not contain living cells. However, this also means they cannot replicate in the soil after application, potentially limiting their long-term effectiveness. Since these biofertilizers are typically liquid-based, they share many of the benefits and drawbacks of liquid formulations, including vulnerability to environmental conditions that may affect compound stability (Tewari et al., 2020). Another major challenge is their variable composition, as the concentration of metabolites, enzymes, and bioactive compounds depends on microbial growth conditions. These fluctuations make large-scale production and consistency more challenging (Fadiji et al., 2024).

3.3.4 Encapsulation Technology

To overcome the limitations of traditional solid and liquid formulations, research has explored the encapsulation of microbial strains as a promising alternative. Encapsulation techniques include macroencapsulation, microencapsulation, and nanoencapsulation, each offering different levels of protection and controlled release for bioinoculants. Macroencapsulation involves beads ranging from millimeters to centimeters in size, microencapsulation includes particles between 1 and 1000 μm, and nanoencapsulation consists of structures smaller than 1 μm (de Vos et al., 2014). These encapsulation techniques embed microbial strains within a permeable polymeric matrix, shielding them from environmental stress and mechanical damage while ensuring a controlled and consistent liberation of biofertilizers (Khan et al., 2023).

Despite its advantages, encapsulation technology presents several challenges. Macro- and microencapsulated formulations may degrade over time, exposing microbial cells to environmental stressors and reducing their viability. Additionally, the polymers used in encapsulation could pose environmental risks due to their continuous release into ecosystems, requiring studies on their potential impacts on human and animal health (Pateiro et al., 2021).

Nanoencapsulation often incorporates silica, silver, boron, zinc, chitosan, or carbon nanoparticles, raising further concerns about their long-term effects on soil, water, and microbial communities. The interactions of these nanoparticles with rhizosphere microorganisms and their potential toxicity require further investigation. Additionally, nanoencapsulated particles may aggregate, reducing their stability and effectiveness during storage (De Gregorio et al., 2017; El-Fakharany, 2021; Singh & Gurjar, 2022).

Despite advancements in biofertilizer formulations, significant challenges remain in optimizing their production and large-scale implementation. Economic factors are particularly important, as high production costs, especially for complex formulations like nano-encapsulated biofertilizers, pose financial barriers to commercialization. Additionally, the difficulty of scaling up biofertilizer production while maintaining quality and consistency further limits their widespread adoption.

Addressing these economic and industrial challenges is essential for improving the accessibility and feasibility of biofertilizers in modern agriculture. Continued research, investment in cost-effective production methods, and advancements in formulation technologies will be key to overcoming these barriers and ensuring biofertilizers become a viable alternative to conventional fertilizers.

3.4 Economic Barriers

Biofertilizers face economic challenges compared to conventional fertilizers. Achieving an equilibrium between agricultural productivity and sustainability is imperative, as fertilizers enhance soil fertility and crop yields at a minimal cost (Tang et al., 2025).

While biofertilizers contribute to agricultural sustainability, their large-scale adoption is hampered by economic barriers that limit their implementation. This production requires the growth of microorganisms on a commercial scale, with high costs for optimization in bioreactors, processing, and ensuring stability in the field (Ferreyra-Suarez et al., 2024). In addition, the distribution infrastructure is inadequate, and regulations are unclear, which could be corrected with government support and investment incentives (Raimi et al., 2021). Logistical and knowledge inequalities also persist, promoting the use of chemical fertilizers, which are more accessible and familiar to farmers (Joshi & Gauraha, 2022).

These economic and infrastructural challenges, combined with long-standing reliance on synthetic fertilizers, have slowed the adoption of biofertilizers. The market remains dominated by synthetic fertilizers, partly due to the perception that biofertilizers are less effective, have a shorter shelf life, and require specific storage conditions. Their higher cost also raises doubts about their viability (Ortiz & Sansinenea, 2023).

Although synthetic fertilizers continue to dominate the market, some countries have implemented policies and innovations to expand the use of biofertilizers, showing varied approaches to overcoming economic and regulatory barriers. In Asia, China has developed biofertilizers with nitrogen-fixing algae to improve crops such as *Panax ginseng* by reducing root rot and promoting beneficial bacteria (Wang et al., 2015; Dong et al., 2019). India has the most advanced regulatory framework, including those on the Essential List and regulating them under the Commodities Act of 1955 (Kumar et al., 2024; Santos et al., 2024). In Latin America, Brazil leads the sector with annual consumption of 60 000–70 000 tons and major producers such as IFB and Embrafós (Saritha & Prasad Tollamadugu, 2019). In the United States, companies such as Lallemand Plant Care and Agro Bio USA contribute to a market that reached $3.55 billion in 2024, with a projected growth of 26% by 2026 (Santos et al., 2024).

Overcoming economic barriers requires raising awareness and empirical evidence of their effectiveness, especially for retail farmers (Asad & Ahmed, 2024). Economic support through subsidies, affordable financing, and private sector participation is also essential (Raimi et al., 2021). Investment in research would optimize formulations and extend shelf life to improve storage and transport (Miranda et al., 2024).

In summary, biofertilizers represent a viable solution for sustainable agriculture. Although their adoption faces economic challenges, strategies such as financial incentives, research, and education can facilitate their integration, benefiting agricultural productivity and long-term soil health.

3.5 Regulatory Challenges in Biofertilizer Commercialization

Biofertilizers offer a sustainable alternative for agricultural production; however, their commercialization remains limited due to methodological, environmental, and regulatory constraints (Allouzi et al., 2022; Laishram et al., 2025). One of the main obstacles is the complexity and stringency of legislative and regulatory frameworks, which often hinder their widespread adoption. To foster the widespread adoption of biofertilizers, international and country-specific regulations must be revised and adapted to better support the development and application of biofertilizers (Laishram et al., 2025).

3.5.1 Regulatory Frameworks in Developed Countries

Several developed nations, including Australia, China, India, the United States, and the United Kingdom, have established comprehensive legislative frameworks for biofertilizers (Malusá & Vassilev, 2014; Raimi et al., 2020). International quality control programs in Australia, New Zealand, and South Africa oversee microbial inoculant products to ensure their safety and efficacy (Lupwayi et al., 2000). China has implemented legal quality standards for biofertilizers based on eight key criteria: live cell count, carbon and water content, pH, granule size (for solid products), appearance, contamination levels, and validity (Suh et al., 2006; Malusá & Vassilev, 2014).

India, similarly, enforces seven quality parameters, including physical form, microbial viability, contamination, pH, granule size, water content, and efficiency, under the Fertilizer Control Order (FCO) of 1985, issued by the Ministry of Agriculture (Ministry of Agriculture, Government of India, 2009; Malusá & Vassilev, 2014). Recognizing the need for improved production standards, India amended the FCO in 2005 to formally integrate biofertilizers into its legal framework (Reddy, 2017; Praveen & Singh, 2019). The country has also established specific quality standards for various microbial inoculants, including *Rhizobium, Azotobacter, Azospirillum*, phosphate-solubilizing bacteria, and mycorrhizal fungi (Malusá & Vassilev, 2014). These standards, known as Bureau of Indian Standards, define quality requirements for *Rhizobium* (IS 8268:2001), *Azotobacter* (IS 9138:2002), *Azospirillum* (IS 14806:2000), and phosphate-solubilizing bacteria (IS 14807:2000) (Agarwal et al., 2021).

However, despite these regulations, the enforcement of product quality is largely left to producers, not only in India but also in countries such as the United States and the United Kingdom (Smith, 1992). In the United States, biofertilizer use in organic farming falls under the USDA National Organic Program, while microbial pesticides are regulated by Environmental Protection Agency (Malusá & Vassilev, 2014).

3.5.2 Global Disparities in Biofertilizer Regulations

No universal regulatory standards exist for biofertilizers due to significant differences in quality control requirements among countries (Lupwayi et al., 2000). Some nations, such as Brazil, Canada, France, and Uruguay, have legislation that empowers regulatory authorities to oversee biofertilizer production and distribution (Agarwal et al., 2021). Others, including Australia, New Zealand, and South Africa, implement quality control programs in which producers actively ensure compliance (Lupwayi et al., 2000). In Spain, the Andalusian regional government permits biofertilizers through its administrative framework (Malusá & Vassilev, 2014). Meanwhile, Italy regulates mycorrhizal inoculants as soil-enhancing products under Legislative Decree #75, requiring them to be produced under sterile conditions to meet quality standards (Malusá & Vassilev, 2014).

One of the main drivers of stringent quality management is the inconsistency in biofertilizer performance. In some cases, biofertilizers fail to meet manufacturers' claims due to the lack of standardized evaluation methods (Lesueur et al., 2016). These issues highlight the importance of regulatory agencies establishing clear guidelines and enforcing quality standards to ensure product reliability and efficacy (Mahanty et al., 2017).

3.5.3 The Challenge of GM Biofertilizers

Another emerging issue in biofertilizer regulation is using GM microorganisms to enhance efficiency. While GM inoculants have the potential to improve nutrient uptake, stress resistance, and overall effectiveness, obtaining regulatory approval remains a significant challenge. The primary concern is convincing regulatory bodies and the public of their safety, which has sparked ethical and environmental debates (Ma et al., 2011;

Laishram et al., 2025). Additionally, legislative restrictions in many countries limit the large-scale application of GM microorganisms, further hindering their integration into agricultural systems (Qiu et al., 2019).

Expanding biofertilizer adoption requires more assertive communication and collaboration among researchers, industry stakeholders, policymakers, and farmers (Laishram et al., 2025). Key barriers to industry growth include inconsistent product quality, limited awareness of biofertilizers, and low adoption rates among farmers (Malusá & Vassilev, 2014). Addressing these challenges through improved regulatory frameworks, scientific innovation, and farmer education will be essential to promote biofertilizers as a viable and sustainable alternative in modern agriculture.

3.6 Social and Cultural Acceptance

Biofertilizers have the potential to enhance sustainable agricultural production significantly, but their widespread adoption depends on overcoming regulatory, social, and cultural barriers (Qiu et al., 2019). Their successful implementation requires the involvement of regulatory agencies, policymakers, and collaboration between the public and private sectors, with field performance data playing a crucial role (Mitter et al., 2021). Effective communication between the scientific and industrial communities is essential for addressing producers' needs while meeting consumers' ethical, social, and economic expectations. Moreover, farmers play a pivotal role in advancing biofertilizer development, as their observations and feedback on field applications contribute to scientific validation and broader social and economic acceptance (Mahmud et al., 2021).

A significant challenge to biofertilizer adoption is the gap between researchers and farmers, compounded by limited awareness and inadequate information dissemination regarding their advantages over chemical fertilizers. Raising awareness about their efficacy is crucial to encouraging farmers to transition from conventional fertilizers to biofertilizers. Building farmer confidence requires improving product quality, implementing robust quality assessments, and providing regular training programs (Adeleke et al., 2022). Since biofertilizers can yield variable results due to differences in soil types, crop varieties, seasonal variations, and environmental conditions, continuous evaluation is necessary to ensure their reliability (Debnath et al., 2019; Adeleke et al., 2022).

In developed and developing regions, including parts of Central and Latin America and Southeast Asia, biofertilizers are increasingly integrated into agricultural practices (Tsoi et al., 2019). Historically, farmers have utilized microbial consortia through traditional practices such as transporting productive soil from one location to another (Mougel et al., 2006). This method remains widely used in tropical regions of Central and South America and in southeastern Georgia (Mahmud et al., 2021).

Beyond farmer adoption, consumer perception must also be considered. Ensuring biofertilizer quality and product safety is essential before presenting them to consumers (Mitter et al., 2021). Public acceptance may be influenced by increased awareness of the adverse effects of chemical fertilizers, such as soil degradation, nitrate emissions, and governmental regulations aimed at promoting sustainable practices (Mehra, 2020). Therefore, to achieve greater acceptance of biofertilizers, efforts must focus on raising awareness of their benefits, maintaining high-quality standards, and continuously improving formulations to meet the demands of both farmers and consumers (O'Callaghan et al., 2022).

3.7 Research and Development

Just as ecological, technical, and economic barriers hinder the standardized and practical application of biofertilizers in the field, continuous innovation in research and development is crucial to overcoming these challenges. However, the research and development process faces multiple obstacles, from microbial isolation to large-scale production, distribution, and field application.

One of the primary challenges lies in strain selection. Despite the vast microbial diversity in the rhizosphere, over 90% of microbial species remain uncultured due to their complex ecological requirements (Bodor et al., 2020). Traditional laboratory cultivation methods introduce additional biases, as commonly used media, such as LB or KB, fail to replicate the intricate plant-associated environments. This limitation often excludes strains with significant agricultural potential (Mitter et al., 2021). Implementing plant-based nutrient formulations during microbial isolation and cultivation may improve the identification of strains better suited for biofertilizer development.

Modern metaomic tools, such as metataxonomics and metaproteomics, have advanced our understanding of the soil microbiome. For instance, studies have shown that rhizosphere microbes can enhance *Arabidopsis thaliana* resistance to *Botrytis cinerea* through plant hormone production (Sham et al., 2019). Similarly, nitrogen-fixing proteins have been detected in the rhizosphere of *Rehmannia glutinosa*, emphasizing the functional roles of microbial interactions in nutrient cycling (Wu et al., 2011). These insights reveal the untapped potential of native rhizosphere communities that could guide the selection of strains for commercial use.

Enhancing the effectiveness of biofertilizers requires the development of formulations with microbial strains adapted to specific crops. Optimizing production by adjusting substrates and component proportions can improve their stability and efficiency in the field (Barin et al., 2022). Current research focuses on increasing biomass production, nutrient availability, and stress resistance to maximize plant growth, regulate phytohormones, and mitigate the effects of pathogens and abiotic stress (Booali et al., 2024). Moreover, genetic engineering is employed to develop microbial strains with enhanced nutrient uptake, stress tolerance, and pathogen resistance, further strengthening plant resilience and productivity (Adetunji et al., 2022).

However, once candidate strains are selected for commercial application, a significant challenge compromising biofertilizer effectiveness is the development of stable bioformulations that can reach the field in optimal conditions. Achieving this requires selecting carrier materials that protect microbial strains from environmental factors while preserving their viability and shelf life (Arora & Mishra, 2016). Investing in research to enhance formulation stability is crucial, as successfully implementing biofertilizers could significantly improve global food production while maintaining environmental sustainability.

Addressing these challenges requires collaboration between academia and industry, integrating interdisciplinary approaches that bring together chemistry, materials science, biology, ecology, and agronomy experts.

3.8 Conclusions

Biofertilizers represent a promising alternative to synthetic fertilizers, offering benefits such as improved plant nutrition, enhanced stress resistance, and reduced environmental impact. However, their widespread adoption is hindered by biological, technical, regulatory, economic, and social barriers. Factors such as microbial interactions within the rhizosphere, competition with native microbiota, and agrochemical residues can limit their effectiveness. Additionally, formulation, storage, and large-scale production challenges must be addressed to ensure their viability in agricultural systems. Regulatory inconsistencies, high production costs, and limited farmer awareness further slow adoption. Advancing research, optimizing formulation technologies, and fostering collaboration among researchers, industry, and policymakers are essential to overcoming these challenges and establishing biofertilizers as a sustainable solution for modern agriculture.

References

Adeleke, B.S., Fadiji, A.E., Ayilara, M.S., Igiehon, O.N., Nwachukwu, B.C., & Babalola, O.O. 2022. Strategies to enhance the use of endophytes as bioinoculants in agriculture. *Horticulturae* 8(6), 498. https://doi.org/10.3390/horticulturae8060498

Adetunji, C.O., Anani, O.A., Olaniyan, O.T., Bodunrinde, R.E., Osemwegie, O.O., & Ubi, B.E. 2022. Sustainability of biofertilizers and other allied products from genetically modified microorganisms. In: *Biomass, Biofuels, Biochemicals* (pp. 363–393). Elsevier. https://doi.org/10.1016/B978-0-323-89855-3.00003-0

Agarwal, P.K., Gupta, R., & Sharma, S. 2021. Biofertilizers: a key player in sustainable agriculture. In: *Advances in Biofertilizers and Biopesticides* (pp. 283–298). Wiley. https://doi.org/10.1002/9781119724995.ch14

Ahemad, M., & Khan, M.S. 2009. Effect of insecticide-tolerant and plant growth-promoting Mesorhizobium on the performance of chickpea grown in insecticide stressed alluvial soils. *Journal of Crop Science and Biotechnology* 12, 217–226.

Ahemad, M., & Khan, M.S. 2011. Effect of pesticides on plant growth promoting traits of greengram-symbiont, *Bradyrhizobium* sp. strain MRM6. *Bulletin of Environmental Contamination and Toxicology* 86, 384–388.

Ahmad, Z., Niyazi, S., Firdoos, A., Wang, C., Manzoor, M.A., Ramakrishnan, M., Upadhyay, A., & Ding, Y. 2024. Enhancing plant resilience: nanotech solutions for sustainable agriculture. *Heliyon* 10(23), e40735. https://doi.org/10.1016/j.heliyon.2024.e40735

Aliste, M., Hernández, V., El Aatik, A., Pérez-Lucas, G., Fenoll, J., & Navarro, S. 2024. Coupled bio-solar photocatalytic treatment for reclamation of water polluted with pharmaceutical and pesticide residues: impact on tomato irrigation. *Ecotoxicology and Environmental Safety* 287, 117291. https://doi.org/10.1016/j.ecoenv.2024.117291

Allison, S.D., & Martiny, J.B. 2008. Resistance, resilience, and redundancy in microbial communities. *Proceedings of the National Academy of Sciences of the United States of America* 105(supplement_1), 11512–11519.

Allouzi, M.M.A., Allouzi, S.M.A., Keng, Z.X., Supramaniam, C.V., Singh, A., & Chong, S. 2022. Liquid biofertilizers as a sustainable solution for agriculture. *Heliyon* 8(12). https://doi.org/10.1016/j.heliyon.2022.e12609

Amin, N., Sinha, R.P., & Kannaujiya, V.K. 2024. Effects of ultraviolet and photosynthetically active radiation on morphogenesis, antioxidants and photoprotective defense mechanism in a hot-spring cyanobacterium *Nostoc* sp. strain VKB02. *Research in Microbiology* 175, 104180.

Angelini, J., Silvina, G., Taurian, T., Ibáñez, F., Tonelli, M.L., Valetti, L., Anzuay, M.S., Ludueña, L., Muñoz, V., & Fabra, A. 2013. The effects of pesticides on bacterial nitrogen fixers in peanut-growing area. *Archives of Microbiology* 195, 683–692.

Arora, N.K., & Mishra, J. 2016. Prospecting the roles of metabolites and additives in future bioformulations for sustainable agriculture. *Applied Soil Ecology* 107, 405–407.

Asad, M., & Ahmed, S.Y. 2024. Utilizing biofertilizer for achieving sustainable agriculture and rural development strategy towards vision 2040, Oman. *Sustainability* 16(10), 4015. https://doi.org/10.3390/su16104015

Balkrishna, A., Kumari, A., Kumar, A., Arya, V., Chauhan, A., Upadhyay, N.K., Guleria, I., Amarowicz, R., Kumar, D., & Kuca, K. 2024. Biosensors for detection of pesticide residue, mycotoxins and heavy metals in fruits and vegetables: a concise review. *Microchemical Journal* 205, 111292.

Barin, M., Asadzadeh, F., Hosseini, M., Hammer, E.C., Vetukuri, R.R., & Vahedi, R. 2022. Optimization of biofertilizer formulation for phosphorus solubilizing by *Pseudomonas fluorescens* Ur21 via response surface methodology. *Processes* 10(4), 650. https://doi.org/10.3390/pr10040650

Berninger, T., González López, Ó., Bejarano, A., Preininger, C., & Sessitsch, A. 2018. Maintenance and assessment of cell viability in formulation of non-sporulating bacterial inoculants. *Microbial Biotechnology* 11(2), 277–301.

Bertin, S., Yates, K., & Petrie, B. 2020. Enantiospecific behaviour of chiral drugs in soil. *Environmental Pollution* 262, 114364. https://doi.org/10.1016/j.envpol.2020.114364

Bodor, A., Bounedjoum, N., Vincze, G.E., Erdeiné Kis, Á., Laczi, K., Bende, G., Szilágyi, Á., Kovács, T., Perei, K., & Rákhely, G. 2020. Challenges of unculturable bacteria: environmental perspectives. *Reviews in Environmental Science and Bio/Technology* 19, 1–22.

Booali, S., Zoufan, P., & Zare Bavani, M.R. 2024. Effect of biofertilizer containing *Thiobacillus* bacteria along with different levels of chemical sulfur fertilizer on growth response and photochemical efficiency of small radish plants (*Raphanus sativus* L. var. shushtari) under greenhouse conditions. *Scientia Horticulturae* 327, 112835. https://doi.org/10.1016/j.scienta.2023.112835

Buta, M., Hubeny, J., Zieliński, W., Harnisz, M., & Korzeniewska, E. 2021. Sewage sludge in agriculture – the effects of selected chemical pollutants and emerging genetic resistance determinants on the quality of soil and crops – a review. *Ecotoxicology and Environmental Safety* 214, 112070. https://doi.org/10.1016/j.ecoenv.2021.112070

Chatterjee, A., Singh, S., Agrawal, C., Yadav, S., Rai, R., & Rai, L.C. 2017. Role of algae as a biofertilizer. In: *Algal Green Chemistry* (pp. 189–200). Elsevier. https://doi.org/10.1016/B978-0-444-63784-0.00010-2

da Costa, P.B., van Elsas, J.D., Mallon, C., dos Anjos Borges, L.G., & Passaglia, L.M.P. 2020. Efficiency of probiotic traits in plant inoculation is determined by environmental constrains. *Soil Biology and Biochemistry* 148, 107893.

De Gregorio, P.R., Michavila, G., Ricciardi Muller, L., de Souza Borges, C., Pomares, M.F., Saccol de Sá, E.L., Pereira, C., & Vincent, P.A. 2017. Beneficial rhizobacteria immobilized in nanofibers for potential application as soybean seed bioinoculants. *PloS One* 12(5), e0176930.

de Vos, P., Lazarjani, H.A., Poncelet, D., & Faas, M.M. 2014. Polymers in cell encapsulation from an enveloped cell perspective. *Advanced Drug Delivery Reviews* 67, 15–34.

Debnath, S., Rawat, D., Mukherjee, A.K., Adhikary, S., & Kundu, R. 2019. Applications and constraints of plant beneficial microorganisms in agriculture. In *Biostimulants in plant science*. IntechOpen. https://doi.org/10.5772/intechopen.89190

Dong, L., Li, Y., Xu, J., Yang, J., Wei, G., Shen, L., Ding, W., & Chen, S. 2019. Biofertilizers regulate the soil microbial community and enhance *Panax ginseng* yields. *Chinese Medicine* 14(1), 20. https://doi.org/10.1186/s13020-019-0241-1

Drouin, P., Sellami, M., Prevost, D., Fortin, J., & Antoun, H. 2010. Tolerance to agricultural pesticides of strains belonging to four genera of Rhizobiaceae. *Journal of Environmental Science and Health Part B* 45(8), 757–765.

El-Fakharany, E.M. 2021. Nanoformulation approach for improved stability and efficiency of lactoperoxidase. *Preparative Biochemistry & Biotechnology* 51(7), 629–641.

Elumalai, P., Gao, X., Parthipan, P., Luo, J., & Cui, J. 2025. Agrochemical pollution: a serious threat to environmental health. *Current Opinion in Environmental Science & Health* 43, 100597.

Fadiji, A.E., Xiong, C., Egidi, E., & Singh, B.K. 2024. Formulation challenges associated with microbial biofertilizers in sustainable agriculture and paths forward. *Journal of Sustainable Agriculture and Environment* 3(3), e70006.

Ferreyra-Suarez, D., García-Depraect, O., & Castro-Muñoz, R. 2024. A review on fungal-based biopesticides and biofertilizers production. *Ecotoxicology and Environmental Safety* 283, 116945. https://doi.org/10.1016/j.ecoenv.2024.116945

Francioli, D., Schulz, E., Lentendu, G., Wubet, T., Buscot, F., & Reitz, T. 2016. Mineral vs. organic amendments: microbial community structure, activity and abundance of agriculturally relevant microbes are driven by long-term fertilization strategies. *Frontiers in Microbiology* 7, 1446.

González-Mancilla, A., Almaraz-Suárez, J.J., Ferrera-Cerrato, R., Rodríguez-Guzmán, M.P., & Taboada-Gaytán, O.R. 2024. Photosynthetic activity and growth of poblano pepper biofertilized with plant growth promoting rhizobacteria and arbuscular mycorrhizal fungi. *Current Research in Microbial Sciences* 7, 100269. https://doi.org/10.1016/j.crmicr.2024.100269

Gworek, B., Kijeńska, M., Wrzosek, J., & Graniewska, M. 2021. Pharmaceuticals in the soil and plant environment: a review. *Water Air & Soil Pollution* 232(4), 145. https://doi.org/10.1007/s11270-020-04954-8

Häder, D.P., Williamson, C.E., Wängberg, S.Å., Rautio, M., Rose, K.C., Gao, K., Helbling, E.W., Sinha, R.P., & Worrest, R. 2015. Effects of UV radiation on aquatic ecosystems and interactions with other environmental factors. *Photochemical & Photobiological Sciences* 14(1), 108–126.

Haroun, M., Xie, S., Awadelkareem, W., Wang, J., & Qian, X. 2023. Influence of biofertilizer on heavy metal bioremediation and enzyme activities in the soil to revealing the potential for sustainable soil restoration. *Scientific Reports* 13(1), 20684.

He, Y., Zhang, J., Gao, C., Su, J., Chen, J., & Shen, J. 2013. Regression analysis of dynamics of insecticide resistance in field populations of *Chilo suppressalis* (Lepidoptera: Crambidae) during 2002–2011 in China. *Journal of Economic Entomology* 106(4), 1832–1837.

Higueras-Valdivia, M., Silva-Castro, G.A., Paniagua-López, M., Romero-Freire, A., & García-Romera, I. 2025. Mitigation of heavy metal soil contamination: a novel strategy with mycorrhizal fungi and biotransformed olive residue. *Journal of Hazardous Materials Advances* 17, 100570.

Hu, H., Li, P., Tong, Y.W., Li, J., & He, Y. 2024. Reutilization recovered nutrients from wastewater by synthesizing eco-friendly slow-release hydrogel fertilizer for sustainable vegetable production. *Chemical Engineering Journal* 499, 155943. https://doi.org/10.1016/j.cej.2024.155943

Hussein, S.H., Qurbani, K., Ahmed, S.K., Tawfeeq, W., & Hassan, M. 2024. Bioremediation of heavy metals in contaminated environments using Comamonas species: a narrative review. *Bioresource Technology Reports* 25, 101711.

Joshi, S.K., & Gauraha, A.K. 2022. Global biofertilizer market: emerging trends and opportunities. In: *Trends of Applied Microbiology for Sustainable Economy* (pp. 689–697). Elsevier. https://doi.org/10.1016/B978-0-323-91595-3.00024-0

Karnwal, A., Shrivastava, S., Al-Tawaha, A.R.M.S., Kumar, G., Kumar, A., & Kumar, A. 2024. PGPR-mediated breakthroughs in plant stress tolerance for sustainable farming. *Journal of Plant Growth Regulation* 43(9), 2955–2971.

Khan, A., Kanwal, F., Shahzad, M., Naz, S., Jalil, S., & Zhang, G. 2024. Interactions of arsenic and phosphorus in their uptake and transportation in plants: advances and prospective research on the mechanisms and approaches for alleviating arsenic stress. *Journal of Integrative Agriculture*. https://doi.org/10.1016/j.jia.2024.07.022

Khan, A., Singh, A.V., Gautam, S.S., Agarwal, A., Punetha, A., Upadhayay, V.K., Kukreti, B., Bundela, V., Jugran, A.K., & Goel, R. 2023. Microbial bioformulation: a microbial assisted biostimulating fertilization technique for sustainable agriculture. *Frontiers in Plant Science* 14, 1270039.

Khare, E., & Arora, N.K. 2011. Dual activity of pyocyanin from *Pseudomonas aeruginosa* – antibiotic against phytopathogen and signal molecule for biofilm development by rhizobia. *Canadian Journal of Microbiology* 57(9), 708–713.

Kovacs, E.D., Kovacs, M.H., Barcelo, D., & Paulo, P. 2024. Nonsteroidal anti-inflammatory drugs impact the microbial community in three different soil types – a laboratory experiment. *Case Studies in Chemical and Environmental Engineering* 10, 100833. https://doi.org/10.1016/j.cscee.2024.100833

Kremer, R.J., & Means, N.E. 2009. Interactions of glyphosate and glyphosate-resistant crops with rhizosphere microorganisms. *European Journal of Agronomy* 31(3), 153–161.

Kröber, M., Wibberg, D., Grosch, R., Eikmeyer, F., Verwaaijen, B., Chowdhury, S.P., Hartmann, A., Pühler, A., & Schlüter, A. 2014. Effect of the strain *Bacillus amyloliquefaciens* FZB42 on the microbial community in the rhizosphere of lettuce under field conditions analyzed by whole metagenome sequencing. *Frontiers in Microbiology* 5, 252.

Kuila, D., & Ghosh, S. 2022. Aspectos, problems and utilization of Arbuscular Mycorrhizal (AM) application as bio-fertilizer in sustainable agriculture. *Current Research in Microbial Sciences* 3, 100107.

Kumar, A., Saharan, B.S., Parshad, J., Gera, R., Choudhary, J., & Yadav, R. 2024. Revolutionizing Indian agriculture: the imperative of advanced biofertilizer technologies for sustainability. *Discover Agriculture* 2(1), 24. https://doi.org/10.1007/s44279-024-00037-y

Kurniawan, S.B., Ramli, N.N., Said, N.S.M., Alias, J., Imron, M.F., Abdullah, S.R.S., Othman, A.R., Purwanti, I.F., & Hasan, H.A. 2022. Practical limitations of bioaugmentation in treating heavy metal contaminated soil and role of plant growth promoting bacteria in phytoremediation as a promising alternative approach. *Heliyon* 8(4), e08995.

Laishram, B., Devi, O.R., Dutta, R., Senthilkumar, T., Goyal, G., Paliwal, D.K., & Rasool, A. 2025. Plant-microbe interactions: PGPM as microbial inoculants/biofertilizers for sustaining crop productivity and soil fertility. *Current Research in Microbial Sciences* 8, 100333. https://doi.org/10.1016/j.crmicr.2024.100333

Le Page, G., Gunnarsson, L., Snape, J., & Tyler, C.R. 2017. Integrating human and environmental health in antibiotic risk assessment: a critical analysis of protection goals, species sensitivity and antimicrobial resistance. *Environment International* 109, 155–169.

Lesueur, D., Deaker, R., Herrmann, L., Bräu, L., & Jansa, J. 2016. The production and potential of biofertilizers to improve crop yields. In *Bioformulations: For Sustainable Agriculture* (pp. 71–92). New Delhi: Springer India. . https://doi.org/10.1007/978-81-322-2779-3_4

Li, B., Qiao, Y., & Yao, R. 2023. What promote farmers to adopt green agricultural fertilizers? Evidence from 8 provinces in China. *Journal of Cleaner Production* 426, 139123.

Lupwayi, N.Z., Olsen, P.E., Sande, E.S., Keyser, H.H., Collins, M.M., Singleton, P.W., & Rice, W.A. 2000. Inoculant quality and its evaluation. *Field Crops Research* 65(2-3), 259–270.

Ma, Y., Prasad, M.N.V., Rajkumar, M., & Freitas, H.J.B.A. 2011. Plant growth promoting rhizobacteria and endophytes accelerate phytoremediation of metalliferous soils. *Biotechnology Advances* 29(2), 248–258. https://doi.org/10.1016/j.biotechadv.2010.12.001

Mahanty, T., Bhattacharjee, S., Goswami, M., Bhattacharyya, P.N., Das, B., Ghosh, A., & Tribedi, P. 2017. Biofertilizers: a potential approach for sustainable agriculture development. *Environmental Science and Pollution Research* 24(7), 6441–6455. https://doi.org/10.1007/s11356-016-8104-0

Mahmud, K., Missaoui, A., Lee, K., Ghimire, B., Presley, H.W., & Makaju, S. 2021. Rhizosphere microbiome manipulation for sustainable crop production. *Current Plant Biology* 27, 100210. https://doi.org/10.1016/j.cpb.2021.100210

Mallick, S., Pradhan, T., & Das, S. 2025. Bacterial biomineralization of heavy metals and its influencing factors for metal bioremediation. *Journal of Environmental Management* 373, 123977.

Malusá, E., & Vassilev, N. 2014. A contribution to sustainable agriculture by biofertilization and biocontrol of plant diseases and pests. *Applied Microbiology and Biotechnology* 98(15), 6599–6613. https://doi.org/10.1007/s00253-014-5828-y

Mateos-Naranjo, E., García-López, J.V., Flores-Duarte, N.J., Romano-Rodríguez, E., Rodríguez-Llorente, I.D., Pérez-Romero, J.A., Pajuelo, E., & Redondo-Gómez, S. 2025. Development of a PGPB-based biofertilizer to optimize strawberry cultivation in semiarid regions: screening, validation and scaling up to commercial production. *Scientia Horticulturae* 340, 113929. https://doi.org/10.1016/j.scienta.2024.113929

Mawarda, P.C., Le Roux, X., Van Elsas, J.D., & Salles, J.F. 2020. Deliberate introduction of invisible invaders: a critical appraisal of the impact of microbial inoculants on soil microbial communities. *Soil Biology and Biochemistry* 148, 107874.

Mehra, A. 2020. Biofertilizers market worth $4.5 billion by 2026. *MarketsandMarkets*. https://www.marketsandmarkets.com/PressReleases/compound-biofertilizerscustomized-fertilizers.asp (accessed 30 January 2025)

Mejía-Guerra, P.A., Salas-Sanjuán, M.C., Martínez-Gallardo, M.R., Jurado, M.M., Estrella-González, M.J., López-González, J.A., Suárez-Estrella, F., & López, M.J. 2025. Microbial biofertilizers and biopesticides from vermicompost tea and rhizosphere of organic soilless melon crop: in vitro assessment. *Biocatalysis and Agricultural Biotechnology* 63, 103453. https://doi.org/10.1016/j.bcab.2024.103453

Menacherry, S.P.M., Kočárek, M., Kacerova, T., Kotíková, Z., Kačer, P., & Kodešová, R. 2022. The impact of initial concentration of selected pharmaceuticals on their early stage of dissipation in soils. *Journal of Soils and Sediments* 22(2), 522–535. https://doi.org/10.1007/s11368-021-03095-7

Ministry of Agriculture, Government of India. 2009. Biofertilizers and organic fertilizers covered in fertilizer (Control) Order, 1985 (as amended, March 2006 and November 2009). *Official Gazette*, 3 November.

Miranda, A.M., Hernandez-Tenorio, F., Villalta, F., Vargas, G.J., & Sáez, A.A. 2024. Advances in the development of biofertilizers and biostimulants from microalgae. *Biology* 13(3), 199. https://doi.org/10.3390/biology13030199

Mishra, J., Singh, R., & Arora, N.K. 2017. Alleviation of heavy metal stress in plants and remediation of soil by rhizosphere microorganisms. *Frontiers in Microbiology* 8, 1706.

Mitter, E.K., Tosi, M., Obregón, D., Dunfield, K.E., & Germida, J.J. 2021. Rethinking crop nutrition in times of modern microbiology: innovative biofertilizer technologies. *Frontiers in Sustainable Food Systems* 5, 606815. https://doi.org/10.3389/fsufs.2021.606815

Mougel, C., Offre, P., Ranjard, L., Corberand, T., Gamalero, E., Robin, C., & Lemanceau, P. 2006. Dynamic of the genetic structure of bacterial and fungal communities in soil during biofumigation as assessed by microbial population DNA fingerprinting. *New Phytologist* 170(1), 165–175. https://doi.org/10.1111/j.1469-8137.2006.01650.x

O'Callaghan, M., Ballard, R. A., & Wright, D. 2022. Soil microbial inoculants for sustainable agriculture: limitations and opportunities. *Soil Use and Management* 38(3), 1340–1369. https://doi.org/10.1111/sum.12811

Omar, I., O'neill, T.M., & Rossall, S. 2006. Biological control of Fusarium crown and root rot of tomato with antagonistic bacteria and integrated control when combined with the fungicide carbendazim. *Plant Pathology* 55(1), 92–99.

Ortiz, A., & Sansinenea, E. 2023. Microbial-based biopesticides. In: *Development and Commercialization of Biopesticides* (pp. 103–118). Elsevier. https://doi.org/10.1016/B978-0-323-95290-3.00020-0

Pateiro, M., Gómez, B., Munekata, P.E., Barba, F.J., Putnik, P., Kovačević, D.B., & Lorenzo, J.M. 2021. Nanoencapsulation of promising bioactive compounds to improve their absorption, stability, functionality and the appearance of the final food products. *Molecules* 26(6), 1547.

Pawłowska, B., & Biczak, R. 2024. Drugs in the environment – impact on plants: a review. *Environmental Toxicology and Pharmacology* 111, 104557. https://doi.org/10.1016/j.etap.2024.104557

Piromyou, P., Noisangiam, R., Uchiyama, H., Tittabutr, P., Boonkerd, N., & Teaumroong, N. 2013. Indigenous microbial community structure in rhizosphere of Chinese kale as affected by plant growth-promoting rhizobacteria inoculation. *Pedosphere* 23(5), 577–592.

Pirttilä, A.M., Mohammad Parast Tabas, H., Baruah, N., & Koskimäki, J.J. 2021. Biofertilizers and biocontrol agents for agriculture: how to identify and develop new potent microbial strains and traits. *Microorganisms* 9(4), 817.

Pourhosseini, S.H., Azizi, A., Sadat Seyedi, F., & Hadian, J. 2024. Bio-fertilizer as a pathway to minimize nitrate leaching from chemical fertilizer in high yield peppermint production. *Journal of Cleaner Production* 468, 143100. https://doi.org/10.1016/j.jclepro.2024.143100

Praveen, K.V., & Singh, A. 2019. Realizing the potential of a low-cost technology to enhance crop yields: evidence from a meta-analysis of biofertilizers in India. *Agricultural Economics Research Review* 32, 77–91. https://doi.org/10.5958/0974-0279.2019.00018.1

Qiu, Z., Egidi, E., Liu, H., Kaur, S., & Singh, B.K. 2019. New frontiers in agricultural microbiome research and its role in sustainable agriculture. *Biotechnology Advances* 37(6), 107371. https://doi.org/10.1016/j.biotechadv.2019.03.010

Rai, S., Mago, Y., Aggarwal, G., Yadav, A., & Tewari, S. 2024. Liquid bioformulation: a trending approach towards achieving sustainable agriculture. *Molecular Biotechnology* 66(10), 2725–2750.

Raimi, A., Roopnarain, A., & Adeleke, R. 2021. Biofertilizer production in Africa: current status, factors impeding adoption and strategies for success. *Scientific African* 11, e00694. https://doi.org/10.1016/j.sciaf.2021.e00694

Raimi, A., Roopnarain, A., Chirima, G. J., & Adeleke, R. 2020. Insights into the microbial composition and potential efficiency of selected commercial biofertilisers. *Heliyon* 6(7). https://doi.org/10.1016/j.heliyon.2020.e04342

Rathjen, J.R., Ryder, M.H., Riley, I.T., Lai, T.V., & Denton, M.D. 2020. Impact of seed-applied pesticides on rhizobial survival and legume nodulation. *Journal of Applied Microbiology* 129(2), 389–399.

Reddy, A. 2017. *Impact of Paramparagat Krishi Vikas Yojana*. National Institute of Agricultural Extension Management (MANAGE), Hyderabad.

Revellin, C., Hartmann, A., Solanas, S., & Topp, E. 2018. Long-term exposure of agricultural soil to veterinary antibiotics changes the population structure of symbiotic nitrogen-fixing rhizobacteria occupying nodules of soybeans (Glycine max). *Applied and Environmental Microbiology* 84(9), e00109-18.

Rodrigues, T.F., Bender, F.R., Sanzovo, A.W.S., Ferreira, E., Nogueira, M.A., & Hungria, M. 2020. Impact of pesticides in properties of *Bradyrhizobium* spp. and in the symbiotic performance with soybean. *World Journal of Microbiology and Biotechnology* 36(11), 172.

Santos, F., Melkani, S., Oliveira-Paiva, C., Bini, D., Pavuluri, K., Gatiboni, L., Mahmud, A., Torres, M., McLamore, E., & Bhadha, J.H. 2024. Biofertilizer use in the United States: definition, regulation, and prospects. *Applied Microbiology and Biotechnology* 108(1), 511. https://doi.org/10.1007/s00253-024-13347-4

Santos, M.S., Rondina, A.B., Nogueira, M.A., & Hungria, M. 2020. Compatibility of Azospirillum brasilense with pesticides used for treatment of maize seeds. *International Journal of Microbiology* 2020(1), 8833879.

Saritha, M., & Prasad Tollamadugu, N.V.K.V. 2019. The status of research and application of biofertilizers and biopesticides: global scenario. In: *Recent Developments in Applied*

Microbiology and Biochemistry (pp. 195–207). Elsevier. https://doi.org/10.1016/B978-0-12-816328-3.00015-5

Sham, A., Al-Ashram, H., Whitley, K., Iratni, R., El-Tarabily, K.A., & AbuQamar, S.F. 2019. Metatranscriptomic analysis of multiple environmental stresses identifies RAP2. 4 gene associated with *Arabidopsis* immunity to *Botrytis cinerea*. *Scientific Reports* 9(1), 17010.

Sharma, I., Sharma, S., Sharma, V., Singh, A.K., Sharma, A., Kumar, A., Singh, J., & Sharma, A. 2024. PGPR-enabled bioremediation of pesticide and heavy metal-contaminated soil: a review of recent advances and emerging challenges. *Chemosphere* 362, 142678.

Sharma, K., & Kumar, P. 2020. Mitigating the effect of biofertilizers on morphological and biochemical level in pearl millet grown under mercury toxicity. *Journal of Pharmacognosy and Phytochemistry* 9(4), 955–961.

Singh, A., & Kostova, I. 2024. Health effects of heavy metal contaminants Vis-à-Vis microbial response in their bioremediation. *Inorganica Chimica Acta* 568, 122068.

Singh, D., & Gurjar, B.R. 2022. Nanotechnology for agricultural applications: facts, issues, knowledge gaps, and challenges in environmental risk assessment. *Journal of Environmental Management* 322, 116033.

Smith, R.S. 1992. Inoculant formulation and delivery systems. *Canadian Journal of Microbiology* 38(6), 485–492. https://doi.org/10.1139/M92-080

Solanki, M.K., Joshi, N.C., Singh, P.K., Singh, S.K., Santoyo, G., de Azevedo, L.C.B., & Kumar, A. 2024. From concept to reality: transforming agriculture through innovative rhizosphere engineering for plant health and productivity. *Microbiological Research* 279, 127553.

Suh, J.S., Jiarong, P., & Toan, P.V. 2006. Quality control of biofertilizers. In: *Biofertilizers Manual* (pp. 112–115). Forum for Nuclear Cooperation in Asia.

Tang, X., Li, J.H., Yang, J.Y., Xiang, Z.C., He, Y.Y., Huang, Y.Z., Zhou, N., Luo, W., & Zhou, Z. 2025. Agricultural sustainability: biochar and bio-based polyurethane coupling coating to prepare novel controlled-release fertilizers. *Industrial Crops and Products* 223, 120296. https://doi.org/10.1016/j.indcrop.2024.120296

Tardy, V., Mathieu, O., Lévêque, J., Terrat, S., Chabbi, A., Lemanceau, P., Ranjard, L., & Maron, P.A. 2014. Stability of soil microbial structure and activity depends on microbial diversity. *Environmental Microbiology Reports* 6(2), 173–183.

Tewari, S., Pooniya, V., & Sharma, S. 2020. Next generation bioformulation prepared by amalgamating Bradyrhizobium, cell free culture supernatant, and exopolysaccharides enhances the indigenous rhizospheric rhizobial population, nodulation, and productivity of pigeon pea. *Applied Soil Ecology* 147, 103363.

Trabelsi, D., Mengoni, A., Ben Ammar, H., & Mhamdi, R. 2011. Effect of on-field inoculation of *Phaseolus vulgaris* with rhizobia on soil bacterial communities. *FEMS Microbiology Ecology* 77(1), 211–222.

Tsoi, R., Dai, Z., & Zhang, Z. 2019. Emerging strategies for engineering microbial communities. *Biotechnology Advances* 37(6), 107372. https://doi.org/10.1016/j.biotechadv.2019.03.011

Vafa, Z.N., Sohrabi, Y., Mirzaghaderi, G., Heidari, G., Rizwan, M., & Sayyed, R.Z. 2024. Effect of bio-fertilizers and seaweed extract on growth and yield of wheat (*Triticum aestivum* L.) under different irrigation regimes: two-year field study. *Chemosphere* 364, 143068. https://doi.org/10.1016/j.chemosphere.2024.143068

Vyas, P., & Bansal, R. 2023. Challenges in the compatibility of microbial inoculants with agrochemicals. In: S. Kaur, V. Dwibedi, P.K. Sahu, & G.S. Kocher (Eds.), *Metabolomics,*

Proteomes and Gene Editing Approaches in Biofertilizer Industry. Singapore: Springer. https://doi.org/10.1007/978-981-99-3561-1_8

Wang, R., Peng, B., & Huang, K. 2015. The research progress of CO2 sequestration by algal biofertilizer in China. *Journal of CO2 Utilization* 11, 67–70. https://doi.org/10.1016/j.jcou.2015.01.007

Wang, Z., Chen, Z., Kowalchuk, G.A., Xu, Z., Fu, X., & Kuramae, E.E. 2021. Succession of the resident soil microbial community in response to periodic inoculations. *Applied and Environmental Microbiology* 87(9), e00046-21.

Wang, Z., Solanki, M.K., Kumar, A., Solanki, A.C., Pang, F., Ba, Z.X., Niu, J.Q., & Ren, Z.X. 2023. Promoting plant resilience against stress by engineering root microenvironment with Streptomyces inoculants. *Microbiological Research* 277, 127509.

Wang, Z., Fu, X., & Kuramae, E.E. 2024. Insight into farming native microbiome by bioinoculant in soil-plant system. *Microbiological Research* 285, 127776.

Wu, L., Wang, H., Zhang, Z., Lin, R., Zhang, Z., & Lin, W. 2011. Comparative metaproteomic analysis on consecutively *Rehmannia glutinosa*-monocultured rhizosphere soil. *PloS One* 6(5), e20611.

Xiong, W., Guo, S., Jousset, A., Zhao, Q., Wu, H., Li, R., Kowalchuk, G.A., & Shen, Q. 2017. Bio-fertilizer application induces soil suppressiveness against Fusarium wilt disease by reshaping the soil microbiome. *Soil Biology and Biochemistry* 114, 238–247.

Yadav, R., Bhullar, M.S., Kaur, S., Kaur, T., & Jhala, A.J. 2017. Weed control in conventional soybean with pendimethalin followed by imazethapyr+ imazamox/quizalofop-p-ethyl. *Canadian Journal of Plant Science* 97(4), 654–664.

Zhang, N.N., Sun, Y.M., Li, L., Wang, E.T., Chen, W.X., & Yuan, H.L. 2010. Effects of intercropping and inoculation with Rhizobium on yield and bacterial community in the rhizosphere of broad bean (*Vicia faba* L.). *Biology and Fertility of Soils* 46, 625–639.

Zhang, Y., Gao, X., Shen, Z., Zhu, C., Jiao, Z., Li, R., & Shen, Q. 2019. Pre-colonization of PGPR triggers rhizosphere microbiota succession associated with crop yield enhancement. *Plant and Soil* 439, 553–567.

4

The Regulation of Nutrient Uptake, Transportation, and Physiological Functions of Microbial Consortia in Rice Crop Under Abiotic Stresses

Mario Alberto Soltero-Rios

Departamento de Genética y Biología Molecular, Centro de Investigación y de Estudios Avanzados del IPN, Ciudad de México, México

Corresponding author: mario.soltero@cinvestav.mx

4.1 Introduction

Rice (*Oryza sativa* L.) is one of the most widely cultivated and consumed crops worldwide due to its central role in the diets of many cultures and its frequent use as a raw material in the food industry. It is estimated that at least half of the global population relies on rice as their primary source of dietary intake. This occurs because rice is a highly domesticated crop that serves as a significant source of carbohydrates, proteins, vitamins, and minerals (Sarma et al., 2023). However, a growing concern has emerged: as the global population continues to expand rapidly (projected to reach 9.9 billion by 2050), current rice production will be insufficient. To meet future demand, global rice output must increase by at least 1% annually (Azad et al., 2024). Nevertheless, the issue is not only about cultivating more; the current philosophy of sustainability emphasizes producing more with existing crops by reducing yield losses rather than simply increasing cultivation area. Among the primary causes of such losses is abiotic stress, which has become increasingly prevalent due to environmental changes. Events such as drought, salinity, extreme heat, and cold now affect vast agricultural areas, disrupting plant nutritional processes and reducing food production, impacting an estimated 70% of all crops (Akram et al., 2019). Rice is no exception, as these stressors significantly impair its growth, viability, and yield potential, resulting in an average yield reduction of approximately 32%. In response to this situation, several strategies have been explored to enhance plant tolerance to abiotic stress (Azad et al., 2024). One particularly promising approach is the exploitation of plant–microbe interactions. Microbial consortia associated with rice (and virtually all plants) have been shown to play a crucial role in regulating the plant's nutritional dynamics under stress conditions, ultimately contributing to improved stress tolerance. This is achieved through a variety of molecular mechanisms that microbes trigger as part of their symbiotic relationship with the plant (Negi et al., 2024).

Microbial Consortia in Plant Science and Sustainable Agriculture, First Edition. Edited by Ajay Tomar, et al.

This chapter explores how microbial consortia regulate diverse molecular and physiological aspects of rice plants to enhance their adaptability to abiotic stress. It also provides a general overview of current scientific knowledge surrounding these interactions and discusses recent research efforts aimed at translating this knowledge into practical technologies that can mitigate crop yield losses caused by environmental stress.

4.2 Root System in Rice and Essential Nutrients

4.2.1 Root Structure

It is important to begin by establishing that in the absorption of nutrients by rice (and many other plants) the determining factor in initiating and carrying out this process is the architecture of the root system because it plays a fundamental role in the plant's ability to explore the soil and absorb water and essential nutrients, which is crucial for this plant, as it is a semiaquatic species (Yang et al., 2004a).

Rice has fibrous root systems, which means it is composed of multiple, functionally distinct root types, and each one contributes in specific ways to the overall capacity of the plant to absorb water and nutrients. These root types include the primary root, seminal roots, crown roots, and lateral roots, each with specialized developmental origins, anatomical features, and physiological roles (Figure 4.1) (Verma et al., 2022b).

The primary root, also known as the radicle, is the first root to emerge from the seed and plays a crucial role during the early stages of seedling development (Meng et al., 2019). The primary root is critical for early seedling establishment. In young rice seedlings, it grows for 7–10 days after germination and then stops. Its function is to explore the soil in search of nutrients and consequently establish the signaling pathways that regulate plant growth in response to the presence of nutrients (Verma et al., 2022b).

The seminal root of rice emerges from the seed shortly after germination; it is common to confuse it with the primary root, but the difference lies in that the primary root arises from the embryo as the radicle while the seminal roots arise from the embryonic axis but develop slightly later as additional roots; its function therefore is to complement the absorption during the first days, especially when the primary root is still developing and the crown roots have not been established. It has even been observed that they have an important role in the adaptability of the plant (Kawai et al., 2022a; Zou et al., 2021).

Crown roots are considered a very important and distinct component of root architecture compared to those mentioned earlier. First, they are postembryonic, emerging four to five days after germination. Their main attribute is that

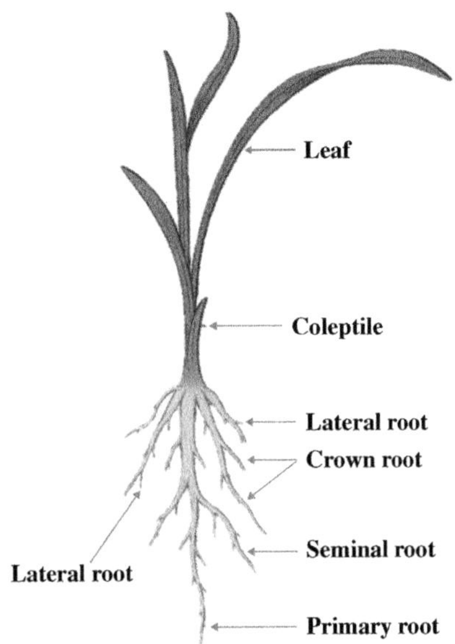

Figure 4.1 Rice root system.

they are dominant in the root system, as they have the largest number of signaling sites for macro- and micronutrients. They also have an anchoring function, and, very importantly for the context of this chapter, most of the symbiotic interactions with microorganisms take place in this type of root (He et al., 2024; Luo et al., 2022).

Finally, it is important to explain the lateral roots, which arise from the crown and seminal roots. Their main role is to expand the absorptive surface area of the root system. To this end, they are divided into two important groups: large laterals and short laterals, both of which perform nutrient uptake but in different layers of the space where the root system develops and with varying nutrient specificity (Huei Kao, 2013; Kawai et al., 2022b).

4.2.2 Essential Nutrients

Rice requires various macro- and micronutrients for its optimal development and, consequently, better productivity. When these nutrients are available, they can be absorbed through the root system, allowing them to generate specific signals that trigger various metabolic pathways in the plant and therefore contribute to specific physiological processes for each nutrient, all of which are essential for plant development. In Table 4.1, the main essential nutrients for rice are detailed, along with their role in plant physiology (Ye et al., 2019).

It is necessary to clarify that there are many other trace nutrients that have key and particular effects on the development of rice; however, those mentioned previously turn out to be the main ones.

4.3 Mechanisms of Nutrient Uptake and Transport in Rice

Once the close relationship between the nutrients present in the environment and the root system capable of capturing them has been established, it is relevant to understand the mechanisms or processes that allow said capture (nutrient uptake) as well as the processes capable of distributing the nutrients to the different sections of the plant (transport) where different metabolic processes are carried out, all involved in the physiology of rice.

4.3.1 Nutrient Uptake

The first process in absorption is that by which nutrients can be internalized into the vascular cylinder of the root and subsequently enter the cells that make up this tissue and others. Nutrients move radially through epidermis, exodermis, cortex, endodermis, and pericycle. To this end, three distinct pathways have been described and managed: apoplastic, symplastic, and transcellular. These pathways are usually combined (Robe & Barberon, 2023).

The apoplastic pathway is responsible for the movement of solutes through the apoplast, that is, between plant cells and their cell walls. In this process, nutrients do not cross the plasmatic membranes, so their diffusion is passive and therefore does not require an energy input. Because cations are favored, this pathway is highly efficient in terms of flow volume, as it allows rapid transit (Robe & Barberon, 2023).

The other two pathways (symplastic and transcellular) require ions to cross plasma membranes at least twice. The symplastic pathway refers to the movement of nutrients through the cytoplasm of root cells, passing from one cell to another via plasmodesmata.

Table 4.1 Key nutrients required by rice.

Name	Absorbed as	Function	Common absorption site	References
		Macronutrients		
Nitrogen (N)	NO_3^-/NH_4^+	Primary component of amino acids, nucleic acids, and chlorophyll	Primary, seminal, and crown roots	Shrestha et al. (2020), Sun et al. (2016)
Phosphorus (P)	$H_2PO_4^-/HPO_4^{2-}$	Energy transfer (ATP), root development	Primary and lateral roots	Dissanayaka et al. (2018), Zheng et al. (2016)
Potassium (K)	K^+	Osmotic regulation, stomatal opening, enzyme activation, root development	Seminal and crown roots	Lian et al. (2023), Shrestha et al. (2020)
Calcium (Ca)	Ca^{2+}	Cell signaling, cell elongation	Seminal root	Wu and Wang (2012), Yamada et al. (2015)
Magnesium (Mg)	Mg^{2+}	Enzyme cofactor, photosynthesis (central component of chlorophyll)	Primary root	Kobayashi et al. (2013), Meng et al. (2023)
		Micronutrients		
Iron (Fe)	Fe^2	Electron transport, chlorophyll synthesis, respiratory enzymes	Lateral roots and root hairs	Shrestha et al. (2020), Zhang et al. (2020)
Manganese (Mn)	Mn^{2+}	Enzyme activator, photosynthesis	Lateral roots	Shrestha et al. (2020)
Zinc (Zn)	Zn^{2+}	Hormonal regulation, auxin synthesis	Root hairs	Kandil et al. (2022)
Molybdenum (Mo)	MoO_4^{2-}	Cofactor in the enzyme nitrate reductase	Crown roots	Li et al. (2018)
Chlorine (Cl)	Cl^-	Ionic balance	Crown roots	Guo et al. (2024)

This requires the participation of specific transport proteins, such as cotransporters, ion channels, and facilitated transporters. This obviously means that an energy input is required (active diffusion), but, at the same time, this makes it a controlled and selective pathway with the capacity for redistribution, qualities that the apoplastic pathway lacks (Keller, 2020).

In the transcellular pathway, nutrients pass through each cell independently, by a series of successive endocytosis and exocytosis events, or protein-mediated transport. In each cell, the nutrient enters the cytoplasm by crossing the plasma membrane, moves through the cytosol, and then exits again to the next cell. This pathway is the most energy-intensive, but it also exerts the greatest control over nutrient absorption in the plant. Therefore, its activity is relevant in emergency or stressful situations, such as severe nutritional deficiency or the absorption of complex nutrients (Barberon & Geldner, 2014).

It is important to mention that these pathways do not act independently but rather act in combination to respond to the dynamics of stimuli that the root system receives from the environment, where said dynamics are regulated by at least two key aspects: the control of how many nutrients are internalized and their accumulation.

4.3.2 Nutrient Transport

Once nutrients have passed from the environment into the root cells as part of the absorption process, the next crucial step is transport, because it is responsible for carrying and distributing nutrients to the different tissues and areas of the plant for their storage or their use in a controlled manner, allowing for crucial metabolic processes in rice physiology. Nutrient transport is dynamic and varies depending on the stage of development, the nutritional status of the plant, the demand of the growing organism, and environmental conditions such as waterlogging (White, 2012).

There are two main forms of transport: by xylem and by phloem. There is also transport resulting from the interaction between the two aforementioned forms. Xylem is the main tissue that conducts water and nutrients absorbed by the roots. Flow in the xylem is unidirectional, and upward, from the root to the aerial organs (leaves, stems, panicles), driven mainly by leaf transpiration. Movement in the xylem depends on cohesion tension, which explains how water and nutrients rise (Dinneny & Yanofsky, 2004). Transpiration at the stomata generates a suction force that moves molecules through columns from the roots. The adhesion of water to the xylem walls prevents the disruption of the flow. This mechanism is passive but effective (Venturas et al., 2017).

On the other hand, the phloem is responsible for the transport of photosynthetic assimilates and nutrients. This dynamic, bidirectional transport adapts to the physiological demands of growing organs. Phloem transport is described by the pressure flow model, where phloem load in mature leaves (nutrients and photosynthetic assimilates) is actively loaded into companion cells, generating a pressure that pushes them in solution toward organs such as roots, grains, or shoots. This process occurs independently of transpiration, allowing compounds to be transported in multiple directions simultaneously (Jensen, 2018).

It is clear that there is a complex and regulated system between nutrient absorption and transport, which does not go in one direction since it is a constant feedback between cells found in the root, leaves, and stem (White, 2012). The important thing to understand is that despite being a complex process with multiple factors involved (many known and many others unknown), transporters turn out to be key for this review since they are the most evident molecular actuators, being the bridge between the absorption and transport processes, mainly due to their functions of internalizing or externalizing nutrients and other molecules from one cell type to another (Gill et al., 2021). Transporters are essentially proteins capable of interacting with nutrients or other molecules. There are specific transporters for each of them. Specific transporters and their specific functions in nutrient management will be discussed later.

4.4 Microbial Consortia and Their Role in Rice Nutrition

It is well known that many microbial genera maintain a significant symbiotic relationship with plants, and several fundamental biological processes require the presence of both partners. Rice is no exception; in fact, its interactions could be considered more varied, since, as a crop that grows in flooded soils, a very special ecological niche is established and it is a microbial consortia composed of bacteria, archaea, fungi, and actinomycetes, capable of interacting with the rhizosphere, endosphere, and phyllosphere (Doni et al., 2022). Broadly speaking, microbial consortia promote increased nutrient availability, stimulate plant growth, and mediate responses to various types of stress. It is important to delve deeper into the role of microbial consortia in rice nutrition, especially by classifying each consortium by the region in which it develops and the different microbial genera involved (Mia et al., 2023).

4.4.1 Rhizosphere Microbes

In the understanding that the rhizosphere refers to the area of soil within 1–10 mm surrounding the plant roots, the microorganisms associated with it are evidently characterized by generating effects on the plant roots and since this is where nutrient absorption occurs, rhizospheric microorganisms turn out to be those that have the greatest participation in the aspect of nutrient availability (Misu et al., 2025).

Under standard crop management practices, the rice rhizosphere microbiome is dominated by bacterial genera, such as *Azospirillum* spp., *Herbaspirillum* spp., *Rhizobium* spp., and *Pseudomonas* spp. These bacteria are mainly responsible for fixing nitrogen and solubilizing phosphate in the rhizosphere, improving the amount that the plant can access. There are also other phyla like Firmicutes (*Bacillus* spp.), Actinobacteria (*Streptomyces* spp.), and Bacteroidetes (*Flavobacterium* spp.) with the same kind of activities in nutrient supply (Dong et al., 2021).

However, there are other microorganisms with relevant activities; for example, fungi from the phylum Glomeromycota and Ascomycota that have symbiotic activity for the uptake of phosphorus, copper, and zinc. The relevance of the participation of Cyanobacteria and Chlorophyta in nitrogen fixation and organic matter degradation has even been observed (Misu et al., 2025).

4.4.2 Phyllosphere Microbes

Unlike what was seen previously, the microorganisms of the phyllosphere have a very different activity, but still concerning the nutrition of the plant, since they participate in the management of nutrients already stored in aerial parts of the plant (especially leaves and stems). In this case, bacteria belonging to the phyla Proteobacteria, Actinobacteria, and Firmicutes are the most recognized, as are fungi from the phylum Ascomycota. In general, their activities include nutrient recycling, redistribution, and assimilation. They are also capable of performing biological processes that are not strictly relevant to nutrition (such as biocontrol). Finally, they have stress response activity (with a nutritional focus), which will be addressed later (Danso Ofori et al., 2024; Doni et al., 2022).

4.4.3 Endosphere Microbes

These types of microorganisms are characterized by colonizing internal plant tissues without causing disease; on the contrary, they fulfill symbiotic roles. They can be found in virtually any plant tissue or organ and have very varied functions. In fact, the same phyla mentioned earlier are present in this type of microorganism, with the exception of distinct genera or species. In general, they contribute to nutrient acquisition, hormonal balance, and stress response (Wang et al., 2021).

The functional roles of microbial communities vary depending on their spatial niche, as shown in Figure 4.2, where their main contributions to nutrient cycling and stress adaptation in rice are summarized. There is another important aspect in which rhizosphere, endosphere, and phyllosphere microorganisms participate, which has not been explored as much but has recently been considered a novel and relevant approach. This is the effect of microorganisms on the expression or repression of genes related to nutrient absorption and transport. In this case, the most studied aspect is the regulation of genes such as OsPT2, OsHAK5, and OsIRT1, which express transporters responsible for interacting with different nutrients (Wu & Cheng, 2014).

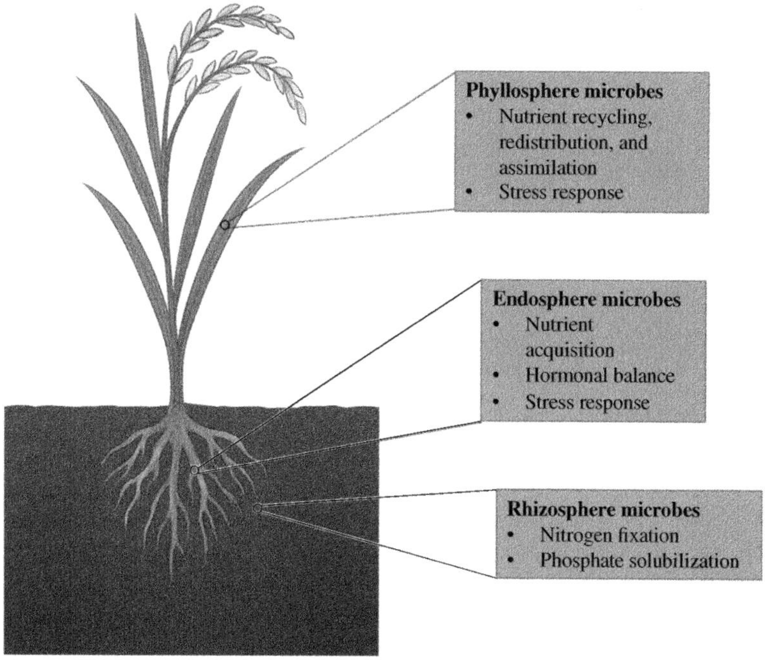

Figure 4.2 Interaction of microbial consortia with different zones of the rice plant.

4.5 Abiotic Stress in Rice Crops: Effects and Physiological Responses

Abiotic stress is understood as any non-living agent that causes damage or undesired stimuli to the plant (in this case, rice) and is typically related to environmental and climate factors. Abiotic stressors have been reported to be one of the main causes of yield losses in rice crops, evidently meaning that these stressors have significantly negative effects on plant physiology. Some of the most common and relevant agents in rice care and their effects are discussed in the following subsections (Mahmood-ur-Rahman et al., 2019).

4.5.1 Drought Stress

It is one of the major abiotic stressors that interfere with normal rice physiological processes, resulting in massive crop losses. It decreases leaf water potential, reduces stomatal conductance and photosynthetic rate, and causes impaired nutrient uptake and translocation. As a physiological consequence, stomatal closure occurs, which is regulated by increased abscisic acid (ABA) and an accumulation of osmolytes. The plant's response to this stress is to upregulate genes that respond to the production of ABA and the transcription factor dehydration-responsive element-binding (DREB), which is responsible for promoting the expression of stress response genes (Ma et al., 2024).

4.5.2 Salinity Stress

This is an accumulation and potential toxicity due to Na^+ and Cl^- ions, which causes a breakdown in the K^+/Na^+ balance and therefore compromises the integrity of the membrane. As a result, the genes of the Salt Overly Sensitive (SOS) pathway are induced, the high-affinity K^+ transporters (HKT) and sodium/hydrogen exchanger (NHX) transporters are activated, and antioxidant mechanisms are upregulated. This causes a decrease in transpiration and enhances ABA signaling.

4.5.3 Flooding Stress

Excessive water accumulation in the environment leads to oxygen deprivation and, therefore, inhibits aerobic respiration, allowing the accumulation of toxic metabolites. In response, genes such as SUB1A and SNORKEL1/2 are induced. These immersion-responsive genes are very similar in function to another gene activated under these conditions, an ethylene-responsive transcription factor (ERF). The result of all this is a shift to anaerobic respiration (ethanol and lactate are produced), promoting ethylene-mediated shoot elongation or growth quiescence (Hasan et al., 2024).

4.5.4 Heat Stress

Temperatures above the optimum for rice development (greater than 34 °C) can generate stress that causes membrane protein denaturation and destabilization, in addition to reducing pollen viability. In response, heat shock proteins (HSPs) and induction of heat-responsive transcription factors (HSFs) are activated. This generates an increase in transpiration, a heat shock response, and changes in leaf orientation to mitigate stress (Ren et al., 2023).

4.5.5 Cold Stress

Temperatures lower than the optimum for rice development (less than 8 °C) can generate stress that causes impaired germination, chlorosis, slow metabolism, and growth inhibition. This causes an upregulation of cold-responsive (COR) genes and C-repeat binding factor/dehydration-responsive element binding (CBF/DREB)-mediated pathways and an induction of antioxidant enzymes, which reduces membrane fluidity and impairs enzyme function, as well as accumulation of osmoprotectants (Shahzad et al., 2024).

4.5.6 Nutrient Deficiency Stress

As mentioned earlier in this chapter, fewer nutrients result in an inhibition of the normal rate of development. The response to this is the gene expression of high-affinity nutrient transporter proteins such as OsPTs, OsAMTs, and OsZIPs, and the activation of signaling pathways linked to the aforementioned genes or others. Therefore, this should lead to enhanced root growth and a reallocation of nutrients from older to younger tissues (Cai et al., 2024).

There are various agents that interfere with the normal development and activity of rice, but, at the same time, there are plant response systems to these agents. Interestingly, these systems are dynamic, meaning they are activated or deactivated only when the plant requires them. The objective is always the same: to provide a state of resistance to adverse conditions that allows the plant to develop or not perish. Understanding these mechanisms is key to developing better agricultural techniques.

4.6 Response of Microbial Consortia to Abiotic Stress Conditions

Until now, the mechanisms of abiotic stress response in plants have been addressed from the plant's own perspective. However, contrary to earlier assumptions, these responses are not exclusive to the plant. Microbial consortia associated with plants also perceive stress signals and generate specific molecular responses. These microbial signals can trigger unique adaptive mechanisms that the plant alone is unable to initiate, or they can synergize with plant-derived pathways to amplify the overall stress response. In both scenarios, a constant feature is observed: a dynamic, microbe-mediated process that contributes to plant resilience and adaptability (Hossain et al., 2022; Singh et al., 2023). Therefore, the objective of the following section is to highlight the critical role of microbial consortia in enhancing plant adaptability under abiotic stress conditions.

In general, it can be established that the primary mechanism through which microorganisms promote plant adaptability under abiotic stress is the production of a wide range of bioactive compounds, including phytohormones, osmolytes, and antioxidants. More recently, attention has turned to the emerging role of nucleic acids in this process, where microRNAs (miRNAs), small interfering RNAs (siRNAs), and extracellular DNA and RNA have been identified as fundamental signaling molecules. These molecules collectively function as mediators of interkingdom communication, activating diverse signaling cascades within the host plant. Importantly, different types of abiotic stress are addressed by distinct microorganisms, each eliciting unique physiological and molecular effects on the plant. This complexity underscores the need for a more in-depth examination of the specificity and functional outcomes of microbial responses to environmental stressors (Ding et al., 2024; Khraiwesh et al., 2012).

4.6.1 Drought Stress

Rhizospheric bacteria from the genera *Pseudomonas, Bacillus, Rhizobium*, and *Azospirillum* are known for their ability to promote plant growth under drought stress. This growth promotion is closely associated with enhanced root development, which in turn improves water uptake. A key mechanism underlying this effect is the microbial production of phytohormones such as indole-3-acetic acid (IAA). Additionally, these bacteria can secrete exopolysaccharides (EPSs) that improve soil structure and water retention, effects that occur independently of the plant's intrinsic physiology. On the other hand, mycorrhizal fungi enhance the plant's ability to access water by binding soil particles into aggregates, thereby improving soil porosity and reducing water evaporation. These fungi also modulate hormonal pathways that activate the plant's antioxidant defenses under drought conditions. Furthermore, certain fungal and bacterial species are capable of producing volatile organic compounds (VOCs) that similarly regulate plant growth responses (Gu et al., 2024; Shaffique et al., 2022).

Finally, the adaptive mechanisms described earlier are also influenced by gene expression patterns modulated by microorganisms. In this context, microbially induced epigenetic modifications can alter the expression of stress-responsive genes, such as those involved in antioxidant pathways, thereby contributing to the plant's resilience to drought (Singh et al., 2023).

4.6.2 Salinity Stress

In this context, it is most commonly observed that microorganisms exhibit mechanisms such as osmolyte production, extracellular protease synthesis, and activation of Na^+/H^+ antiporters. They may also induce the biosynthesis of phytohormones such as auxins, cytokinins, and gibberellins. Under saline stress conditions, the plant hormone ABA is upregulated, promoting the accumulation of osmolytes in root vacuoles and enhancing the uptake of Ca^{2+} and K^+, which are critical for maintaining cellular homeostasis (Liu et al., 2022).

Since salt stress negatively affects the uptake of nitrogen, phosphorus, potassium, and water, microbial consortia associated with the rhizosphere play a fundamental role by acidifying the root environment and chelating key minerals, thereby enhancing their bioavailability. This is achieved through the synthesis of specific microbial metabolites, such as organic acids and siderophores, which solubilize nutrients and facilitate their uptake by the plant, while also preventing their precipitation or accumulation in toxic ionic forms that would otherwise intensify salt stress. Moreover, these microorganisms can activate plant signaling pathways that lead to the upregulation of antioxidant genes, including those encoding peroxidase, superoxide dismutase (SOD), catalase (CAT), nitrate reductase (NR), and glutathione reductase (GR) enzymes that are essential for alleviating oxidative damage associated with salt stress (Acharya et al., 2024; Singh et al., 2023).

4.6.3 Flooding Stress

Certain facultative and strict anaerobic microorganisms play a central role in the response to this type of stress by activating metabolic pathways such as fermentation (via *adhE* and *ldh* genes) and denitrification (via *nar, nir,* and *nor* genes), which help modulate the redox potential of the rhizosphere. This facilitates the mobilization of nutrients such as iron and manganese, supporting plant metabolism under hypoxic or anoxic conditions (Martínez-Arias et al., 2022).

Additionally, microbial activity leads to the release of metabolites such as ethylene, lactic acid, and VOCs, which promote the formation of aerenchyma and adventitious roots, as well as the activation of hormonal pathways through the action of microbially induced transcription factors, notably the ERFs. Finally, in response to the accumulation of reactive oxygen species (ROS) caused by this stress, certain endophytic microorganisms stimulate the secretion of antioxidant compounds and the activation of plant genes encoding antioxidant enzymes, SOD, CAT, and GR (Tyagi et al., 2024).

4.6.4 Heat and Cold Stress

In this case, microbial consortia are capable of conferring thermotolerance to plants by promoting the expression of HSPs, such as groES, groEL, dnaK, htpG, and clpB, which act as molecular chaperones that stabilize protein folding processes compromised by heat stress. These microbes also produce EPS that buffer temperature fluctuations in the rhizosphere (Dastogeer et al., 2022).

Additionally, the synthesis of phytohormones plays a central role. For instance, IAA and cytokinins contribute to heat stress signaling by regulating the expression of transcriptional elements such as the heat shock factors. Other microbes interact with Mitogen-Activated Protein Kinase (MAPK) signaling cascades and DREB/HSF transcription factors, which in turn regulate the expression and translation of HSP70, HSP90, and various antioxidant enzymes in rice (Singh et al., 2023).

Moreover, these effects are often accompanied by the induction of epigenetic modifications (such as DNA methylation and histone acetylation) that modulate the transcriptional activity of genes associated with thermotolerance (Verma et al., 2022a).

In the case of cold stress, most of the mechanisms remain similar; signaling pathways are affected, phytohormones are produced, and transcription factors are activated. However, the main difference lies in the identity of the molecules expressed in response to microbial consortia. Specifically, this involves the gene expression of cold regulated proteins (such as OsCOR410), the accumulation of cryoprotective compounds like trehalose and EPSs, and the activation of signaling pathways involving CBF and DREB transcription factors, which in turn induce the expression of COR genes (Muralibabu et al., 2024).

As expected, these responses are driven by a diverse range of microorganisms, with notable representatives from the genera *Bacillus, Pseudomonas*, and *Thermus*, as well as various fungal species (Singh et al., 2023). The complexity of plant–microbe interactions under abiotic stress conditions is further exemplified by specific microbial strains that confer adaptive advantages to rice plants. These functional associations are summarized in Table 4.2.

Table 4.2 Specific mechanisms by which various microbial strains mitigate the impact of abiotic stress in rice.

Strain	Type of stress treated	Action	Effect on rice	References
		Bacteria		
Pseudomonas putida	Drought	Boosts IAA synthesis	Increases shoot and root biomass and water content	Chieb and Gachomo (2023)
Bacillus subtilis NM-2	Drought	Synthesizes 1-aminocyclopropane-1-carboxylic acid (ACC), deaminase activity, and phosphate-solubilizing activity	Improves root length and surface; accumulates osmoprotectant	Mahreen et al. (2023)
Brevibacterium frigoritolerans W19	Salinity	Modulation of SOD1, CATa, NHX1, and PAL1 proteins	Regulates ionic balance, reduces ROS, and stabilizes membranes	Gupta et al. (2023)
Pantoea stewartii A	Salinity	Modulation of OsSOS1, OsPIN1, and OsCIPK15	Promotes osmoprotective profile and improves ionic balance	Yang et al. (2024b)
Rhizobium sp. IIRR N1	Heat	Expression of OsHsp90, OsHsp100, and silicon transporters (OsLsi1 and OsLsi2)	Improves chlorophyll content, electrolyte leakage, and relative water content	Chaganti et al. (2023)
Gluconacetobacter diazotrophicus	Heat	Expression of heat shock proteins, hormone-related genes (OsIAA6), and silicon transporters (OsLsi1 and OsLsi2)	Improves chlorophyll content and decreases electrolyte leakage	Chaganti et al. (2023)
Kosakonia sp. CIR2	Cold	Synthesis of ACC and participation in the activation of por DREB1/CBF	Reduces phenological delay and oxidative damage; protects cell membranes and maintains growth	De Souza et al. (2021)
Pseudomonas mosselii	Cold	Increases in proline and reduced glutathione levels; modulation of cold tolerance gene expression	Decreases electrolyte leakage, indicating increased membrane integrity	Balasjin et al. (2024)

Strain	Type of stress treated	Action	Effect on rice	References
		Fungi		
Serendipita indica	Drought	Regulates the expression of the P5CS gene (proline accumulation) and other genes related to total antioxidant capacity. Also regulates miR396 and miR159, related to the production of ABA	Improves water and nutrient uptake and increases chlorophyll synthesis	Boorboori and Zhang (2022)
Glomus etunicatum *Glomus geosporum*	Salinity	Increased proline as an osmoprotector. Regulation of anthocyanins as compounds associated with antioxidant capacity	Stable proteins and cell membranes	Tisarum et al. (2020)
Funneliformis mosseae	Heat	Modulation of genes related to hormonal pathways and OsMGD2, OsPAP23, OsSPX3, and OsNIA1 proteins related to heat stress	Increases root biomass; improves root architecture and photosynthetic efficiency	Guigard et al. (2023)
Piriformospora indica	Drought and cold	Strengthens the plant's antioxidant and osmotic regulation mechanisms and modulates the expression of antifreeze proteins (CBF and COR factors)	Enlarges the root zone of plants and enhances their photosynthetic efficiency	Shi et al. (2024)

4.7 Microbial Mechanisms for Enhancing Nutrient Availability Under Abiotic Stress

It is clear that different abiotic stressors generate a range of adverse effects in rice, from DNA damage to physiological alterations affecting stems, roots, and leaves. While each of these impacts is relevant for understanding how stress influences plant development and productivity, the specific effect of interest in this chapter is the one related to the reduced ability of rice to absorb and utilize nutrients under stress conditions. During such events, the plant is exposed to an altered soil chemistry that is often suboptimal, as the stressor likely impairs ion mobility and compromises root function (Sarma et al., 2023).

Under these conditions, a key factor comes into play, not only helping to mitigate the effects of the stressor through various mechanisms from the onset of stress but also facilitating specific processes related to nutrient uptake. This factor is the action of microbial consortia. The following section will examine the microbial mechanisms (many already mentioned) but in a nutrient uptake context with the aim of understanding their relevance to plant adaptability due to efficient nutrient acquisition under abiotic stress and its subsequent relief (Singh et al., 2023).

In all cases of stress agents, a common microbial mechanism is the production of EPSs, which, among their various effects, contribute to soil moisture retention. A particularly relevant indirect effect is that, by increasing soil moisture, nutrient diffusion toward the root system is enhanced, facilitating nutrient absorption. Additionally, the microbial production of osmolytes and VOCs has been previously discussed, and, in this context, they improve root hydraulic conductivity, further facilitating nutrient transport within the plant (Almeida et al., 2023; Wei et al., 2024). Finally, the microbial production of auxins and cytokinins is also highly relevant, as these phytohormones stimulate root growth and branching, thereby increasing the absorptive surface area and enhancing the plant's overall capacity to absorb a greater quantity and diversity of nutrients (Mishra et al., 2022).

In heat and drought stress, certain microbial consortia are capable of stimulating the expression of genes such as *HAK*, *AKT*, and *AMT* (nutrient transporters), so the uptake and transport of potassium and ammonium in the plant are improved, leading to better reserves of two of the most critical nutrients for plant metabolism. Evidently, these are not the only transporters upregulated under microbial influence (Parasar et al., 2024).

On the other hand, under salt stress, the expression of genes such as *NHX1* and *HKT1* is induced, leading to the restoration of key ionic homeostasis essential for nutrient absorption. These genes facilitate the extrusion of certain ions and the retention of others according to the plant's physiological needs (Gupta et al., 2023). A similar nutritional effect is observed following cold stress, but it is mediated through the induction of COR genes such as CBF/DREB by microbial consortia, which help maintain nutrient uptake efficiency under low-temperature conditions (Shi et al., 2024). Microorganisms thus possess the capacity to trigger multiple responses, among which enhanced nutrient absorption stands out as a crucial factor in plant resilience to various types of stress as shown in Figure 4.3. This establishes a key conceptual bridge between stress tolerance and nutrient dynamics, mediated by microbial consortia.

Furthermore, it is important to emphasize that these benefits manifest across multiple levels: genetic, molecular, biochemical, cellular, physiological, and even ecological.

Thanks to a wide range of mechanisms, including organic acid production, secretion of enzymes and other proteins, and nutrient fixation within the environment. Therefore, gaining a deep understanding of these interactions is essential for constructing a comprehensive view of how rice plants withstand environmental challenges and for designing strategies to promote or enhance this resilience (Munir et al., 2022).

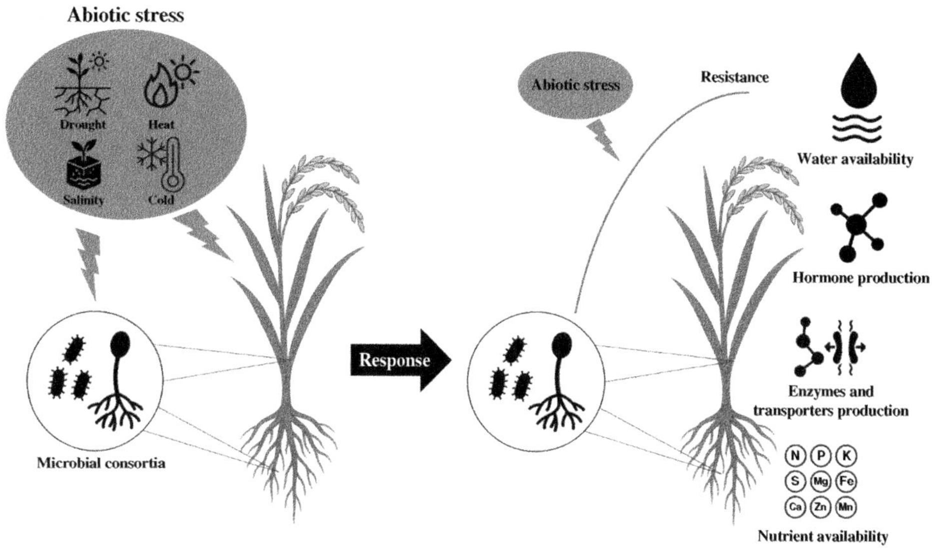

Figure 4.3 Effect of microbial consortia on nutrient uptake in response to abiotic stress.

4.8 Perspectives and Challenges in the Research of Microbial Consortia and Their Regulation Under Stress Conditions

Based on the aforementioned research, it becomes clear that current knowledge about microbial consortia and their dynamics in stress resistance (particularly in rice and other plants) is substantial yet still limited when compared to the full scope of this phenomenon. Although numerous signaling and response mechanisms are known, ranging from the genetic to the physiological, their full potential remains unexplored, and certainly not all existing mechanisms have been identified. Nonetheless, the current understanding has made it possible to establish and propose strategies based on the use of microbial consortia to improve plant productivity under adverse conditions, with promising results that have driven the development of foundational science aimed at uncovering new mechanisms (Misu et al., 2025).

Therefore, future perspectives point toward a deeper investigation of plant gene expression modulated by microbial consortia within the context of functional genomics. The integration of omics sciences such as metagenomics, transcriptomics, proteomics, and metabolomics offers a broad view of the signaling pathways involved and how they might be intentionally manipulated (Kapazoglou et al., 2025). This approach would also help identify microbial strains with specific effects, enabling the design of tailored synthetic consortia by combining nonconventional but effective strains for defined purposes (Che et al., 2024). Furthermore, the potential of genome editing technologies like CRISPR-Cas could be considered for developing artificial consortia that promote abiotic stress resistance in rice and other crops (Duncker et al., 2021). Another perspective is that editing microorganisms or searching for novel strains is not the only path forward; native or well-known strains present in the rice rhizosphere may also generate desirable effects when they are appropriately stimulated. Thus, advancing our understanding of microbial-induced signaling pathways and resistance mechanisms could allow for the activation of specific stress tolerance responses by the application of certain phytohormones, organic acids, or other key molecules (Santoyo et al., 2021).

Given the growing role of computational modeling in all scientific disciplines, it is reasonable to envision its application in this field as well. Integrated models could be developed to combine the knowledge and variables mentioned above, ultimately providing tailored strategies against specific stressors. These models could guide the design of next-generation microbial consortia capable of colonizing, adapting to, and benefiting a wide range of crops, ultimately contributing to enhanced food production (Duncker et al., 2021).

While these perspectives are encouraging and synergize well with current technological advances, there are clear challenges to overcome to realize them. First, the immense microbial diversity and complexity of microbial communities demand time and effort to fully understand species behavior and community interactions underlying resistance mechanisms. Functional redundancy, horizontal gene transfer, and strain-specific effects further complicate efforts to predict or harness microbial behavior (Jawed et al., 2019). Additionally, the context-dependent nature of microbial activity, shaped by plant genotype, soil type, and environmental conditions, means that there is no single, universal response to stress. Understanding the variability in stimulus-response dynamics is therefore essential to reflect the true nature of plant–microbe interactions in real-world settings (Liu et al., 2019).

Regulatory frameworks also pose a major challenge. Since this research touches agricultural systems, any emerging technologies based on microbial consortia must be evaluated and approved by national and international authorities in agriculture and environmental protection. The current regulatory landscape may not fully accommodate these innovations, and new policies may need to be established. Moreover, genome editing remains a sensitive issue, requiring careful risk assessment, particularly when nonnative microbial consortia are introduced into fields, which could cause even subtle ecological imbalances (Nunes, 2012).

Ultimately, the greatest challenge (and the culmination of overcoming all others) will be to establish a strong link between scientific discovery and real-world application, where knowledge about microbial consortia is not only generated but also translated into accessible, effective technologies that farmers and the general population are willing to adopt. Though this transition involves a complex intellectual and practical process, it holds the potential to significantly advance current strategies for improving rice and other crop production (Misu et al., 2025).

4.9 Conclusions

The growing demand for food, combined with the increasing frequency and severity of abiotic stressors due to climate change, presents a significant global challenge for enhancing rice production and ensuring a sufficient food supply. Given that current agricultural strategies are proving ineffective for these conditions, the role of microbial consortia as agents of abiotic stress resistance has gained increasing interest and has shown promising results in addressing this issue. As discussed throughout the chapter, microbial communities are capable of stimulating gene expression related to the production of transporters, enzymes, and a variety of proteins involved in key metabolic processes. These, in turn, trigger physiological effects that strengthen the plant's ability to withstand environmental stressors. Hence, understanding the specific effects of each microbial strain, their localization (rhizosphere, phyllosphere, and endosphere), and their relationship with the molecular and physiological mechanisms that enable nutrient uptake becomes essential. All this knowledge is vital for establishing new strategies in sustainable agriculture, as the generation of novel insights in this field offers encouraging perspectives for achieving this goal. Potential technologies derived from this research include bioinoculants, synthetic microbial consortia, and precision agriculture tools based on computational models. And as always, the guiding premise is that realizing their full potential will require deeper exploration of molecular mechanisms, context-specific effects, and the safe, scalable application of these technologies in real-world agricultural systems. Ultimately, this will be essential for designing climate-resilient and highly productive cropping systems, thereby contributing to global food security amid growing environmental challenges.

References

Acharya, B.R., Gill, S.P., Kaundal, A., & Sandhu, D. 2024. Strategies for combating plant salinity stress: The potential of plant growth-promoting microorganisms. *Frontiers in Plant Science* 15, 1406913. https://doi.org/10.3389/fpls.2024.1406913

Akram, R., Fahad, S., Masood, N., Rasool, A., Ijaz, M., Ihsan, M.Z., Maqbool, M.M., Ahmad, S., Hussain, S., Ahmed, M., Kaleem, S., Sultana, S.R., Mubeen, M., Saud, S., Kamran, M., & Nasim, W. 2019. Plant growth and morphological changes in rice under abiotic stress. In: *Advances in Rice Research for Abiotic Stress Tolerance* (pp. 69–85). Elsevier. https://doi.org/10.1016/B978-0-12-814332-2.00004-6

Almeida, O.A.C., De Araujo, N.O., Mulato, A.T.N., Persinoti, G.F., Sforça, M.L., Calderan-Rodrigues, M.J., & Oliveira, J.V.D.C. 2023. Bacterial volatile organic compounds (VOCs) promote growth and induce metabolic changes in rice. *Frontiers in Plant Science* 13, 1056082. https://doi.org/10.3389/fpls.2022.1056082

Azad, M., Tohidfar, M., Ghanbari Moheb Seraj, R., Mehralian, M., & Esmaeilzadeh-Salestani, K. 2024. Identification of responsive genes to multiple abiotic stresses in rice (*Oryza sativa*): a meta-analysis of transcriptomics data. *Scientific Reports* 14(1), 5463. https://doi.org/10.1038/s41598-024-54623-7

Balasjin, N.M., Maki, J.S., & Schläppi, M.R. 2024. *Pseudomonas mosselii* improves cold tolerance of Asian rice (*Oryza sativa* L.) in a genotype-dependent manner by increasing proline in *japonica* and reduced glutathione in *indica* varieties. *Canadian Journal of Microbiology* 70(1), 15–31. https://doi.org/10.1139/cjm-2023-0030

Barberon, M., & Geldner, N. 2014. Radial transport of nutrients: the plant root as a polarized epithelium. *Plant Physiology* 166(2), 528–537. https://doi.org/10.1104/pp.114.246124

Boorboori, M.R., & Zhang, H.-Y. 2022. The role of *Serendipita indica* (*Piriformospora indica*) in improving plant resistance to drought and salinity stresses. *Biology* 11(7), 952. https://doi.org/10.3390/biology11070952

Cai, Q., Zou, Y., Zhong, N., & Jia, G. 2024. Effects of nutritional stress on soil fertility and antioxidant enzymes of rice in different growth periods. *Frontiers in Plant Science* 15, 1471682. https://doi.org/10.3389/fpls.2024.1471682

Chaganti, C., Phule, A.S., Chandran, L.P., Sonth, B., Kavuru, V.P.B., Govindannagari, R., & Sundaram, R.M. 2023. Silicate solubilizing and plant growth promoting bacteria interact with biogenic silica to impart heat stress tolerance in rice by modulating physiology and gene expression. *Frontiers in Microbiology* 14, 1168415. https://doi.org/10.3389/fmicb.2023.1168415

Che, S., Xu, Y., Qin, X., Tian, S., Wang, J., Zhou, X., Cao, Z., Wang, D., Wu, M., Wu, Z., Yang, M., Wu, L., & Yang, X. 2024. Building microbial consortia to enhance straw degradation, phosphorus solubilization, and soil fertility for rice growth. *Microbial Cell Factories* 23(1), 232. https://doi.org/10.1186/s12934-024-02503-0

Chieb, M., & Gachomo, E.W. 2023. The role of plant growth promoting rhizobacteria in plant drought stress responses. *BMC Plant Biology* 23(1), 407. https://doi.org/10.1186/s12870-023-04403-8

Danso Ofori, A., Su, W., Zheng, T., Datsomor, O., Titriku, J.K., Xiang, X., Kandhro, A.G., Ahmed, M.I., Mawuli, E.W., Awuah, R.T., & Zheng, A. 2024. Roles of phyllosphere microbes in rice health and productivity. *Plants* 13(23), 3268. https://doi.org/10.3390/plants13233268

Dastogeer, K.M.G., Zahan, Mst.I., Rhaman, M.S., Sarker, M.S.A., & Chakraborty, A. 2022. Microbe-mediated thermotolerance in plants and pertinent mechanisms – a meta-analysis and review. *Frontiers in Microbiology* 13, 833566. https://doi.org/10.3389/fmicb.2022.833566

De Souza, E.M., Lamb, T.I., Lamb, T.A., Dos Santos Silva, A., Da Fré De Carvalho, S., Nyland, V., Lopes, M.C.B., Grohs, M., Marconatto, L., Dos Anjos Borges, L.G., Giongo, A., Granada, C.E., & Sperotto, R.A. 2021. Rhizospheric soil from rice paddy presents isolable bacteria able to induce cold tolerance in rice plants. *Journal of Soil Science and Plant Nutrition* 21(3), 1993–2006. https://doi.org/10.1007/s42729-021-00496-y

Ding, T., Li, W., Li, F., Ren, M., & Wang, W. 2024. MicroRNAs: key regulators in plant responses to abiotic and biotic stresses via endogenous and cross-kingdom mechanisms. *International Journal of Molecular Sciences* 25(2), 1154. https://doi.org/10.3390/ijms25021154

Dinneny, J.R., & Yanofsky, M.F. 2004. Vascular patterning: xylem or phloem? *Current Biology* 14(3), R112–R114. https://doi.org/10.1016/j.cub.2004.01.017

Dissanayaka, D.M.S.B., Plaxton, W.C., Lambers, H., Siebers, M., Marambe, B., & Wasaki, J. 2018. Molecular mechanisms underpinning phosphorus-use efficiency in rice. *Plant, Cell & Environment* 41(7), 1483–1496. https://doi.org/10.1111/pce.13191

Dong, H., Fan, S., Sun, H., Chen, C., Wang, A., Jiang, L., & Ma, D. 2021. Rhizosphere-associated microbiomes of rice (*Oryza sativa* L.) under the effect of increased nitrogen fertilization. *Frontiers in Microbiology* 12, 730506. https://doi.org/10.3389/fmicb.2021.730506

Doni, F., Suhaimi, N.S.M., Mispan, M.S., Fathurrahman, F., Marzuki, B.M., Kusmoro, J., & Uphoff, N. 2022. Microbial contributions for rice production: from conventional crop management to the use of "Omics" technologies. *International Journal of Molecular Sciences* 23(2), 737. https://doi.org/10.3390/ijms23020737

Duncker, K.E., Holmes, Z.A., & You, L. 2021. Engineered microbial consortia: strategies and applications. *Microbial Cell Factories* 20(1), 211. https://doi.org/10.1186/s12934-021-01699-9

Gill, R.A., Ahmar, S., Ali, B., Saleem, M.H., Khan, M.U., Zhou, W., & Liu, S. 2021. The role of membrane transporters in plant growth and development and abiotic stress tolerance.

International Journal of Molecular Sciences 22(23), 12792. https://doi.org/10.3390/ijms222312792

Gu, Z., Hu, C., Gan, Y., Zhou, J., Tian, G., & Gao, L. 2024. Role of microbes in alleviating crop drought stress: a review. *Plants* 13(3), 384. https://doi.org/10.3390/plants13030384

Guigard, L., Jobert, L., Busset, N., Moulin, L., & Czernic, P. 2023. Symbiotic compatibility between rice cultivars and arbuscular mycorrhizal fungi genotypes affects rice growth and mycorrhiza-induced resistance. *Frontiers in Plant Science* 14, 1278990. https://doi.org/10.3389/fpls.2023.1278990

Guo, J., Ge, C., Wang, G., & Zhou, D. 2024. Mechanisms of chloride to promote the uptake and accumulation of cadmium in rice (*Oryza sativa* L.). *Science of the Total Environment* 926, 172046. https://doi.org/10.1016/j.scitotenv.2024.172046

Gupta, A., Tiwari, R.K., Shukla, R., Singh, A.N., & Sahu, P.K. 2023. Salinity alleviator bacteria in rice (*Oryza sativa* L.), their colonization efficacy, and synergism with melatonin. *Frontiers in Plant Science* 13, 1060287. https://doi.org/10.3389/fpls.2022.1060287

Hasan, Md.M., Rahman, M.A., Corpas, F.J., Rahman, Md.M., Jahan, M.S., Liu, X., Ghimire, S., Alabdallah, N.M., Wassem, M., Alharbi, B.M., Raza, A., & Fang, X. 2024. Salt stress tolerance in rice (*Oryza sativa* L.): a proteomic overview of recent advances and future prospects. *Plant Stress* 11, 100307. https://doi.org/10.1016/j.stress.2023.100307

He, J., Dai, H., Zhang, X., & Wang, E. 2024. Mycorrhizal signals promote root development dependent on LysM-receptor like kinases in rice. *New Crops* 1, 100009. https://doi.org/10.1016/j.ncrops.2023.12.004

Hossain, A., Ali, M.E., Maitra, S., Bhadra, P., Rahman, Md.M.E., Ali, S., & Aftab, T. 2022. The role of soil microorganisms in plant adaptation to abiotic stresses: current scenario and future perspectives. In: *Plant Perspectives to Global Climate Changes* (pp. 233–278). Elsevier. https://doi.org/10.1016/B978-0-323-85665-2.00001-7

Huei Kao, C. 2013. Role of rice heme oxygenase in lateral root formation. *Plant Signaling & Behavior* 8(10), e25766. https://doi.org/10.4161/psb.25766

Jawed, K., Yazdani, S.S., & Koffas, M.Ag. 2019. Advances in the development and application of microbial consortia for metabolic engineering. *Metabolic Engineering Communications* 9, e00095. https://doi.org/10.1016/j.mec.2019.e00095

Jensen, K.H. 2018. Phloem physics: mechanisms, constraints, and perspectives. *Current Opinion in Plant Biology* 43, 96–100. https://doi.org/10.1016/j.pbi.2018.03.005

Kandil, E.E., El-Banna, A.A.A., Tabl, D.M.M., Mackled, M.I., Ghareeb, R.Y., Al-Huqail, A.A., Ali, H.M., Jebril, J., & Abdelsalam, N.R. 2022. Zinc nutrition responses to agronomic and yield traits, kernel quality, and pollen viability in rice (*Oryza sativa* L.). *Frontiers in Plant Science* 13, 791066. https://doi.org/10.3389/fpls.2022.791066

Kapazoglou, A., Tani, E., Papasotiropoulos, V., Letsiou, S., Gerakari, M., Abraham, E., & Bebeli, P.J. 2025. Enhancing abiotic stress resilience in Mediterranean woody perennial fruit crops: genetic, epigenetic, and microbial molecular perspectives in the face of climate change. *International Journal of Molecular Sciences* 26(7), 3160. https://doi.org/10.3390/ijms26073160

Kawai, T., Chen, Y., Takahashi, H., Inukai, Y., & Siddique, K.H.M. 2022a. Rice genotypes express compensatory root growth with altered root distributions in response to root cutting. *Frontiers in Plant Science* 13, 830577. https://doi.org/10.3389/fpls.2022.830577

Kawai, T., Shibata, K., Akahoshi, R., Nishiuchi, S., Takahashi, H., Nakazono, M., Kojima, T., Nosaka-Takahashi, M., Sato, Y., Toyoda, A., Lucob-Agustin, N., Kano-Nakata, M.,

Suralta, R.R., Niones, J.M., Chen, Y., Siddique, K.H.M., Yamauchi, A., & Inukai, Y. 2022b. *WUSCHEL-related homeobox* family genes in rice control lateral root primordium size. *Proceedings of the National Academy of Sciences* 119(1), e2101846119. https://doi.org/10.1073/pnas.2101846119

Keller, M. 2020. Water relations and nutrient uptake. In: *The Science of Grapevines* (pp. 105–127). Elsevier. https://doi.org/10.1016/B978-0-12-816365-8.00003-8

Khraiwesh, B., Zhu, J.-K., & Zhu, J. 2012. Role of miRNAs and siRNAs in biotic and abiotic stress responses of plants. *Biochimica et Biophysica Acta (BBA) – Gene Regulatory Mechanisms* 1819(2), 137–148. https://doi.org/10.1016/j.bbagrm.2011.05.001

Kobayashi, N.I., Iwata, N., Saito, T., Suzuki, H., Iwata, R., Tanoi, K., & Nakanishi, T.M. 2013. Different magnesium uptake and transport activity along the rice root axis revealed by ^{28}Mg tracer experiments*. *Soil Science and Plant Nutrition* 59(2), 149–155. https://doi.org/10.1080/00380768.2012.751554

Li, Y., Jin, Q., Yang, D., & Cui, J. 2018. Molybdenum sulfide induce growth enhancement effect of rice (*Oryza sativa* L.) through regulating the synthesis of chlorophyll and the expression of aquaporin gene. *Journal of Agricultural and Food Chemistry* 66(16), 4013–4021. https://doi.org/10.1021/acs.jafc.7b05940

Lian, W., Geng, A., Wang, Y., Liu, M., Zhang, Y., Wang, X., & Chen, G. 2023. The molecular mechanism of potassium absorption, transport, and utilization in rice. *International Journal of Molecular Sciences* 24(23), 16682. https://doi.org/10.3390/ijms242316682

Liu, F., Mao, J., Lu, T., & Hua, Q. 2019. Synthetic, context-dependent microbial consortium of predator and prey. *ACS Synthetic Biology* 8(8), 1713–1722. https://doi.org/10.1021/acssynbio.9b00110

Liu, Y., Xun, W., Chen, L., Xu, Z., Zhang, N., Feng, H., Zhang, Q., & Zhang, R. 2022. Rhizosphere microbes enhance plant salt tolerance: toward crop production in saline soil. *Computational and Structural Biotechnology Journal* 20, 6543–6551. https://doi.org/10.1016/j.csbj.2022.11.046

Luo, L., Zhu, M., Jia, L., Xie, Y., Wang, Z., & Xuan, W. 2022. Ammonium transporters cooperatively regulate rice crown root formation responding to ammonium nitrogen. *Journal of Experimental Botany* 73(11), 3671–3685. https://doi.org/10.1093/jxb/erac059

Ma, Y., Tang, M., Wang, M., Yu, Y., & Ruan, B. 2024. Advances in understanding drought stress responses in rice: molecular mechanisms of ABA signaling and breeding prospects. *Genes* 15(12), 1529. https://doi.org/10.3390/genes15121529

Mahmood-ur-Rahman, Ijaz, M., Qamar, S., Bukhari, S.A., & Malik, K. 2019. Abiotic stress signaling in rice crop. In: *Advances in Rice Research for Abiotic Stress Tolerance* (pp. 551–569). Elsevier. https://doi.org/10.1016/B978-0-12-814332-2.00027-7

Mahreen, N., Yasmin, S., Asif, M., Yahya, M., Ejaz, K., Mehboob-ur-Rahman, Yousaf, S., Amin, I., Zulfiqar, S., Imran, A., Khaliq, S., & Arif, M. 2023. Mitigation of water scarcity with sustained growth of Rice by plant growth promoting bacteria. *Frontiers in Plant Science* 14, 1081537. https://doi.org/10.3389/fpls.2023.1081537

Martínez-Arias, C., Witzell, J., Solla, A., Martin, J.A., & Rodríguez-Calcerrada, J. 2022. Beneficial and pathogenic plant-microbe interactions during flooding stress. *Plant, Cell & Environment* 45(10), 2875–2897. https://doi.org/10.1111/pce.14403

Meng, F., Xiang, D., Zhu, J., Li, Y., & Mao, C. 2019. Molecular mechanisms of root development in rice. *Rice* 12(1), 1. https://doi.org/10.1186/s12284-018-0262-x

Meng, X., Bai, S., Wang, S., Pan, Y., Chen, K., Xie, K., Wang, M., & Guo, S. 2023. The sensitivity of photosynthesis to magnesium deficiency differs between rice (*Oryza sativa* L.) and cucumber (*Cucumis sativus* L.). *Frontiers in Plant Science* 14, 1164866. https://doi.org/10.3389/fpls.2023.1164866

Mia, M.A.B., Momotaj, A., & Islam, T. 2023. Consortia of probiotic bacteria and their potentials for sustainable rice production. In: D.K. Maheshwari, & S. Dheeman (Eds.), *Sustainable Agrobiology* (Vol. 43, pp. 151–176). Springer Nature Singapore. https://doi.org/10.1007/978-981-19-9570-5_8

Mishra, D.C., Arora, D., Budhlakoti, N., Solanke, A.U., Mithra, S.V.A.C., Kumar, A., Pandey, P.S., Srivastava, S., Kumar, S., Farooqi, M.S., Lal, S.B., Rai, A., & Chaturvedi, K.K. 2022. Identification of potential cytokinin responsive key genes in rice treated with Trans-Zeatin through systems biology approach. *Frontiers in Genetics* 12, 780599. https://doi.org/10.3389/fgene.2021.780599

Misu, I.J., Kayess, Md.O., Siddiqui, Md.N., Gupta, D.R., Islam, M.N., & Islam, T. 2025. Microbiome engineering for sustainable rice production: strategies for biofertilization, stress tolerance, and climate resilience. *Microorganisms* 13(2), 233. https://doi.org/10.3390/microorganisms13020233

Munir, N., Hanif, M., Abideen, Z., Sohail, M., El-Keblawy, A., Radicetti, E., Mancinelli, R., & Haider, G. 2022. Mechanisms and strategies of plant microbiome interactions to mitigate abiotic stresses. *Agronomy* 12(9), 2069. https://doi.org/10.3390/agronomy12092069

Muralibabu, A.S., Manikandan, K.B., Kim, S.-H., & Subramanian, P. 2024. Strategies of cold-adapted PGPB to elicit cold-stress acclimatory responses in plants. In: *Beneficial Microbes for Sustainable Agriculture Under Stress Conditions* (pp. 241–261). Elsevier. https://doi.org/10.1016/B978-0-443-13193-6.00013-0

Negi, R., Sharma, B., Parastesh, F., Kaur, S., Khan, S.S., Kour, D., Singh, S., Rai, A.K., Rustagi, S., Yadav, N., Shreaz, S., Madouh, T.A., & Yadav, A.N. 2024. Microbial consortia mediated regulation of plant defense: a promising tool for sustaining crops protection. *Physiological and Molecular Plant Pathology* 134, 102393. https://doi.org/10.1016/j.pmpp.2024.102393

Nunes, C.A. 2012. Biological control of postharvest diseases of fruit. *European Journal of Plant Pathology* 133(1), 181–196. https://doi.org/10.1007/s10658-011-9919-7

Parasar, B.J., Kashyap, S., Sharma, I., Marme, S.D., Das, P., & Agarwala, N. 2024. Microbe mediated alleviation of drought and heat stress in plants- current understanding and future prospects. *Discover Plants* 1(1), 26. https://doi.org/10.1007/s44372-024-00022-1

Ren, H., Bao, J., Gao, Z., Sun, D., Zheng, S., & Bai, J. 2023. How rice adapts to high temperatures. *Frontiers in Plant Science* 14, 1137923. https://doi.org/10.3389/fpls.2023.1137923

Robe, K., & Barberon, M. 2023. Nutrient carriers at the heart of plant nutrition and sensing. *Current Opinion in Plant Biology* 74, 102376. https://doi.org/10.1016/j.pbi.2023.102376

Santoyo, G., Guzmán-Guzmán, P., Parra-Cota, F.I., Santos-Villalobos, S.D.L., Orozco-Mosqueda, Ma.D.C., & Glick, B.R. 2021. Plant growth stimulation by microbial consortia. *Agronomy* 11(2), 219. https://doi.org/10.3390/agronomy11020219

Sarma, B., Kashtoh, H., Lama Tamang, T., Bhattacharyya, P.N., Mohanta, Y.K., & Baek, K.-H. 2023. Abiotic stress in rice: visiting the physiological response and its tolerance mechanisms. *Plants* 12(23), 3948. https://doi.org/10.3390/plants12233948

Shaffique, S., Khan, M.A., Imran, M., Kang, S.-M., Park, Y.-S., Wani, S.H., & Lee, I.-J. 2022. Research progress in the field of microbial mitigation of drought stress in plants. *Frontiers in Plant Science* 13, 870626. https://doi.org/10.3389/fpls.2022.870626

Shahzad, N., Nabi, H.G., Qiao, L., & Li, W. 2024. The molecular mechanism of cold-stress tolerance: cold responsive genes and their mechanisms in rice (*Oryza sativa* L.). *Biology* 13(6), 442. https://doi.org/10.3390/biology13060442

Shi, F., Zhu, S., Li, H., Zhang, B., Liu, J., & Song, F. 2024. Further enhancement of cold tolerance in rice seedlings by *Piriformospora indica* collaborating with plant growth-promoting bacteria: evidence from the antioxidant defense, osmoregulation, photosynthesis, and related genes. *Plant Stress* 14, 100656. https://doi.org/10.1016/j.stress.2024.100656

Shrestha, J., Kandel, M., Subedi, S., & Shah, K.K. 2020. Role of nutrients in rice (*Oryza sativa* L.): a review. *Agrica* 9(1), 53. https://doi.org/10.5958/2394-448X.2020.00008.5

Singh, A., Mazahar, S., Chapadgaonkar, S.S., Giri, P., & Shourie, A. 2023. Phyto-microbiome to mitigate abiotic stress in crop plants. *Frontiers in Microbiology* 14, 1210890. https://doi.org/10.3389/fmicb.2023.1210890

Sun, J., Ye, M., Peng, S., & Li, Y. 2016. Nitrogen can improve the rapid response of photosynthesis to changing irradiance in rice (*Oryza sativa* L.) plants. *Scientific Reports* 6(1), 31305. https://doi.org/10.1038/srep31305

Tisarum, R., Theerawitaya, C., Samphumphuang, T., Polispitak, K., Thongpoem, P., Singh, H.P., & Cha-um, S. 2020. Alleviation of salt stress in upland rice (*Oryza sativa* L. ssp. Indica cv. Leum Pua) using arbuscular mycorrhizal fungi inoculation. *Frontiers in Plant Science* 11, 348. https://doi.org/10.3389/fpls.2020.00348

Tyagi, A., Ali, S., Mir, R.A., Sharma, S., Arpita, K., Almalki, M.A., & Mir, Z.A. 2024. Uncovering the effect of waterlogging stress on plant microbiome and disease development: current knowledge and future perspectives. *Frontiers in Plant Science* 15, 1407789. https://doi.org/10.3389/fpls.2024.1407789

Venturas, M.D., Sperry, J.S., & Hacke, U.G. 2017. Plant xylem hydraulics: what we understand, current research, and future challenges. *Journal of Integrative Plant Biology* 59(6), 356–389. https://doi.org/10.1111/jipb.12534

Verma, N., Giri, S.K., Singh, G., Gill, R., & Kumar, A. 2022a. Epigenetic regulation of heat and cold stress responses in crop plants. *Plant Gene* 29, 100351. https://doi.org/10.1016/j.plgene.2022.100351

Verma, P.K., Verma, S., & Pandey, N. 2022b. Root system architecture in rice: impacts of genes, phytohormones and root microbiota. *3 Biotech* 12(9), 239. https://doi.org/10.1007/s13205-022-03299-9

Wang, P., Kong, X., Chen, H., Xiao, Y., Liu, H., Li, X., Zhang, Z., Tan, X., Wang, D., Jin, D., Deng, Y., & Cernava, T. 2021. Exploration of intrinsic microbial community modulators in the rice endosphere indicates a key role of distinct bacterial taxa across different cultivars. *Frontiers in Microbiology* 12, 629852. https://doi.org/10.3389/fmicb.2021.629852

Wei, H., Li, Y., Wei, L., Peng, S., Zhang, B., Xu, D., & Cheng, X. 2024. Exploring the mechanism of exopolysaccharides in mitigating cadmium toxicity in rice through analyzing the changes of antioxidant system. *Journal of Hazardous Materials* 461, 132678. https://doi.org/10.1016/j.jhazmat.2023.132678

White, P.J. 2012. Long-distance transport in the xylem and phloem. In: *Marschner's Mineral Nutrition of Higher Plants* (pp. 49–70). Elsevier. https://doi.org/10.1016/B978-0-12-384905-2.00003-0

Wu, G.Q., & Wang, S.M. 2012. Calcium regulates K^+/Na^+ homeostasis in rice (*Oryza sativa* L.) under saline conditions. *Plant, Soil and Environment* 58(3), 121–127. https://doi.org/10.17221/374/2011-PSE

Wu, W., & Cheng, S. 2014. Root genetic research, an opportunity and challenge to rice improvement. *Field Crops Research* 165, 111–124. https://doi.org/10.1016/j.fcr.2014.04.013

Yamada, N., Theerawitaya, C., Kageyama, H., Cha-um, S., & Takabe, T. 2015. Expression of developmentally regulated plasma membrane polypeptide (DREPP2) in rice root tip and interaction with Ca2+/CaM complex and microtubule. *Protoplasma* 252(6), 1519–1527. https://doi.org/10.1007/s00709-015-0781-x

Yang, C., Yang, L., Yang, Y., & Ouyang, Z. 2004a. Rice root growth and nutrient uptake as influenced by organic manure in continuously and alternately flooded paddy soils. *Agricultural Water Management* 70(1), 67–81. https://doi.org/10.1016/j.agwat.2004.05.003

Yang, X., Yuan, R., Yang, S., Dai, Z., Di, N., Yang, H., He, Z., & Wei, M. 2024b. A salt-tolerant growth-promoting phyllosphere microbial combination from mangrove plants and its mechanism for promoting salt tolerance in rice. *Microbiome* 12(1), 270. https://doi.org/10.1186/s40168-024-01969-9

Ye, T., Li, Y., Zhang, J., Hou, W., Zhou, W., Lu, J., Xing, Y., & Li, X. 2019. Nitrogen, phosphorus, and potassium fertilization affects the flowering time of rice (*Oryza sativa* L.). *Global Ecology and Conservation* 20, e00753. https://doi.org/10.1016/j.gecco.2019.e00753

Zhang, H., Li, Y., Pu, M., Xu, P., Liang, G., & Yu, D. 2020. *Oryza sativa* Positive Regulator of Iron Deficiency Response 2 (OsPRI2) and OsPRI3 are involved in the maintenance of Fe homeostasis. *Plant, Cell & Environment* 43(1), 261–274. https://doi.org/10.1111/pce.13655

Zheng, H., Pan, X., Deng, Y., Wu, H., Liu, P., & Li, X. 2016. AtOPR3 specifically inhibits primary root growth in Arabidopsis under phosphate deficiency. *Scientific Reports* 6(1), 24778. https://doi.org/10.1038/srep24778

Zou, X., Liu, L., Hu, Z., Wang, X., Zhu, Y., Zhang, J., Li, X., Kang, Z., Lin, Y., & Yin, C. 2021. Salt-induced inhibition of rice seminal root growth is mediated by ethylene-jasmonate interaction. *Journal of Experimental Botany* 72(15), 5656–5672. https://doi.org/10.1093/jxb/erab206

5

Plant Pathogen-specific PR Proteins Produced in Response to PGPR

*Selene Aguilera-Aguirre, María Fernanda Olmedo-Díaz, Miriam del Carmen Bañuelos-González, Gema Cárdenas-Flores, Ulises Miguel López-García, Rosa Isela Ortiz-Basurto, and Martina Alejandra Chacón-López**

Laboratorio Integral de Investigación en Alimentos/Departamento de Ingeniería Química y Bioquímica, Tecnológico Nacional de México-Instituto Tecnológico de Tepic, Tepic, Nayarit, México

* Corresponding author: mchacon@tepic.tecnm.mx

5.1 Introduction

Plants, being sessile organisms, have developed diverse and sophisticated defense mechanisms to recognize and overcome pathogen infections. Among these strategies, the production of pathogenesis-related (PR) proteins is one of the most extensively studied, which can exert direct antimicrobial activity or modulate defense responses through the activation of signaling pathways. The production of these proteins is regulated by the perception of specific environmental signals, including the detection of pathogen-associated molecular patterns (PAMPs) and interactions with pathogen effectors (Menéndez et al., 2020).

PR proteins are classified into 17 families, each with distinct roles in plant immunity, such as pathogen cell wall degradation, enzyme inhibition, or the production of toxic compounds. Their production is regulated by physical factors (UV light, temperature, and mechanical wounding), chemical factors (salicylic acid [SA] and jasmonate), and biological factors (interaction with beneficial microorganisms). Among the latter, plant growth-promoting rhizobacteria (PGPRs) have gained significant interest for their capacity to trigger systemic resistance in plants, reducing the need for chemical pesticides to keep them free of diseases (Van Loon & Van Strien, 1999; Menéndez et al., 2020).

PGPR-induced resistance is based on the induction of hormonal pathways that enhance the plant's defensive response against future attacks. This phenomenon, known as "priming," has been reported to be effective against a widespread range of pathogens. Additionally, PGPRs facilitate nutrient uptake and improve soil health, making them a key tool for sustainable agriculture (Chandran et al., 2021).

Microbial Consortia in Plant Science and Sustainable Agriculture, First Edition. Edited by Ajay Tomar, et al.

Given the growing concern over the environmental impact of agrochemicals, PGPR application represents a viable alternative to enhance agricultural productivity while reducing the need for synthetic chemicals. Combining different PGPR strains has shown promising results in strengthening plant resistance and disease management (Zhu et al., 2022). Harnessing these beneficial organisms is a key strategy for increasing agricultural productivity and caring for the environment. In this regard, reviewing the plant defense mechanism and how the interaction of PGPR can activate it is helpful. This study explores the role of PR proteins in plant defense, their regulation, and the significance of PGPR in systemic resistance induction.

5.2 Recognition Systems in Plant Immunity

Plants are frequently exposed to a wide range of microbes, including bacteria, fungi, viruses, oomycetes, and nematodes. The interactions between these organisms and plants can vary, ranging from mutualistic to parasitic or pathogenic. Organisms involved in the latter two types of interactions are generally classified as pathogens (Newton et al., 2010). Pathogens compromise plant survival and productivity by disrupting essential physiological processes, reducing growth, and causing disease symptoms that can lead to significant yield losses. To defend themselves, they have developed a sophisticated defense system capable of detecting and responding to these threats. The vegetal defense relies on specialized recognition mechanisms that allow plants to perceive pathogen-derived signals and activate appropriate responses. This implies that plants have evolved adaptive strategies to identify and counteract diverse types of threats posed by pathogens. To achieve this, they rely on two primary recognition systems:

1. **Pattern-triggered Immunity (PTI):** This defense mechanism is mediated by the identification of PAMPs, which are preserved structural motifs shared by different groups of microorganisms. These PAMPs are recognized by pattern recognition receptors (PRRs), which are situated on the plasma membrane and initiate a signaling cascade that includes an oxidative burst induced by reactive oxygen species (ROS), the activation of kinases, and the induction of defense-related genes, including those that encode PR proteins (Henry et al., 2012; Bigeard et al., 2015).
2. **Effector-triggered Immunity (ETI):** Some pathogens have evolved effectors capable of suppressing PTI. In reply, plants expressed resistance (R) proteins capable of specifically identifying pathogen-encoded effectors, triggering a more robust and amplified defense response known as ETI (Remick et al., 2023).

These recognition systems represent the initial stage of plant immune responses, allowing plants to detect and counteract potential threats rapidly and effectively. A key aspect of this defense is the activation of defense-related genes, among which PR proteins play a crucial role. These proteins are a diverse group of molecules that contribute to plant immunity through various mechanisms, including antimicrobial activity, cell wall reinforcement, and signaling functions that enhance plant resistance (Figure 5.1) (Ali et al., 2018).

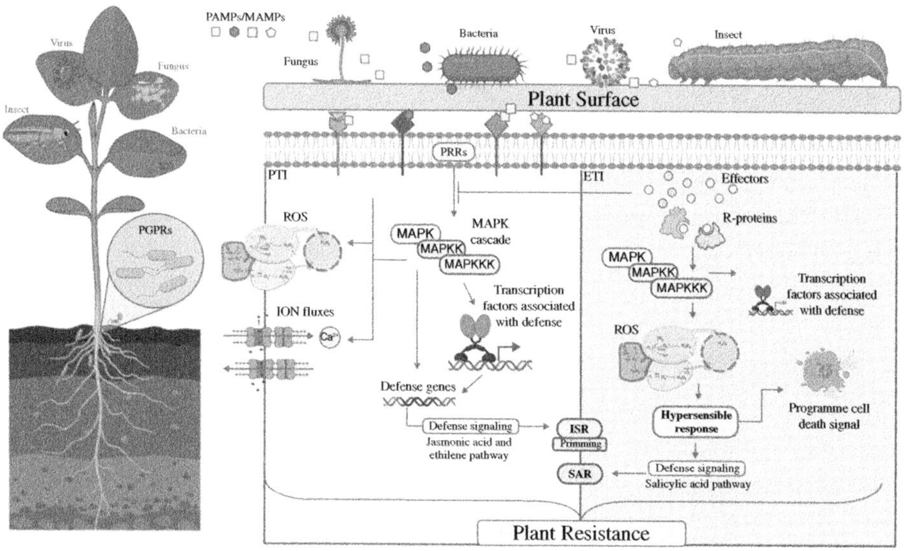

Figure 5.1 PTI and ETI in plants. The first line of plant defense is PTI, which is triggered by PAMP recognition through membrane proteins named PRRs. Several PTI signaling events occur, such as the activation of mitogen-activated protein kinase (MAPK) cascades, an influx of Ca^{2+} into the cytosol, and the production of ROS, activating defense-associated transcription factors (TFs), likewise activating defense genes. Many pathogens may be able to suppress PTI by producing virulence-associated molecules called effectors. PR proteins recognize these effectors to trigger ETI, a second line of defense in plants. In parallel, systemic immunity can be initiated as the induced systemic response (ISR) or the systemic acquired response (SAR), which will depend on the type of molecules perceived and the activated PR proteins. Jasmonic acid (JA) and ethylene (ET) are key phytohormones that regulate ISR, while salicylic acid (SA) regulates SAR.

Both plants' immune mechanisms, PTI or ETI, involve PR proteins responding to pathogens' attacks. The following are the characteristics of PR proteins, their classification, and their role in the plant defense system.

5.3 PR Proteins

PR proteins constitute a family of proteins expressed in plants in response to pathogens and other stress factors. These proteins were first identified in tobacco plants following infection with the tobacco mosaic virus (TMV) but have since been identified in numerous other plant species. Their primary function is to protect plants through direct mechanisms, such as antimicrobial activity, or indirect mechanisms, including the activation of defense signaling pathways. PR proteins are usually low-molecular-weight proteins, from 6 to 43 kDa. They exhibit thermostability, hydrolase resistance, and solubility in highly acidic conditions (pH < 3) (Ali et al., 2018).

The expression of these proteins can be regulated through various pathways, with hormonal signaling playing a key role. Certain phytohormones regulate defense responses, including SA, JA, and ET. A prominent example is the accumulation of SA, which is related to the triggering of systemic acquired resistance (SAR) and the induction of PR proteins such as NPR1. NPR1, in turn, promotes the transcription of PR genes, thereby enhancing plant defense against biotrophic pathogens. Conversely, JA and ET are involved in induced systemic resistance (ISR), which can also lead to PR protein expression.

Another factor in PR gene induction is the presence of elicitors – molecules recognized by the plant that trigger defense responses. The interaction with these elicitors activates signaling pathways that culminate in the induction of defense-related genes, including those encoding PR proteins (Thakur & Sohal, 2013; Kamle et al., 2020).

5.3.1 PR Protein Classification

PR proteins are organized into 17 distinct families (Table 5.1) based on their structural characteristics and functional roles in plant defense. Each PR family exhibits specific biochemical properties and mechanisms of action, which are crucial for plant responses against different types of pathogens.

The classification of PR proteins provides a basis for understanding their diverse roles in plant defense. Yet, their production is not constitutive and is instead regulated by various internal and external factors. The induction of these proteins can be triggered by a wide range of stimuli, including physical, chemical, and biological signals, which activate specific defense pathways in response to environmental challenges (Menéndez et al., 2020).

Table 5.1 Classification of PR proteins.

PR proteins	Origin	Function	Properties	References
PR-1	*Nicotiana tabacum*	Antifungal	Its expression is associated with the accumulation of salicylic acid in infected plants.	Niderman et al. (1995)
PR-2	*N. tabacum*	β-1,3-glucanase	They degrade the fungal cell wall. They participate in systemic defense signaling.	Van Loon and Van Strien (1999)
PR-3	*N. tabacum*	Chitinase type I, II, IV, V, VI, VII	They break down chitin polymers present in fungal cell walls.	Van Loon and Van Strien (1999)
PR-4	*N. tabacum*	Chitinase type I, II	They are involved in direct defense against fungal infections.	Van Loon and Van Strien (1999)
PR-5	*N. tabacum*	Thaumatin-like	They alter the integrity of cell membranes in pathogens. They have an osmotic effect that leads to the cellular death of the invader.	Van Loon and Van Strien (1999)

PR proteins	Origin	Function	Properties	References
PR-6	*Solanum lycopersicum*	Proteinase inhibitor	They inhibit pathogen proteases, interfering with their ability to digest plant proteins. They generate reactive oxygen species that damage microorganisms.	Van Loon and Van Strien (1999)
PR-7	*Solanum lycopersicum*	Endoproteinase	They are involved in long-term defense responses.	Van Loon and Van Strien (1999)
PR-8	*Cucumis sativus*	Chitinase type III	They are involved in direct defense against fungal infections.	Van Loon and Van Strien (1999)
PR-9	*N. tabacum*	Peroxidase	They reinforce cell walls.	Van Loon and Van Strien (1999)
PR-10	*Petroselinum crispum*	Ribonuclease-like	They possess ribonuclease activity, affecting pathogen replication.	Van Loon and Van Strien (1999)
PR-11	*N. tabacum*	Chitinase type I	They are involved in direct defense against fungal infections.	Van Loon and Van Strien (1999)
R-12	*Raphanus raphanistrum*	Defensin	They inhibit microbial growth.	Terras et al. (1995)
PR-13	*Arabidopsis thaliana*	Thionin	They lyse pathogen cell membranes.	Epple et al. (1995)
PR-14	*Hordeum vulgare*	Lipid transfer protein	They modulate membrane permeability.	García-Olmedo et al. (1995)
PR-15	*H. vulgare*	Oxalate oxidase	Their gene expression shows a response to fungal infections.	Zhang et al. (1995)
PR-16	*H. vulgare*	Oxidase-like	They strengthen the defense response in plants.	Wei et al. (1998)
PR-17	*N. tabacum*	Antifungal and antiviral	Activation can lead to a prolonged defense response.	Okushima et al. (2000)

5.3.2 Factors That Promote PR Protein Production

PR proteins are induced as part of plant defense strategies against various threats. Their production is regulated by multiple physical, chemical, and biological factors that trigger specific signaling pathways. These factors activate genes associated with plant defense, promoting a rapid and effective response against pathogens, pests, and abiotic stress (Menéndez et al., 2020).

Among the physical factors, ultraviolet (UV) radiation, specifically UV-B (280–315 nm), can lead to the accumulation of ROS within plant cells, potentially triggering oxidative stress in plants. These molecules act as signaling intermediates that trigger the expression of PR genes, enhancing the plant's defensive capacity (Bigeard et al., 2015). Temperature fluctuations, including extreme heat and cold stress, also contribute to PR protein induction. Thermal stress promotes the production of SA, which in turn activates PR genes involved in pathogen defense (Wang et al., 2006). Additionally, mechanical wounding caused by environmental factors or herbivory elicits both local and systemic defense responses. One key response is the activation of PR proteins such as protease inhibitors (PR-6), which interfere with herbivore digestion and provide a protective advantage to the plant (Ryan, 1990).

Chemical factors also play a significant role in PR protein induction. Various chemical elicitors, including SA, JA, and ET, regulate plant defense pathways by activating specific sets of PR genes. SA is closely linked to SAR, while JA and ET mediate responses against insect herbivory and necrotrophic pathogens (Glazebrook, 2005). In addition to natural signaling molecules, synthetic compounds such as benzothiadiazole, a functional analog of SA, can stimulate PR gene expression even in the absence of a pathogen (Friedrich et al., 1996). Abscisic acid also plays a role in stress response regulation, modulating PR protein production when plants face multiple stress conditions simultaneously (Yasuda et al., 2008). Furthermore, nutrient availability, particularly calcium (Ca^{2+}) and potassium (K^{+}), influences PR gene activation. Calcium, for instance, serves as a secondary messenger in signal transduction pathways, linking external stimuli to the activation of plant defense responses (Reddy & Reddy, 2004).

Biological factors play a crucial role in regulating PR protein production. Pathogenic microorganisms and pests that attack plants can trigger PR protein accumulation through the activation of immune signaling pathways. When PRRs detect PAMPs, defense responses are initiated, leading to the expression of PR proteins as part of the plant's immune strategy (Boller & Felix, 2009). However, not only can pathogenic organisms induce PR protein production, but beneficial microbes, including bacteria and fungi, can also be recognized by plant cells and activate defense mechanisms. Arbuscular mycorrhizal fungi contribute to plant immunity by enhancing nutrient uptake and stimulating PR gene expression. For instance, *Rhizophagus irregularis* has been shown to induce the accumulation of defensins and peroxidases (PODs) (PR-9) in colonized roots, offering additional protection against potential threats (Pozo & Azcón-Aguilar, 2007). Likewise, PGPRs enhance plant immunity through the production of signaling molecules such as siderophores and lipopolysaccharides (LPSs), which activate defense responses. PGPRs also induce systemic resistance (ISR), a defense mechanism mediated by JA and ET that leads to the induction of specific PR proteins (Van Loon et al., 1998). Certain bacterial species, such as *Pseudomonas fluorescens*, are known to upregulate chitinases (PR-3) and β-1,3-glucanases (PR-2), while *Bacillus subtilis* stimulates the accumulation of defensins (PR-12) and thionins (PR-13), further strengthening plant resistance (Van Wees et al., 1999; Kloepper et al., 2004).

The ability of PGPRs to stimulate PR protein production highlights their potential as key players in plant defense. Understanding their diversity, characteristics, and interactions with plants provides a foundation for recognizing their role in improving resistance and overall plant health.

5.4 Plant Growth-promoting Rhizobacteria

Plant roots act as a primary site for the accumulation of photosynthetic products, releasing organic compounds into the surrounding soil. This localized carbon enrichment fosters the development of diverse microbial communities that interact closely with plants, like plant growth-promoting rhizobacteria (Wang et al., 2023).

PGPRs are a cluster of bacteria found in the soil, specifically in the rhizosphere. The term PGPR was coined by Kloepper and Schroth (1978) and refers to those microbes that inhabit the rhizosphere, specifically colonizing plant roots, and promote the enhancement of plant growth and development (Kloepper & Schroth, 1978; Ahmad et al., 2008). PGPRs can be categorized into two clusters: symbiotic bacteria that reside within plants and actively share metabolites with them and free-living bacteria external to plant cells (Gray & Smith, 2005).

Among the key characteristics reported as ideal for PGPR, it is essential that these bacteria not only effectively promote plant growth and enhance resistance to diseases, but also possess traits that enable their survival and adaptation to the environments where they will be applied. This includes tolerance to various stresses such as heat, desiccation, radiation, and chemical agents. Additionally, PGPRs should be compatible with other rhizobacteria while maintaining high rhizosphere competence and competitive saprophytic ability. Another critical factor is their capacity for mass reproduction, broad-spectrum activity, and safety for human health and the environment (Jeyarajan & Nakkeeran, 2000; Reddy, 2013).

Their role in enhancing plant immunity drives the growing interest in PGPR research. Understanding how these bacteria contribute to plant defense has become a key study area, particularly as an alternative to chemical pesticides and synthetic fertilizers. Recent research has focused on unraveling the molecular and biochemical mechanisms by which PGPRs induce resistance, aiming to harness these interactions for sustainability (Iqbal et al., 2024).

5.4.1 Mechanism of the Production of PR Proteins by PGPRs

As previously mentioned, PGPRs not only possess the capacity to enhance plant growth through improved nutrient accessibility but also exhibit the characteristic of inducing resistance to various types of stresses (Bashan & De-Bashan, 2005). This resistance is not only manifested locally but also extends to distant parts of the plant, providing systemic protection against subsequent pathogen attacks (Van Loon, 2007; Chandrasekaran & Chun, 2016; Van Hulten et al., 2006; Chung et al., 2015).

PGPR mechanisms can be categorized into two types: direct and indirect. Direct mechanisms encompass biofertilization, root growth enhancement, rhizoremediation, and plant stress management. In contrast, the indirect mechanism involves biological control, where rhizobacteria enhance plant growth by mitigating disease impacts through antibiosis, systemic resistance induction, and resource rivalry and ecological niches (Egamberdieva & Lugtenberg, 2014).

ISR by PGPR has been reported in various plant species, proving a broad spectrum of activity against different pathogens. An example is *Bacillus* species, which have conferred resistance to bacterial, fungal, and viral pathogens in different plant hosts (Gopal et al., 2013).

ISR is characterized by the activation of a protective state known as priming, in which induced or elicited plants respond more quickly to future pathogen attacks, generally resulting in a more efficient level of protection. This resistance is regulated by a signaling network in which JA and ET hormones have a fundamental role (Vlot et al., 2021; Pieterse et al., 2014; Conrath, 2011). Following resistance elicitation through PGPRs, the accumulation of JA and ET acts as a strong signal to activate the expression of the *NPR-1* gene, which encodes the NPR-1 (Non-Expressor of Pathogenesis-Related Protein 1) protein. This protein has key functions as part of the molecular mechanism that regulates genetic transcription and protein synthesis of PR proteins related to pathogenesis (Jain et al., 2016; Van Wees et al., 2000).

Plants have a first line of defense in the apoplast, which detects the presence of PGPR microbes by particular molecules. This system is highly sensitive and specific, mediated by a set of immune sensors on the plasma membrane referred to as PRRs. These receptors detect microbe-associated molecular patterns (MAMPs) that refer to structural motifs (epitopes) present in various microbial molecules, such as flagellin, lipopolysaccharides (LPSs), siderophores, antibiotics, biosurfactants, exopolysaccharides, and volatile organic compounds (Gopal et al., 2013).

PRRs activate signaling events that lead to changes in membrane polarity, the production of ROS and nitric oxide (NO), as well as the stimulation of mitogen-activated protein kinases and calcium-dependent protein kinases. When MAMPs bind to PRRs, it triggers a shift in TF activity, activating genes responsible for defense mechanisms. This results in various enzymes and metabolites specifically aimed at dealing with pathogen invasion. As a result of the participation of these enzymes, different defense responses have been reported, such as the deposition of callose, lignin, and phenolic compounds. Some enzymes that present high activity in the defense mechanism of plants are phenylalanine ammonia-lyase (PAL), POD, polyphenol oxidase (PPO), lipoxygenase, ß-1,3-glucanase, chitinase, and the production of phytoalexins (Verhagen et al., 2010).

The expression of POD plays a role in creating structural defenses, including the accumulation of lignin and suberin, as well as generating highly toxic environments through the production of ROS and reactive nitrogen species (Passardi et al., 2005; Cavalcanti et al., 2004). Studies have also indicated that POD is associated with the accumulation of phenolic substances within the cell walls of plants (Gross, 1981; Walter, 1992; Graham & Graham, 1991). PPO catalyzes the oxidation of phenolic compounds present in plants, converting them into quinones, which have antimicrobial activity. PPO is believed to be involved in the terminal oxidation of infected tissues (Daayf et al., 1997). In the phenylpropanoid metabolic pathway, PAL is essential for synthesizing defensive compounds. The cell wall's structural integrity is strengthened by the generation of phenolic substances, which increases the plant's resistance to mechanical stress (Daayf et al., 1997).

5.4.2 PGPR Inducers of PR Proteins in Plants

It is now well established that various bacterial genera exhibit PGPR characteristics, including *Agrobacterium*, *Azotobacter*, *Azospirillum*, *Bacillus*, *Pseudomonas*, and *Rhizobium*. However, not all PGPR genera can activate plant defense systems. The genera most associated with inducing immune responses are those frequently used as biocontrol agents, particularly *Pseudomonas* and *Bacillus* (Abdelkhalek et al., 2020; Pieterse et al., 2021; Ge et al., 2022). Nevertheless, to a lesser extent, certain bacterial species have also been reported to possess this ability. Examples include *Paenibacillus, Arthrobacter, Azospirillum*, and *Chryseobacterium* (Verhagen et al., 2004; Samain et al., 2022).

An example of this can be found in certain *Pseudomonas* species, such as *Pseudomonas simiae* and *Pseudomonas capeferrum*. When inoculated in *Arabidopsis* plants, these bacteria induce ISR, leading to the excretion of root metabolites like scopoletin, which inhibits pathogens present in the rhizosphere (Stringlis et al., 2018). Similarly, other species, such as *P. fluorescens* have been reported to not only promote plant growth but also exhibit biocontrol characteristics and activate ISR through the production of phytohormones (Panpatte et al., 2016). Additionally, *Pseudomonas putida* is known to induce ISR in various species, including *Arabidopsis*, tomato, and bean, primarily due to the recognition of flagellin and pseudobactin (Meziane et al., 2005).

Several *Bacillus* species have also demonstrated the ability to induce systemic resistance in plants. For instance, *Bacillus velezensis* GB1 enhanced the activity of defense-related enzymes in pumpkin plants, leading to a significant reduction in the attraction and reproduction of *Bemisia tabaci* (Soliman et al., 2022). Similarly, the filtrate of *Bacillus amyloliquefaciens* TBorg1 applied to tomato plants reduced TMV infection by up to 90%, likely due to the stimulation of phenolic compound pathways and the production of antimicrobial compounds (Abdelkhalek et al., 2022). Additionally, inoculation with *B. subtilis* CBR05 in tomato plants increased the activity of antioxidant enzymes and alleviated the impact of bacterial spot disease caused by *Xanthomonas campestris* pv. vesicatoria (Chandrasekaran & Chun, 2016)

In nature, approximately 99% of microorganisms are part of microbial consortia (Dong et al., 2017). The PGPRs show better benefits to the plants when they are integrated into consortia because they can generate synergistic for crop health, development, and yield (Santos & Olivares, 2021).

5.4.3 PGPR Consortia in the Induction of the Defense System

The combination of different strains of PGPRs can enhance the effectiveness of ISR against various pathogens affecting crops. Several studies have shown that combinations of PGPR generate a more effective ISR response than treating the strains individually (Raupach and Kloepper, 1998). These combinations have the potential to achieve more significant and more extensive colonization of the rhizosphere, providing beneficial effects under various soil conditions and confronting a larger number of pests or pathogens compared to individually applied strains. The improved disease control achieved by combining different microorganisms can be attributed to the varied mechanisms of action of the various agents, which can be equally effective in different climatic zones (Ganeshamoorthi et al., 2008; Thilagavathi et al., 2007).

The functionality and stability of a microbial consortium depend on its interactions and cooperative relationships (Che & Men, 2019). Rhizobacteria within a consortium interact by sharing metabolites and signals, enabling them to synchronize their actions and offer reciprocal advantages. This collaborative effort allows them to execute intricate functions and carry out sequential processes in the utilization of resources (Atkinson et al., 2022).

Quorum sensing regulates interactions among microorganisms and plants in the rhizosphere. This mechanism is utilized by certain bacteria that promote plant growth to regulate the synthesis of substances like phytohormones and siderophores, which enhance plant development (Moreno-Gámez et al., 2023). Moreover, quorum sensing plays a role in biofilm development, which aids microbial colonization of roots and shields microorganisms from environmental stressors. Gaining insights into quorum-sensing mechanisms could enable researchers to increase the positive impacts of microbial biostimulants, thereby enhancing plant vigor and durability in agricultural environments (Nag et al., 2021). Engineering microbial consortia could be a key strategy to ensure rhizobacterial consortia's long-term stability and functionality (Han et al., 2019).

Samain et al. (2022) developed a PGPR microbial consortium using the B2 strain of *Paenibacillus* sp. and the AA strain of *Arthrobacter* sp. This consortium was used to combat *Zymoseptoria tritici* and drought stress. The researchers reported that the consortium provided effective and long-lasting induced systemic resistance in wheat. On the other hand, in a study that evaluated the effect of applying the strains *P. fluorescens* WCS417r and three PGPR strains, *Azospirillum brasilense* Sp7, *P. fluorescens* AUR6, and *Chryseobacterium balustinum* AUR9, on *Arabidopsis thaliana*, it was observed that consortia induced ISR more effectively than individual strains against *Pseudomonas syringae* pv. tomato DC3000 (Verhagen et al., 2004).

Additionally, Ganeshamoorthi et al. (2008) documented the impact of a combination of *P. fluorescens* strains isolate Pf1 and Py15 and *B. subtilis* isolate 16 (Bs16), which protected mulberry plants against *Macrophomina phaseolina*. The induction of defense-related proteins such as peroxidase (PO), PPO, PAL, β-1,3-glucanase, chitinase, and the production of phenolic compounds was evaluated. This was significantly higher in the combined application of the biocontrol agents Pf1-Py15-Bs16, followed by plants treated with Pf1-Bs16, Pf1-Py15, and Py15-Bs16, compared to the individual treatments and the control without treatment.

Then, the application of individual microbial strains or consortia in various crops has emerged as a promising substitute or supplement to conventional agricultural practices, which often depend on synthetic chemical inputs (Sahu et al., 2019). Advances in omics technologies, such as transcriptomics, proteomics, and metabolomics, have provided more profound insights into the complex interactions within microbial consortia. These tools help identify the microorganisms involved, their functional roles, and the metabolites they produce, offering valuable information for optimizing their application in agriculture (Jia et al., 2016; Manzoni et al., 2018). A comprehensive understanding of these interactions has been fundamental in developing effective biocontrol formulations (Meyer & Roberts, 2002).

5.4.4 Application of PGPR Bacteria in Plants and Their Beneficial Effect Against Phytopathogens

The global population continues to grow, increasing demand for agricultural production (Timmusk et al., 2017). However, various diseases caused by phytopathogens and environmental stress threaten both production and crop security. These conditions not only reduce the quality and value of agricultural products but also increase the use of inputs, raise labor costs, and affect the income and livelihood of farmers (Vurro et al., 2010). On the other hand, the intensive use of chemical fertilizers to boost production has led to the deterioration of soil properties, an imbalance in the nutrients present, and the destruction of ecosystems (Bhattacharyya and Jha, 2012; Mal et al., 2014). Annual losses in global crop yields due to phytopathogenic diseases are estimated to exceed 30%, resulting in economic losses of hundreds of billions of dollars (Oerke, 2006; Savary et al., 2019). Although chemical control has been a common tool for managing crop diseases, its excessive use has generated severe environmental impacts. In this context, it is urgent to develop alternative strategies that are both effective and environmentally friendly for controlling these diseases (Deising et al., 2008; Lucas et al., 2015).

The application of PGPR (*Pseudomonas* and *Bacillus* genera, among others) microorganisms in agricultural production reduces the use of pesticides and chemical fertilizers (Li et al., 2020). In this regard, commercial formulations of PGPR strains have been developed for application in various agricultural practices, such as foliar spraying, soil application, root immersion, or seed treatment (Srivastava & Prasad, 2014). In this way, the application of PGPR can boost crop nutrition, increase output, enhance productivity, and foster better disease management. However, its effectiveness in promoting plant growth and development depends on the bacterial strain and the environment in which it is applied. Therefore, it is essential to select strains appropriately to maximize the crop's benefits (Saeed et al., 2021; de Andrade et al., 2023).

Various genera of microorganisms, such as *Pseudomonas, Bacillus, Azotobacter, Rhizobium*, and *Anabaena*, are used as inoculants to improve nutrient absorption (Ju et al., 2020; Zhang et al., 2017). These microorganisms can be applied to the soil, improving its texture, or directly to seeds and plants, acting as biofertilizers or biostimulants, increasing crop productivity (Igiehon et al., 2024) (Table 5.2).

Various strains of *Pseudomonas* promote plant growth through the production of indole-3-acetic acid, phosphate solubilization, synthesis of enzymes and antifungal metabolites, and biofilm formation, among other mechanisms. In one study, it was reported that *P. fluorescens* UM270 was able to promote the growth of tomato plants under saline stress conditions, resulting in an increase in dry weight, chlorophyll content, as well as root and shoot length (Gohil et al., 2020).

Various formulations containing PGPR consortia have been registered and are commercially available, such as Mycostop®, Prestop®, and EndoRoots Soluble® (Mourouzidou et al., 2023; Demoz & Korsten, 2006; Stockwell & Stack, 2007). Some commercially available species include *Pseudomonas, Serratia, Bacillus, Azotobacter, Azospirillum*, and *Enterobacter* (Andrade et al., 2018).

The PR proteins are important molecular actors in the plant's response to biotic and abiotic stresses. As mentioned earlier, the expression of these proteins is signaled by the perception of PGPRs, which are most effective when integrated into consortia. Table 5.2 mentions the effect of some PGPR strains acting as consortia and individuals. Its application promotes sustainable agricultural production because it contributes to the care of the environment and reduces the use of agrochemical products such as fertilizers and pesticides.

Table 5.2 Effect of PGPR application on crops.

Bacteria	Crop	Beneficial effects	References
Klebsiella spp. (MBE02)	Peanut (*Arachis hypogaea* L.)	Growth stimulation	Sharma et al. (2019)
Bacillus megaterium (MU2) *Bacillus licheniformis* (MU8)	Wheat (*Triticum aestivum L.*)	Induces drought tolerance Increased germination rate, earliness, seedling vigor index, fresh weight, and dry weight under irrigation and drought stress	Rashid et al. (2022)
Bacillus subtilis, Bacillus mojavensis, Paenibacillus polymyxa, Bacillus altitudinis, Brevibacillus laterosporus	Tomato (*Solanum lycopersicum* L.) Tomato seeds	Resistance to abiotic stress due to water shortage in tomato plants	Astorga-Eló et al. (2021)
Enterobacter hormaechei (MF957335)	Tomato (*Solanum lycopersicum*)	Increased tomato yields, disease, and salinity tolerance	Ranawat et al. (2021)
Bacillus sp., *B. sonorensis, B. cereus, B. subtilis, Brevibacillus* sp. *B. safensis, B. paramycoides, Bacillus* sp., *B. cereus and B. tequilensis*	Cotton (*Gossypium hirsutum*, FH-142)	Enhancing growth, productivity, and tolerance in a saline environment	Saleem et al. (2021)
Enterobacter HS9 Bacillus G9	*Mucuna pruriens* (L.) DC	Water stress tolerance	Brunetti et al. (2021)
Burkholderia pyromania FJS-3 *Pseudomonas rhodesiae* FJS-7 *Pseudomonas baetica* FJS-16	Tea Fuding Dabai (*Camellia sinensis)* Tobacco K326 (*Nicotiana tabacum)* Pepper Chaotian (*Capsicum annum)*	Promoting plant growth and improving crop quality	Zhang et al. (2024)
Bacillus velezensis (KHH13)	Red lettuce (*Lactuca sativa* cv. LS-006) Cucumber (*Cucumis sativus* cv. Ashin)	Growth promotion in red lettuce Reduction of wilt caused by *Fusarium oxysporum* f. sp. conglutinans on cucumber	Chen et al. (2024)
B. velezensis Strain GB	Squash (*Cucurbita pepo*)	Induction and increase of β-1,3-glucanase, chitinase, polyphenol oxidase, and peroxidase activities Reduction in the number of *Bemisia tabaci* attracted to the plant Reduction in the number of plant eggs and hatching	Soliman et al. (2022)

Bacteria	Crop	Beneficial effects	References
Bacillus amyloliquefaciens Strain TBorg1	Tomato (*Solanum lycopersicum* L.)	Induction of systemic resistance Increased enzyme activities (PPO and POX) Increased polyphenolic pathway transcripts (C4H, HCT, and CHI) and pathogenesis-related genes (PR-1 and PR-5)	Abdelkhalek et al. (2022)
Bacillus sp. strain JS	Tobacco (*N. tabacum* "Xanthi")	Resistance to fungal and oomycete pathogens *Rhizoctonia solani* and *Phytophthora nicotianae* Expression of genes related to PR defense (expression of acidic PR-4 encoding chitinase and acidic PR-9 encoding peroxidase and PR-14 encoding lipid transfer protein)	Kim et al. (2015)
Bacillus thuringiensis GS1	Cucumber plants	Induction of PR proteins and defense-related enzymes to confer resistance to *R. solani* KACC 40111 (RS)	Seo et al. (2012)

5.5 Conclusion

PR proteins are essential in the plant immune response, providing an effective defense against various pathogens. Physical, chemical, and biological factors influence their production. PGPRs are biological stimuli that trigger the PR protein-mediated plant immune response. These rhizobacteria promote plant growth and activate resistance mechanisms that enhance the plant's ability to combat diseases. In many cases, combining different PGPR strains to form consortia has proven to be an effective strategy for optimizing plant protection. Given the environmental impact of agrochemicals, the use of PGPR in agriculture represents a sustainable and promising alternative to improve crop health and reduce reliance on synthetic pesticides and fertilizers.

References

Abdelkhalek, A., Aseel, D. G., Király, L., Künstler, A., Moawad, H., & Al-Askar, A. A. 2022. Induction of systemic resistance to Tobacco mosaic virus in tomato through foliar application of *Bacillus amyloliquefaciens* Strain TBorg1 culture filtrate. *Viruses*, 14(8), 1830. https://doi.org/10.3390/v14081830

Abdelkhalek, A., Behiry, S.I., & Al-Askar, A.A. 2020. *Bacillus velezensis* PEA1 inhibits *Fusarium oxysporum* growth and induces systemic resistance to cucumber mosaic virus. *Agronomy*. 10(9), 1312.

Ahmad, F., Ahmad, I., & Khan, M.S. 2008. Screening of free-living rhizospheric bacteria for their multiple plant growth promoting activities. *Microbiol. Res.* 163, 173–181. https://doi.org/10.1016/j.micres.2006.04.001

Ali, S., Ganai, B.A., Kamili, A.N., Bhat, A.A., Mir, Z.A., Bhat, J.A., Tyagi, A., Islam, S.T., Mushtaq, M., Yadav, P., Rawat, S., & Grover, A. 2018. Pathogenesis-related proteins and peptides as promising tools for engineering plants with multiple stress tolerance. *Microbiol. Res.* 212–213, 29–37. https://doi.org/10.1016/j.micres.2018.04.008

Andrade, G.F., Paniz, F.P., Martins, A.C., Rocha, B.A., da Silva Lobato, A.K., Rodrigues, J.L., Cardoso-Gustavson, P., Masuda, H.P., & Batista, B.L. 2018. Agricultural use of Samarco's spilled mud assessed by rice cultivation: a promising residue use? *Chemosphere*. 193, 892–902. https://doi.org/10.1016/j.chemosphere.2017.11.099

Astorga-Eló, M., Gonzalez, S., Acuña, J.J., Sadowsky, M.J., & Jorquera, M.A. 2021. Rhizobacteria from "flowering desert" events contribute to the mitigation of water scarcity stress during tomato seedling germination and growth. *Sci. Rep.* 11, 1–12. https://doi.org/10.1038/s41598-021-93303-8

Atkinson, E., Tuza, Z., Perrino, G., Stan, G.B., & Ledesma-Amaro, R. 2022. Resource-aware whole-cell model of division of labour in a microbial consortium for complex-substrate degradation. *Microb. Cell. Fact.* 21(1), 115.

Bhattacharyya, P.N., & Jha, D.K. 2012. Plant growth-promoting rhizobacteria (PGPR): emergence in agriculture. *World J. Microbiol. Biotechnol.* 28, 1327–1350. https://doi.org/10.1007/s11274-011-0979-9

Bigeard, J., Colcombet, J., & Hirt, H. 2015. Signaling mechanisms in Pattern-Triggered Immunity (PTI). *Mol. Plant.* 8, 521–539. https://doi.org/10.1016/j.molp.2014.12.022

Boller, T., & Felix, G. 2009. A renaissance of elicitors: perception of microbe-associated molecular patterns and danger signals by pattern-recognition receptors. *Annu. Rev. Plant. Biol.* 60, 379–406. https://doi.org/10.1146/annurev.arplant.57.032905.105346

Brunetti, C., Saleem, A.R., Della Rocca, G., Emiliani, G., De Carlo, A., Balestrini, R., Khalid, A., Mahmood, T., & Centritto, M. 2021. Effects of plant growth-promoting rhizobacteria strains producing ACC deaminase on photosynthesis, isoprene emission, ethylene formation and growth of *Mucuna pruriens* (L.) DC. in response to water deficit. *J. Biotechnol.* 331, 53–62. https://doi.org/10.1016/j.jbiotec.2021.03.008

Cavalcanti, F.R., Oliveira, J.T.A., Martins-Miranda, A.S., Viégas, R.A., & Silveira, J.A.G. 2004. Superoxide dismutase, catalase and peroxidase activities do not confer protection against oxidative damage in salt-stressed cowpea leaves. *New. Phytol.* 163, 563–571. https://doi.org/10.1111/j.1469-8137.2004.01139.x

Chandran, H., Meena, M., & Swapnil, P. 2021. Plant growth-promoting rhizobacteria as a green alternative for sustainable agriculture. *Sustainability*. 13(19), 10986. https://doi.org/10.3390/su131910986

Chandrasekaran, M., & Chun, S.C. 2016. Expression of PR-protein genes and induction of defense-related enzymes by *Bacillus subtilis* CBR05 in tomato (*Solanum lycopersicum*) plants challenged with *Erwinia carotovora* subsp. carotovora. *Biosci. Biotechnol. Biochem.* 80, 2277–2283. https://doi.org/10.1080/09168451.2016.1206811

Che, S., & Men, Y. 2019. Synthetic microbial consortia for biosynthesis and biodegradation: promises and challenges. *J. Ind. Microbiol. Biotechnol.* 46, 1343–1358. https://doi.org/10.1007/s10295-019-02211-4

Chen, T.-Y., Tzean, Y., Chang, T.-D., Wang, X.-R., Yang, C.-M., & Lin, Y.-H. 2024. Characterization of biofertilization and biocontrol potential of *Bacillus velezensis* KHH13 from organic soils. *Agronomy*. 14, 1135. https://doi.org/10.3390/agronomy14061135

Chung, E.J., Hossain, M.T., Khan, A., Kim, K.H., Jeon, C.O., & Chung, Y.R. 2015. *Bacillus oryzicola* sp. Nov., an endophytic bacterium isolated from the roots of rice with antimicrobial, plant growth promoting, and systemic resistance inducing activities in rice. *Plant. Pathol. J. (Faisalabad)* 31, 152–164. https://doi.org/10.5423/PPJ.OA.12.2014.0136

Conrath, U. 2011. Molecular aspects of defence priming. *Trends. Plant. Sci.* 16, 524–531. https://doi.org/10.1016/j.tplants.2011.06.004

Daayf, F., Bel-Rhlid, R., & Bélanger, R.R. 1997. Methyl ester of p-coumaric acid: a phytoalexin-like compound from long English cucumber leaves. *J. Chem. Ecol.* 23, 1517–1526. https://doi.org/10.1023/B:JOEC.0000006419.22408.65

de Andrade, L.A., Santos, C.H.B., Frezarin, E.T., Sales, L.R., & Rigobelo, E.C. 2023. Plant growth-promoting rhizobacteria for sustainable agricultural production. *Microorganisms*. 11, 1088. https://doi.org/10.3390/microorganisms11041088

Deising, H.B., Reimann, S., & Pascholati, S.F. 2008. Mechanisms and significance of fungicide resistance. *Braz. J. Microbiol.* 39, 286–295. https://doi.org/10.1590/S1517-83822008000200017

Demoz, B.T., & Korsten, L. 2006. *Bacillus subtilis* attachment, colonization, and survival on avocado flowers and its mode of action on stem-end rot pathogens. *Biological. Control*. 37, 68–74. https://doi.org/10.1016/j.biocontrol.2005.11.010

Dong, W., Liu, K., Wang, F., Xin, F., Zhang, W., Zhang, M., Wu, H., Ma, J., & Jiang, M. 2017. The metabolic pathway of metamifop degradation by consortium ME-1 and its bacterial community structure. *Biodegradation*. 28, 181–194. https://doi.org/10.1007/s10532-017-9787-8

Egamberdieva, D., & Lugtenberg, B. 2013. Use of plant growth-promoting rhizobacteria to alleviate salinity stress in plants. In Use of Microbes for the Alleviation of Soil Stresses, Volume 1 (pp. 73–96). New York, NY: Springer New York.

Epple, P., Apel, K., & Bohlmann, H. 1995. An *Arabidopsis thaliana* thionin gene is inducible via a signal transduction pathway different from that for pathogenesis-related proteins. *Plant. Physiol.* 109, 813–820. https://doi.org/10.1104/pp.109.3.813

Friedrich, L., Lawton, K., Ruess, W., Masner, P., Specker, N., Rella, M.G., Meier, B., Dincher, S., Staub, T., Uknes, S., Métraux, J., Kessmann, H., & Ryals, J. 1996. A benzothiadiazole derivative induces systemic acquired resistance in tobacco. *Plant. J.* 10, 61–70. https://doi.org/10.1046/j.1365-313X.1996.10010061.x

Ganeshamoorthi, P., Anand, T., Prakasam, V., Bharani, M., Ragupathi, N., & Samiyappan, R. 2008. Plant Growth Promoting Rhizobacterial (PGPR) bioconsortia mediates induction of defense-related proteins against infection of root rot pathogen in mulberry plants. *J. Plant. Interact.* 3, 233–244. https://doi.org/10.1080/17429140802088097

García-Olmedo, F., Molina, A., Segura, A., & Moreno, M. 1995. The defensive role of nonspecific lipid-transfer proteins in plants. *Trends. Microbiol.* 3, 72–74. https://doi.org/10.1016/S0966-842X(00)88879-4

Ge, M., Gong, M., Jiao, Y., Li, Y., Shen, L., Li, B., Wang, Y., Wang, F., Zhang, S., & Yang, J. 2022. *Serratia marcescens*-S3 inhibits *Potato virus* Y by activating ubiquitination of molecular chaperone proteins NbHsc70-2 in *Nicotiana benthamiana*. *Microb. Biotechnol.* 15(4), 1178–1188.

Glazebrook, J. 2005. Contrasting mechanisms of defense against biotrophic and necrotrophic pathogens. *Annu. Rev. Phytopathol.* 43, 205–227. https://doi.org/10.1146/annurev.phyto.43.040204.135923

Gohil, K., Rajput, V., & Dharne, M. 2020. Pan-genomics of *Ochrobactrum* species from clinical and environmental origins reveals distinct populations and possible links. *Genomics*. 112, 3003–3012. https://doi.org/10.1016/j.ygeno.2020.04.030

Gopal, M., Gupta, A., & Thomas, G.V. 2013. Bespoke microbiome therapy to manage plant diseases. *Front. Microbiol.* 4, 355.

Graham, M.Y., & Graham, T.L. 1991. Rapid accumulation of anionic peroxidases and phenolic polymers in soybean cotyledon tissues following treatment with phytophthora megasperma f. sp. glycinea wall glucan. *Plant. Physiol.* 97, 1445–1455. https://doi.org/10.1104/pp.97.4.1445

Gray, E.J., & Smith, D.L. 2005. Intracellular and extracellular PGPR: commonalities and distinctions in the plant-bacterium signaling processes. *Soil. Biol. Biochem.* 37, 395–412. https://doi.org/10.1016/j.soilbio.2004.08.030

Gross, G.G. 1981. The biochemistry of lignification. *Adv. Bot. Res.* 8, 25–63. https://doi.org/10.1016/S0065-2296(08)60032-4

Han, W., He, P., Lin, Y., Shao, L., & Lü, F. 2019. A methanogenic consortium was active and exhibited long-term survival in an extremely acidified thermophilic bioreactor. *Front. Microbiol.* 10, 1–15. https://doi.org/10.3389/fmicb.2019.02757

Henry, G., Thonart, P., & Ongena, M. 2012. PAMPs, MAMPs, DAMPs and others: an update on the diversity of plant immunity elicitors. *Biotechnol. Agron. Soc. Envir.* 16, 257–268.

Igiehon, B.C., Babalola, O.O., & Hassen, A.I. 2024. Rhizosphere competence and applications of plant growth-promoting rhizobacteria in food production – a review. *Sci. Afr.* 23, e02081. https://doi.org/10.1016/j.sciaf.2024.e02081

Iqbal, M.Z., Singh, K., & Chandra, R. 2024. Recent advances of plant growth promoting rhizobacteria (PGPR) for eco-restoration of polluted soil. *Clean. Eng. Technol.* 23, 100845. https://doi.org/10.1016/j.clet.2024.100845

Jain S., Varma, A., Tuteja N., & Choudhary, D.K. 2016. *Microbial-Mediated Induced Systemic Resistance in Plants* (pp. 1–226). Springer. https://doi.org/10.1007/978-981-10-0388-2

Jeyarajan, R., & Nakkeeran, S. 2000. Exploitation of microorganisms and viruses as biocontrol agents for crop disease management. In: R.K. Upadhyay, K.G. Mukerji, & B.P. Chamola (Eds.), *Biocontrol Potential and its Exploitation in Sustainable Agriculture* (pp. 95–116). Springer. https://doi.org/10.1007/978-1-4615-4209-4_8

Jia, X., Liu, C., Song, H., Ding, M., Du, J., Ma, Q., & Yuan, Y. 2016. Design, analysis and application of synthetic microbial consortia. *Synth. Syst. Biotechnol.* 1, 109–117. https://doi.org/10.1016/j.synbio.2016.02.001

Ju, W., Jin, X., Liu, L., Shen, G., Zhao, W., Duan, C., & Fang, L. 2020. Rhizobacteria inoculation benefits nutrient availability for phytostabilization in copper contaminated soil: drivers from bacterial community structures in rhizosphere. *Appl. Soil. Ecol.* 150, 103450. https://doi.org/10.1016/j.apsoil.2019.103450

Kamle, M., Borah, R., Bora, H., Jaiswal, A.K., Singh, R.K., & Kumar, P. 2020. Systemic Acquired Resistance (SAR) and Induced Systemic Resistance (ISR): role and mechanism of action against phytopathogens. In: A.L. Hesham, R. Upadhyay, G. Sharma, C. Manoharachary, & V. Gupta (Eds.), *Fungal Biotechnology and Bioengineering. Fungal Biology* (pp. 457–470). Springer: Cham. https://doi.org/10.1007/978-3-030-41870-0_20

Kim, J.-S., Lee, J., Lee, C.-H., Woo, S.Y., Kang, H., Seo, S.-G., & Kim, S.-H. 2015. Activation of pathogenesis-related genes by the rhizobacterium, *Bacillus* sp. JS, which induces systemic resistance in tobacco plants. *Plant. Pathol. J. (Faisalabad).* 31, 195–201. https://doi.org/10.5423/PPJ.NT.11.2014.0122

Kloepper, J., & Schroth, M.N. 1978. Plant growth-promoting rhizobacteria on radishes. In: *Fourth International Conference on Plant Pathogenic Bacteria Article. Fourth International Conference on Plant Pathogen Bacteria*, Angers, France (Vol. 2, pp. 879–882).

Kloepper, J.W., Ryu, C.-M., & Zhang, S. 2004. Induced systemic resistance and promotion of plant growth by *Bacillus* spp. *Phytopathology.* 94, 1259–1266. https://doi.org/10.1094/PHYTO.2004.94.11.1259

Li, H., Qiu, Y., Yao, T., Ma, Y., Zhang, H., & Yang, X. 2020. Effects of PGPR microbial inoculants on the growth and soil properties of *Avena sativa, Medicago sativa*, and *Cucumis sativus* seedlings. *Soil. Till. Res.* 199, 104577. https://doi.org/10.1016/j.still.2020.104577

Lucas, J.A., Hawkins, N.J., & Fraaije, B.A. 2015. The evolution of fungicide resistance. *Adv. Appl. Microbiol.* 90, 29–92. https://doi.org/10.1016/bs.aambs.2014.09.001

Mal, B., Mahapatra, P., & Mohanty, S. 2014. Effect of diazotrophs and chemical fertilizers on production and economics of Okra (*Abelmoschus esculentus*, L.) cultivars. *Am. J. Plant. Sci.* 5, 168–174. https://doi.org/10.4236/ajps.2014.51022

Manzoni, C., Kia, D.A., Vandrovcova, J., Hardy, J., Wood, N.W., Lewis, P.A., & Ferrari, R. 2018. Genome, transcriptome and proteome: the rise of omics data and their integration in biomedical sciences. *Brief. Bioinform.* 19, 286–302. https://doi.org/10.1093/BIB/BBW114

Menéndez, E., Navarro, J., López, J., & Dalmau, A. 2020. Mecanismos de defensa en plantas. Proteínas relacionadas con la patogenicidad. *Rev. investig. Univ. Le. Cordon. Bleu.* 7, 98–109. https://doi.org/10.36955/RIULCB.2020v7n2.010

Meziane, H., Van Der Sluis, I., Van Loon, L.C., Höfte, M., & Bakker, P.A. 2005. Determinants of *Pseudomonas putida* WCS358 involved in inducing systemic resistance in plants. *Mol. Plant. Pathol.* 6(2), 177–185.

Meyer, S.L.F., & Roberts, D.P. 2002. Combinations of biocontrol agents for management of plant-parasitic nematodes and soilborne plant-pathogenic fungi. *J. Nematol.* 34, 1–8.

Moreno-Gámez, S., Hochberg, M.E., & van Doorn, G.S. 2023. Quorum sensing as a mechanism to harness the wisdom of the crowds. *Nat. Commun.* 14, 1–10. https://doi.org/10.1038/s41467-023-37950-7

Mourouzidou, S., Ntinas, G.K., Tsaballa, A., & Monokrousos, N. 2023. Introducing the power of plant growth promoting microorganisms in soilless systems: a promising alternative for sustainable agriculture. *Sustainability (Switzerland).* 15, 5959. https://doi.org/10.3390/su15075959

Nag, M., Lahiri, D., Mukherjee, D., Banerjee, R., Garai, S., Sarkar, T., Ghosh, S., Dey, A., Ghosh, S., Pattnaik, S., Edinur, H.A., Kari, Z.A., Pati, S., & Ray, R.R. 2021. Functionalized chitosan nanomaterials: a jammer for quorum sensing. *Polymers (Basel).* 13, 1–17. https://doi.org/10.3390/polym13152533

Newton, A.C., Fitt, B.D., Atkins, S.D., Walters, D.R., & Daniell, T.J. 2010. Pathogenesis, parasitism and mutualism in the trophic space of microbe–plant interactions. *Trends. Microbiol.* 18(8), 365–373.

Niderman, T., Genetet, I., Bruyere, T., Gees, R., Stintzi, A., Legrand, M., Fritig, B., & Mosinger, E. 1995. Pathogenesis-related PR-1 proteins are antifungal (isolation and characterization of three 14-kilodalton proteins of tomato and of a basic PR-1 of tobacco with inhibitory activity against phytophthora infestans). *Plant. Physiol.* 108, 17–27. https://doi.org/10.1104/pp.108.1.17

Oerke, E.C. 2006. Crop losses to pests. *J. Agric. Sci.* 144, 31–43. https://doi.org/10.1017/S0021859605005708

Okushima, Y., Koizumi, N., Kusano, T., & Sano, H. 2000. Secreted proteins of tobacco cultured BY2 cells: identification of a new member of pathogenesis-related proteins. *Plant. Mol. Biol.* 42, 479–488. https://doi.org/10.1023/A:1006393326985

Panpatte, D.G., Jhala, Y.K., Shelat, H.N., & Vyas, R.V. 2016. *Pseudomonas fluorescens*: a promising biocontrol agent and PGPR for sustainable agriculture. In: *Microbial Inoculants in Sustainable Agricultural Productivity* (Vol. 1: Research perspectives, pp. 257–270). Springer

Passardi, F., Cosio, C., Penel, C., & Dunand, C. 2005. Peroxidases have more functions than a Swiss army knife. *Plant. Cell. Rep.* 24, 255–265. https://doi.org/10.1007/s00299-005-0972-6

Pieterse, C.M., Berendsen, R.L., de Jonge, R., Stringlis, I.A., Van Dijken, A.J., Van Pelt, J. A., Van Wees, S.C., Yu, K., Zamioudis, C., & Bakker, P.A. 2021. *Pseudomonas simiae* WCS417: star track of a model beneficial rhizobacterium. *Plant. Soil.* 461, 245–263.

Pieterse, C.M.J., Zamioudis, C., Berendsen, R.L., Weller, D.M., Van Wees, S.C.M. & Bakker, P.A.H.M. 2014. Induced systemic resistance by beneficial microbes. *Annu. Rev. Phytopathol.* 52, 347–375. https://doi.org/10.1146/annurev-phyto-082712-102340

Bashan, Y., & De-Bashan, L.E. 2005. Plant growth-promoting plant growth-promoting. *ESE* 1(1), 103–115.

Pozo, M.J., & Azcón-Aguilar, C. 2007. Unraveling mycorrhiza-induced resistance. *Curr. Opin. Plant. Biol.* 10, 393–398. https://doi.org/10.1016/j.pbi.2007.05.004

Ranawat, B., Mishra, S., & Singh, A. 2021. *Enterobacter hormaechei* (MF957335) enhanced yield, disease and salinity tolerance in tomato. *Arch. Microbiol.* 203, 2659–2667. https://doi.org/10.1007/s00203-021-02226-5

Rashid, U., Yasmin, H., Hassan, M.N., Naz, R., Nosheen, A., Sajjad, M., Ilyas, N., Keyani, R., Jabeen, Z., Mumtaz, S., Alyemeni, M.N., & Ahmad, P. 2022. Drought-tolerant *Bacillus megaterium* isolated from semi-arid conditions induces systemic tolerance of wheat under drought conditions. *Plant. Cell. Rep.* 41, 549–569. https://doi.org/10.1007/s00299-020-02640-x

Raupach, G.S., & Kloepper, J.W. 1998. Mixtures of plant growth-promoting rhizobacteria enhance biological control of multiple cucumber pathogens. *Phytopathology.* 88, 1158–1164. https://doi.org/10.1094/PHYTO.1998.88.11.1158

Reddy, P.P. 2013. *Recent Advances in Crop Protection* (pp. 1–268). Springer. https://doi.org/10.1007/978-81-322-0723-8

Reddy, V.S., & Reddy, A.S.N. 2004. Proteomics of calcium-signaling components in plants. *Phytochemistry.* 65, 1745–1776. https://doi.org/10.1016/j.phytochem.2004.04.033

Remick, B.C., Gaidt, M.M., & Vance, R.E. 2023. Effector-triggered immunity. *Annu. Rev. Immunol.* 41, 453–481.

Ryan, C.A. 1990. Protease inhibitors in plants: genes for improving defenses against insects and pathogens. *Annu. Rev. Phytopathol.* 28, 425–449. https://doi.org/10.1146/annurev.py.28.090190.002233

Saeed, Q., Xiukang, W., Haider, F.U., Kučerik, J., Mumtaz, M.Z., Holatko, J., Naseem, M., Kintl, A., Ejaz, M., Naveed, M., Brtnicky, M., & Mustafa, A. 2021. Rhizosphere bacteria in plant growth promotion, biocontrol, and bioremediation of contaminated sites: a comprehensive review of effects and mechanisms. *Int. J. Mol. Sci.* 22, 10529. https://doi.org/10.3390/ijms221910529

Sahu, P.K., Singh, D.P., Prabha, R., Meena, K.K., & Abhilash, P.C. 2019. Connecting microbial capabilities with the soil and plant health: options for agricultural sustainability. *Ecol. Indic.* 105, 601–612. https://doi.org/10.1016/j.ecolind.2018.05.084

Saleem, S., Iqbal, A., Ahmed, F., & Ahmad, M. 2021. Phytobeneficial and salt stress mitigating efficacy of IAA producing salt tolerant strains in *Gossypium hirsutum*. *Saudi. J. Biol. Sci.* 28, 5317–5324. https://doi.org/10.1016/j.sjbs.2021.05.056

Samain, E., Ernenwein, C., Aussenac, T., & Selim, S. 2022. Effective and durable systemic wheat-induced resistance by a plant-growth-promoting rhizobacteria consortium of *Paenibacillus* sp. strain B2 and *Arthrobacter* spp. strain AA against *Zymoseptoria tritici* and drought stress. *Physiol. Mol. Plant. Pathol.* 119, 101830. https://doi.org/10.1016/j.pmpp.2022.101830

Santos, L.F., & Olivares, F.L. 2021. Plant microbiome structure and benefits for sustainable agriculture. *Curr. Plant. Biol.* 26, 100198. https://doi.org/10.1016/j.cpb.2021.100198

Savary, S., Willocquet, L., Pethybridge, S.J., Esker, P., McRoberts, N., & Nelson, A. 2019. The global burden of pathogens and pests on major food crops. *Nat. Ecol. Evol.* 3, 430–439. https://doi.org/10.1038/s41559-018-0793-y

Seo, D.J., Nguyen, D.M.C., Song, Y.S., & Jung, W.J. 2012. Induction of defense response against *Rhizoctonia solani* in cucumber plants by endophytic bacterium *Bacillus thuringiensis* GS1. *J. Microbiol. Biotechnol.* 22, 407–415. https://doi.org/10.4014/jmb.1107.07027

Sharma, S., Chen, C., Navathe, S., Chand, R., & Pandey, S.P. 2019. A halotolerant growth promoting rhizobacteria triggers induced systemic resistance in plants and defends against fungal infection. *Sci. Rep.* 9, 4054. https://doi.org/10.1038/s41598-019-40930-x

Soliman, A., Matar, S., & Abo-Zaid, G. 2022. Production of *Bacillus velezensis* strain GB1 as a biocontrol agent and its impact on *Bemisia tabaci* by inducing systemic resistance in a squash plant. *Horticulturae.* 8, 511. https://doi.org/10.3390/horticulturae8060511

Srivastava, S., & Prasad, V. 2014. Induction of defence responses for biological control of plant diseases. In: *Biological Controls for Preventing Food Deterioration: Strategies for Pre- and Postharvest Management* (pp. 321–339). Wiley Blackwell, Department of Botany, University of Lucknow, Lucknow, India. https://doi.org/10.1002/9781118533024.ch14

Stockwell, V.O., & Stack, J.P. 2007. Using *Pseudomonas* spp. for integrated biological control. *Phytopathology.* 97, 244–249. https://doi.org/10.1094/PHYTO-97-2-0244

Stringlis, I.A., Yu, K.E., Feussner, K., de Jonge, R., Van Bentum, S., Van Verk, M.C., Berendsen, R.L., Bakker, P.A.H.M., Feussner, I., & Pieterse, C.M. 2018. MYB72-dependent coumarin exudation shapes root microbiome assembly to promote plant health. *Proc. Natl. Acad. Sci. USA.* 115(22), E5213–E5222.

Terras, F.R., Eggermont, K., Kovaleva, V., Raikhel, N.V, Osborn, R.W., Kester, A., Rees, S.B., Torrekens, S., Van Leuven, F., & Vanderleyden, J., 1995. Small cysteine-rich antifungal proteins from radish: their role in host defense. *Plant. Cell.* 7, 573–588. https://doi.org/10.1105/tpc.7.5.573

Thakur, M., & Sohal, B.S. 2013. Role of elicitors in inducing resistance in plants against pathogen infection: a review. *ISRN Biochem.* 2013, 762412. https://doi.org/10.1155/2013/762412

Thilagavathi, R., Saravanakumar, D., Ragupathi, N., & Samiyappan, R. 2007. A combination of biocontrol agents improves the management of dry root rot (*Macrophomina phaseolina*) in greengram. *Phytopathol. Mediterr.* 46, 157–167.

Timmusk, S., Behers, L., Muthoni, J., Muraya, A., & Aronsson, A.C., 2017. Perspectives and challenges of microbial application for crop improvement. *Front. Plant. Sci.* 8, 1–10. https://doi.org/10.3389/fpls.2017.00049

Van Hulten, M., Pelser, M., Van Loon, L.C., Pieterse, C.M.J., & Ton, J., 2006. Costs and benefits of priming for defense in *Arabidopsis. Proc. Natl. Acad. Sci. USA.* 103, 5602–5607. https://doi.org/10.1073/pnas.0510213103

Van Loon, L.C., 2007. Plant responses to plant growth-promoting rhizobacteria. *Eur. J. Plant. Pathol.* 119, 243–254. https://doi.org/10.1007/s10658-007-9165-1

Van Loon, L.C., Bakker, P.A.H.M., & Pieterse, C.M.J. 1998. Systemic resistance induced by rhizosphere bacteria. *Annu. Rev. Phytopathol.* 36, 453–483. https://doi.org/10.1146/annurev.phyto.36.1.453

Van Loon, L.C., & Van Strien, E.A. 1999. The families of pathogenesis-related proteins, their activities, and comparative analysis of PR-1 type proteins. *Physiol. Mol. Plant. Pathol.* 55, 85–97. https://doi.org/10.1006/pmpp.1999.0213

Van Wees, S.C.M., De Swart, E.A.M., Van Pelt, J.A., Van Loon, L.C., & Pieterse, C.M.J. 2000. Enhancement of induced disease resistance by simultaneous activation of salicylate- and jasmonate-dependent defense pathways in *Arabidopsis thaliana. Proc. Natl. Acad. Sci. USA.* 97, 8711–8716. https://doi.org/10.1073/pnas.130425197

Van Wees, S.C.M., Luijendijk, M., Smoorenburg, I., Van Loon, L.C., & Pieterse, C.M.J., 1999. Rhizobacteria-mediated induced systemic resistance (ISR *Arabidopsis* is not associated with a direct effect on expression of known defense-related genes but stimulates the expression of the jasmonate-inducible gene Atvsp upon challenge. *Plant. Mol. Biol.* 41, 537–549. https://doi.org/10.1023/A:1006319216982

Verhagen, B.W.M., Glazebrook, J., Zhu, T., Chang, H.S., Van Loon, L.C., & Pieterse, C.M.J. 2004. The transcriptome of rhizobacteria-induced systemic resistance in *Arabidopsis. Mol. Plant Microbe. Interact.* 17, 895–908. https://doi.org/10.1094/MPMI.2004.17.8.895

Verhagen, B.W.M., Trotel-Aziz, P., Couderchet, M., Höfte, M., and Aziz, A. 2010. Pseudomonas spp.-induced systemic resistance to *Botrytis cinerea* is associated with induction and priming of defence responses in grapevine. *J. Exp. Bot.* 61, 249–260. https://doi.org/10.1093/jxb/erp295

Vlot, A.C., Sales, J.H., Lenk, M., Bauer, K., Brambilla, A., Sommer, A., Chen, Y., Wenig, M., & Nayem, S. 2021. Systemic propagation of immunity in plants. *New. Phytol.* 229, 1234–1250. https://doi.org/10.1111/nph.16953

Vurro, M., Bonciani, B., & Vannacci, G. 2010. Emerging infectious diseases of crop plants in developing countries: impact on agriculture and socio-economic consequences. *Food. Secur.* 2, 113–132. https://doi.org/10.1007/s12571-010-0062-7

Walter, M.H., 1992. Regulation of lignification in defense. In: T. Boller, F. Meins. (Eds.), *Genes Involved in Plant Defense. Plant Gene Research* (pp. 327–352). Springer. https://doi.org/10.1007/978-3-7091-6684-0_13

Wang, L.-J., & Li, S.-H. 2006. Salicylic acid-induced heat or cold tolerance in relation to Ca^{2+} homeostasis and antioxidant systems in young grape plants. *Plant. Sci.* 170, 685–694. https://doi.org/10.1016/j.plantsci.2005.09.005

Wang, Q., Wang, D., Agathokleous, E., Cheng, C., Shang, B., & Feng, Z. 2023. Soil microbial community involved in nitrogen cycling in rice fields treated with antiozonant under ambient ozone. *Appl. Environ. Microbiol.* 89, e00180-23. https://doi.org/10.1128/aem.00180-23

Wei, Y., Zhang, Z., Andersen, C.H., Schmelzer, E., Gregersen, P.L., Collinge, D.B., Smedegaard-Petersen, V., & Thordal-Christensen, H. 1998. An epidermis/papilla-specific oxalate oxidase-like protein in the defence response of barley attacked by the powdery mildew fungus. *Plant. Mol. Biol.* 36, 101–112. https://doi.org/10.1023/A:1005955119326

Yasuda, M., Ishikawa, A., Jikumaru, Y., Seki, M., Umezawa, T., Asami, T., Maruyama-Nakashita, A., Kudo, T., Shinozaki, K., Yoshida, S., & Nakashita, H. 2008. Antagonistic interaction between systemic acquired resistance and the abscisic acid–mediated abiotic stress response in *Arabidopsis. Plant. Cell.* 20, 1678–1692. https://doi.org/10.1105/tpc.107.054296

Zhang, R., Vivanco, J.M., & Shen, Q. 2017. The unseen rhizosphere root–soil–microbe interactions for crop production. *Curr. Opin. Microbiol.* 37, 8–14. https://doi.org/10.1016/j.mib.2017.03.008

Zhang, T., Jian, Q., Yao, X., Guan, L., Li, L., Liu, F., Zhang, C., Li, D., Tang, H., & Lu, L. 2024. Plant growth-promoting rhizobacteria (PGPR) improve the growth and quality of several crops. *Heliyon.* 10, e31553. https://doi.org/10.1016/j.heliyon.2024.e31553

Zhang, Z., Collinge, D.B., & Thordal-Christensen, H. 1995. Germin-like oxalate oxidase, a H_2O_2-producing enzyme, accumulates in barley attacked by the powdery mildew fungus. *Plant. J.* 8, 139–145. https://doi.org/10.1046/j.1365-313X.1995.08010139.x

Zhu, L., Huang, J., Lu, X., & Zhou, C. 2022. Development of plant systemic resistance by beneficial rhizobacteria: recognition, initiation, elicitation and regulation. *Front. Plant. Sci.* 13, 952397. https://doi.org/10.3389/fpls.2022.952397

6

Current and Emerging Molecular Techniques for Microbial Consortia to Diagnose Plant Diseases

Dalpat Lal[1], Aparna[2], Sunidhi Tiwari[2], Nidhi Bhatt[3], and Durga Prasad[1]*

[1] *College of Agriculture, Agriculture University, Mandor, Jodhpur, India*
[2] *Department of Agriculture, Jagan Nath University, Chaksu, Jaipur, India*
[3] *Surajmal University Kichha, Udham Singh Nagar, Uttarakhand, India*

* Corresponding author: dalpat032@gmail.com

6.1 Introduction

The main aim of diagnostic science is to enable swift, accurate, and dependable detection of harmful pathogens responsible for disease. Traditional diagnostic techniques often involve microscopy, cultural, and morphological analyses. Although, these methods are fundamental to diagnostics, they are time-consuming, labor-intensive, and heavily dependent on expertise in classical taxonomy, which can introduce variability and inaccuracies (Nilsson et al., 2011). Reliable outcomes often require highly trained specialists in plant pathogen identification to mitigate the limitations of these approaches (Chalupová et al., 2014). To address these problems, the molecular diagnostic methods have been increasingly adopted, providing more efficient tools for identification and classification. Biotechnological techniques, like immunological assays, probes, and polymerase chain reaction (PCR) and its variants, focus on genotypic characteristics rather than phenotypic traits. These techniques offer significant advantages over the traditional techniques, including faster processing times, greater specificity, and improved accuracy. DNA macroarrays, microarrays, and high-throughput sequencing techniques (next-generation sequencing [NGS]), particularly RNA-Seq. These DNA-focused technologies enhance sensitivity and precision, allowing the rapid recognition of disease-causing microbes when DNA is found in low quantities (Capote et al., 2012; Midorikawa et al., 2018; Luchi et al., 2013; Rollins et al., 2016). Additionally, bioinformatics resources play a critical role in pathogen diagnostics. Platforms like GenBank (NCBI), the Nucleotide Sequence Database Collaboration (EBI), and MycoBank provide comprehensive repositories for nucleotide sequences and nomenclature innovations in mycology. These databases support the development of molecular tools for accurately identifying and categorizing plant

Microbial Consortia in Plant Science and Sustainable Agriculture, First Edition. Edited by Ajay Tomar, et al.

pathogen species, including newly emerging pathogens. The increasing prevalence of novel and rapidly spreading fungal diseases poses substantial risks to global agriculture, underscoring the critical need for precise and timely identification of phytopathogens (Hariharan & Prasannath, 2021). Therefore, we have provided an analysis of the current molecular diagnostic methods for plant pathogens, highlighting their benefits and limitations. Furthermore, it evaluates available molecular tools for detecting both established and emerging fungal pathogens across various crops, emphasizing the importance of such technologies in safeguarding agricultural productivity.

6.2 Molecular Techniques

6.2.1 PCR and Its Modification

6.2.1.1 Conventional PCR

It was originally designed as a precise tool for diagnosing bacterial and viral infections. However, it has broadened significantly in scope in plant pathology, also in identifying plant pathogens. Further, the result of PCR largely depends on its components and the quality of the DNA. Where a single set of primers fails to yield reliable or specific results, techniques like the use of DNA probes or nested primers can enhance the accuracy of the assay (Compton, 1991). PCR is applied successfully to detect the *Mycosphaerella musicola* and *Mycosphaerella fijiensis*, the pathogens responsible for diseases of leaf spot in banana and plantation crops (Henson & French, 1993). Despite its unparalleled specificity, PCR remains a technique that is resource-intensive and costly to perform. There are some variants of PCR as discussed in the following subsections.

6.2.1.2 Nested PCR

This is a molecular technique designed to boost the accuracy and detection limits of nucleic acid amplification processes. Yeo and Wong (2002) demonstrated that its sensitivity surpasses that of conventional PCR by up to 1000-fold, particularly for fungal detection. The procedure entails two sequential rounds of amplification: the first employs an outer primer pair to amplify a general DNA region, while the second utilizes the resulting product as a template, applying a different set of internal primers to amplify a more targeted sequence. Further, nested PCR has also proven effective in identifying pathogens in plant diseases, such as *Phytophthora palmivora* and *Phytophthora parasitica*, which infect orchids (Tsai et al., 2006). Similarly, it has also been used to confirm *Phytophthora cactorum* in strawberry diseases as a pathogen (Bhat & Browne, 2010).

6.2.1.3 Multiplex PCR

It can be optimized to enhance efficiency and reduce costs by incorporating multiple primer pairs in a reaction, allowing the simultaneous multiplex detection of DNA sequences from different source sequences. This approach is widely employed in plant pathology, where plants frequently experience simultaneous infections by multiple pathogens. During this process, specific DNA fragments corresponding to pathogenic fungal DNA are amplified and analyzed for identification purposes. Recent developments have introduced the use of padlock probes in multiplex PCR techniques to improve the recognition of fungi. This multiplex approach has been successfully employed to identify fungal pathogens, including *Fusarium oxysporum, Bipolaris cactivora, Phytophthora nicotianae*, and *P. cactorum*, which are significant causal agents of diseases in propagated cactus (Cho et al., 2016).

6.2.1.4 Real-time PCR (RT/qPCR)

Conventional PCR techniques are limited to the amplification of target DNA, which is insufficient to facilitate prompt diagnosis of diseases. In recent times, an advanced practice named RT-PCR has been developed. It enhances traditional PCR by enabling the quantification of DNA along with its amplification (Mackay, 2004). The use of SYBR Green I (fluorescent dyes) or TaqMan probe, a sequence-specific, fluorescent-labeled probe, has allowed real-time monitoring of the amplification (Badali & Nabili, 2012). Fluorescent signals are generated when the dye intercalates into DNA, and these signals increase proportionally with the accumulation of target DNA during each amplification cycle (McCartney et al., 2003). Although cost effective, fluorescent dyes have a limitation due to their nonspecific binding, which can result in false-positive signals, such as those arising from primer–dimer formations. To address this issue, fluorogenic probes with high specificity were introduced (Atkins & Clark, 2004; Bu et al., 2005). These probes are labeled with two kinds of fluorescent dyes: quencher dye on the 3′ end and reporter dye on the 5′ end. In their initial proximity, the quencher suppresses the reporter's fluorescence. During amplification, the exonuclease activity of *Taq* polymerase separates the reporter dye from the quencher, leading to fluorescence emission (Dasmahapatra & Mallet, 2006). Quantitative PCR (qPCR) has been extensively employed to identify and quantify pathogenic fungi, including *Aspergillus versicolor, Cladosporium cladosporioides, Stachybotrys chartarum*, and *Alternaria alternata* (Black, 2009).

6.2.1.5 Serial Analysis of Gene Expression

Serial analysis of gene expression (SAGE) is an advanced, sequence-based technique for profiling gene expression quantitatively. It enables the simultaneous identification of multiple transcripts by sequencing and analyzing short oligonucleotides (15 base pairs/longer), which are then matched to existing genomic data used to identify actively expressed genes (Velculescu et al., 1995). The methodology involves processing two samples that are ligated and labeled with distinct primers, followed by amplification. Later, the primers are removed, resulting in sticky ends that facilitate the formation of concatemers. These concatemers are cloned into a vector and sequenced, with subsequent computational analysis for gene identification. Despite its utility, SAGE has limitations. It requires substantial quantities of mRNA, and the 15 bp tag may sometimes be insufficient for accurate gene identification of complex genomes. This technique was initially employed on barley leaves to detect the *Blumeria graminis* (Dawei & Peng, 2014).

6.2.1.6 DNA Barcoding

Generally, it is utilized for the identification of flora and fauna, aiding in the assessment of biodiversity and the formulation of conservation strategies. In these years, DNA barcoding has been particularly effective in detecting symptomless plants and latent pathogens. Further, Hsieh et al. (2020) employed a set of molecular markers including internal transcribed spacer (ITS), partial β-tubulin, and histone H3 regions to identify *Cercospora, Colletotrichum,* and *Fusarium* in Chinese masona. Additionally, barcoding has been instrumental in metagenomic research, advancing our understanding, including both cultivable and uncultivable microbial groups in ecological, environmental, and health studies (Harrison et al., 2020; Hoshino & Inagaki, 2017; Shahraki et al., 2019). Techniques like PCR and loop-mediated isothermal amplification (LAMP), utilizing barcoding, have greatly enhanced diagnostic capabilities for various diseases.

DNA barcoding relies on the principle that genetic differences are greater between species than within a single species. Fungi, which are difficult to detect through traditional morphological methods, are detectable at the molecular scale via DNA barcoding. Further, it is facilitated by an appropriate marker, is especially useful for understanding the diverse fungus species present in regular ecosystems (Roe et al., 2010).

Further, genetic barcoding techniques for species differentiation offers numerous scientific benefits– (i) species identification at any life stage (ii) aiding in the discovery of new species through phylogenetic analysis of DNA/RNA sequence, (iii) enhancing our understanding of biological diversity, and (iv) encouraging the development of DNA sequencing technologies that support biodiversity research (Savolainen et al., 2005). Trustworthy barcode sequences provide uniform and effective tools for species identification, which are crucial for biodiversity assessments and environmental studies. The choice of an ideal barcode region involves balancing various options to design worldwide molecular identification using markers suitable for PCR.

6.2.2 DNA Microarray

It is also called a DNA/gene or biochip, which is a collection of microscopic DNA fragments immobilized on a solid substrate, typically glass, in a defined arrangement. These DNA fragments, referred to as probes, are designed to specifically hybridize with complementary targets (cDNA sequences) from a sample. The hybridization event is detected and quantified using labeled targets, such as those tagged with fluorophores, silver, or chemiluminescent markers, to evaluate the relative richness of transcription within the taster (Guigó, 2013). The development of DNA microarray technologies has enabled high-throughput detection and simultaneous analysis of various microbes.

For example, qPCR microarray technology utilizing primers and fluorescent probes has been employed to screen for fungal species such as *Rhizoctonia solani, Spongospora subterranea* (targeting ITS regions), *Alternaria solani, A. alternata* (Alt_a1 gene), *Fusarium* spp. (TEF-1α gene), and *Colletotrichum coccodes* (TUB2 gene) using silicon-based microarrays (Nikitin et al., 2018). Additionally, sugar beet root rot pathogens have been identified using array tube microarrays with marker genes such as ITS, TEF-1α, and 16S rDNA, including *Aphanomyces cochlioides, Botrytis cinerea*, and *Penicillium expansum* (Liebe et al., 2016). These techniques are advantageous due to their ability to be produced in batches on standard microscope slides in a rapid, cost-effective, and reproducible manner (Johannes et al., 2020). However, a primary limitation is that the technology can only identify sequences for which probes have been specifically designed (Bumgarner, 2013).

6.2.3 DNA Macroarrays

The development of DNA macroarrays involves designing probes that are unique to each species and made up of 15–30 base oligonucleotides, which are organized in well plates and immobilized onto nylon or nitrocellulose membranes. These probes are hybridized with an amplified targeted sequence and allow for their detection (Clark et al., 1999; Zhang et al., 2008). For instance, a digoxigenin-labeled RNA probe-based oligo-DNA macroarray assay was established to detect microbial populations residing on apple leaf surfaces. This technique utilized 40-base pair oligonucleotide sequences derived from the ITS region of fungal rDNA to identify various fungal species. This technique successfully identified fungal pathogens like *Venturia inaequalis* (scab), *Alternaria mali (Alternaria blotch)*, and *Diplocarpon mali (Marssonina blotch)*, as well as various nonpathogenic fungi and bacteria (He et al., 2012). Moreover, a DNA macroarray was employed to detect 61 fungal species, with 34 pathogens linked to young vine decline in grapevines (Úrbez-Torres et al., 2015).

DNA macroarrays are widely regarded as a dependable robust method for pathogen identification, even in samples containing numerous pathogens (Zhang et al., 2008). However, there are some limitations, including the limited lifespan of the filters, difficulties in annotating new genes with low transcript abundance, and the substantial number of filters needed for hybridization (Gammill & Lee, 2008). Other disadvantages include the inability to quantify pathogens and the challenge of verifying whether the detected genetic material is from viable organisms (Úrbez-Torres et al., 2015).

6.2.4 Fluorescence *In Situ* Hybridization

Fluorescence *in situ* hybridization (FISH) is a widely recognized molecular technique increasingly used for detecting and diagnosing plant diseases. This technique offers a combination of strong specificity of DNA sequence recognition with the sensitivity provided by fluorochrome-based detection systems (Cui et al., 2016; Hijri, 2009). It works by identifying cellular or tissue-specific DNA or RNA sequences, which is achieved using fluorescently labeled nucleic acid probes, applied through either direct or indirect labeling strategies (Shakoori, 2017). Typically, FISH assays involve fluorescently labeled oligonucleotide probes that hybridize with ribosomal RNA in pathogenic cells. These labeled cells are then examined using microscopy for in-depth analysis (Lukumbuzya et al., 2019). When plants are infected by pathogens, pathogen-specific rRNA sequences act as unique markers and can be detected using these techniques (Fang & Ramasamy, 2015). For example, FISH was successfully used to identify Southern blight in tomatoes, caused by the soilborne fungus *Sclerotium rolfsii*. Oligonucleotide probes labeled with cyanine dyes (Cy3 and Cy5) were able to detect concentrations as low as 0.06 pg μL^{-1} of *S. rolfsii* DNA in soil samples (Milner et al., 2019).

FISH offers a range of strengths, including excellent sensitivity, reproducibility, specificity, accuracy, and the capacity to deliver rapid results (Bozorg-Ghalati et al., 2019). Additionally, it gives valuable information on morphology and resolution of pathogens, allowing for identification of specific species within mixed microbial communities (Frickmann et al., 2017). However, challenges such as false positives caused by autofluorescence remain a limitation and can reduce assay specificity (Moter & Göbel, 2000). This promising diagnostic technique enhances the ability to detect and study plant pathogens, particularly within environments with complex mixed specimens.

6.2.5 Next-generation Sequencing

The high-throughput sequencing, a novel strategy for identifying and studying phytopathogens, is a transformative diagnostic approach enabling precise and comprehensive pathogen detection (Chalupowicz et al., 2019). The standard workflow for DNA-based NGS includes several essential stages, starting with DNA isolation, followed by library preparation, high-throughput parallel sequencing, and bioinformatic analysis, and concluding with the annotation of genetic variations or mutations (Qin, 2019). Advanced NGS methodologies, such as massively parallel signature sequencing, pyrosequencing, polony sequencing, and sequencing by oligonucleotide ligation detection, have been developed to enhance sequencing efficiency and accuracy (Rajesh & Jaya, 2017). RNA sequencing has also been developed as a robust tool for studying the transcriptome with unparalleled resolution and dynamic range. Platforms like the Illumina HiSeq system have set a benchmark for RNA-Seq applications, and the compact Illumina MiSeq system has further broadened accessibility for smaller-scale sequencing projects (Kukurba & Montgomery, 2015). RNA-Seq-based NGS is particularly valuable for identifying fungal plant pathogens responsible for new diseases. For example, whole-genome sequencing using Illumina MiSeq revealed a 51.4 Mb genome in *Calonectria pseudonaviculata*, a fungus implicated in *Sarcococca* blight in ornamental plants, identifying unique single-nucleotide polymorphisms (SNPs) in host isolates (Malapi-Wight et al., 2016).

Potgieter et al. (2020) studied how NGS-driven population genomics facilitates the analysis of genetic variations such as SNPs, insertions, deletions, and structural variations, which was demonstrated in the causal organism of yellow rust in wheat.

6.2.6 RNA Interference

RNA interference (RNAi) is a highly tissue-specific method for managing fungal infections and controlling mycotoxin-producing fungi in crops. This innovative approach hinges on several critical factors to achieve effective gene silencing. Key considerations include (i) designing vectors that generate double-stranded RNAs capable of producing small-interfering RNAs targeting specific fungal genes, (ii) ensuring the availability of abundant siRNA molecules at infection sites, (iii) facilitating the uptake of siRNAs by fungal cells, (iv) maintaining siRNA stability (e.g. sufficient half-life), and (v) enhancing the silencing effects through amplification mechanisms (Panwar et al., 2012). RNAi offers a sustainable alternative to conventional disease management strategies, addressing the growing challenge of fungicide resistance among plant pathogens (Ishii & Holloman, 2015). By leveraging the natural cellular defense mechanisms, RNAi technology disrupts fungal gene expression through sequence-specific degradation of target transcripts upon recognition of dsRNA or hairpin RNA (hpRNA) (Nakayashiki et al., 2005). The identification of unique fungal gene targets necessitates the use of cell-specific and dual RNA sequencing data to design dsRNA or hpRNA constructs that precisely inhibit fungal pathogenicity. Once designed, these RNA molecules exploit the cell's innate machinery, particularly siRNAs, to downregulate the expression of target genes, effectively reducing fungal infections. RNAi has been effectively applied in crops like maize to inhibit *Aspergillus flavus* and in barley to combat *Fusarium graminearum* (Masanga et al., 2015; Chen et al., 2006).

6.2.7 CRISPR/Cas Technology

Clustered regularly interspaced short palindromic repeats (CRISPR/Cas) is among the most advanced technologies for targeted genetic modification. Diagnostic methods utilizing Cas nucleases offer significant potential, particularly for applications in detecting plant pathogens and managing diseases. Detection systems based on Cas12a and Cas13a leverage the precise targeting capabilities of these nucleases to identify specific sequences in pathogen genetic materials (Gootenberg et al., 2017; Li et al., 2018; Chen et al., 2018). These methods often incorporate isothermal amplification techniques such as T7 transcription-recombinase polymerase amplification (specific to Cas13a), recombinase polymerase amplification, or LAMP before introducing Cas nuclease. Advanced approaches like SHERLOCK, SHERLOCKv2, HOLMES, and DETECTR employ activated Cas nucleases to degrade and quantify fluorescent reporter oligonucleotides. Compared to traditional nucleic acid detection methods genome editing tools require significantly less amount of genetic material and achieve exceptional sensitivity, with detection limits as low as femtomolar (10^{-15}) to zeptomolar (10^{-21}), depending on the methodology used (Li et al., 2019). Currently, Cas12a-based systems have been successfully employed to detect the rice blast caused by *Magnaporthe oryzae*, which was found to be abundant in infected rice tissues (Zhang et al., 2020; Kang et al., 2021). Some of the plant pathogens have been successfully detected from molecular techniques as discussed in Table 6.1.

Table 6.1 Molecular techniques utilized for detecting various plant pathogens.

Sl. no.	Detection technique	Causative agent	Disease	References
1.	Conventional PCR	*Mycosphaerella musicola*	Banana leaf spots disease	Henson and French (1993)
2.	Nested PCR	*Phytophthora cactorum*	Crown rot in Strawberry	Bhat and Browne (2010)
3.	Multiplex PCR	*Bipolaris cactivora*	Stem rot disease in cactus	Cho et al. (2016)
4.	RT-PCR	*Fusarium graminearum*	*Fusarium* ear blight (FEB) disease	Brown et al. (2011)
5.	SAGE	*Blumeria graminis*	Powdery mildew in barley	Dawei and Peng (2014)
6.	Northern blotting	*Magnaporthe grisea*	Blast disease in rice	Qi and Yang (2002)
7.	Macroarray	*Fusarium solani*	*Fusarium* rot in solanaceous crops	Sato et al. (2010)
8.	RNAi	*Aspergillus flavus*	Corn ear rot	Masanga et al. (2015)
9.	CRISPR/Cas (CAS 12A-based method)	*Magnaporthe oryzae*	Rice blast	Zhang et al. (2020)

6.3 Immunological Methods

An immunoassay is a biochemical technique widely used to quantify the concentration of specific target molecules. This method is based on the specific binding interaction between an analyte or antigen (Ag) and a corresponding antibody (Ab), forming an Ab–Ag complex. The performance of the assay largely depends on the efficiency and kinetics of this complex formation. Following binding, the complex is separated from unbound reagents using physical or chemical separation methods. Detection in immunoassays typically involves the use of labels, such as enzymes or radioisotopes, which are easily measurable and are conjugated to either the antibody or the analyte, facilitating accurate quantification (Darwish, 2006). The identification of the target antigen is essential when developing immunological assays. The antigen selected should be immunogenic enough to cause an antibody response and need to be specific to the pathogen to prevent cross-reactions with closely related pathogens (Richards et al., 2023).

Immunoassay technology that employs monoclonal antibodies offers high specificity for plant virus detection, making it well-suited for large-scale screening of plant samples and on-site pathogen identification, as demonstrated by methods such as tissue print enzyme-linked immunosorbent assay (ELISA) and lateral flow (LF) devices (Khatar et al., 2017). While the polyclonal antisera contain antibodies targeting multiple epitopes (Werres & Steffens, 1994), the problem of cross-reactivity can arise. The use and advancement of immunological methods have captivated significant attention and interest due to their capability to precisely and sensitively identify specific targets (Zhao et al., 2024). The main objective of immunodiagnostic assay is to quantify the specific molecular interaction between the antibody and its target antigen (Kokoskova & Janse, 2009).

The immunotechniques, including ELISA, multiplex immune-detection, and immunofluorescence, are also commonly employed. However, immunology-based kits tend to be less cost-effective than PCR assays and often suffer from limited stability over time (Martinelli et al., 2015). The most common immunoassays used are ELISA and lateral flow immunoassay (LFIA). Some of the pathogens detected from various immunological techniques are mentioned in Table 6.2.

6.3.1 Enzyme-linked Immunosorbent Assay

During the 1970s, it was introduced as a method for identifying plant viruses in aqueous solution. This method relies on the specific binding of target effectors from pathogens to antibodies conjugated with an enzyme (Fang & Ramasamy, 2015). ELISAs are the most commonly used immunodiagnostic technique as they are cost-effective and suitable for high-throughput testing. These assays employ both monoclonal and polyclonal antibodies, wherein monoclonal antibodies enhance assay sensitivity and specificity, thereby rendering ELISA a highly versatile tool for pathogen detection, while polyclonal antibodies are regarded as advantageous for the detection of diverse pathogen variants but can sometimes result in cross-reactivity. There are different ELISA protocols, primarily categorized into four core types: (i) direct, (ii) indirect, (iii) sandwich, and (iv) competitive ELISA. In these methods, antigens or antibodies are labeled with enzymes known as enzyme immunoassay (Sakamoto et al., 2017), with commonly used enzyme substrate including alkaline

Table 6.2 Various immunological methods used to detect plant pathogens.

Sr. no.	Technique used	Crop	Pathogen	References
1.	Lateral flow immunoassay	Cucumber and tobacco	Cucumber mosaic cucumovirus and tobacco mosaic tobamovirus	Tsuda et al. (1992)
		Potato	Potato virus x	Drygin et al. (2012)
		Maize	*Pantoea stewartia* subsp. *Stewartia*	Zhang et al. (2014)
		Maize	*P. stewartia* subsp. *Stewartia*	Feng et al. (2015)
			Xanthomonas campestris pv. *Musacearum*	Hodgett et al. (2015)
		Potato	*Pectobacterium atrosepticum* (potato blackleg)	Safenkova et al. (2015)
		Brassica	*Alternaria brassicae*	Wakeham et al. (2016)
		Prunus	*Xanthomonas arboricola* pv. *pruni*	Lo′pez-Soriano et al. (2017)
		Potato	Leaf roll virus of potato	Panferov et al. (2017)
		Grapes	Grapevine leafroll-associated virus	Byzova et al. (2018)
		Barley	Yellow dwarf virus of barley	Kim et al. (2022)
		Sugarcane	Mosaic virus of sugarcane	Thangavelu et al. (2022)
		Maize	Chlorotic mottle virus of maize	Lei et al. (2023)
2.	ELISA	Soybean	Soybean mosaic virus	Hill et al. (1981)
		Bean	Yellow mosaic virus of bean	Alkhalaf et al. (2009)
		Broad bean	Broad bean strain virus Red clover mottle virus	Musil and Gallo (1993)
	TAS-ELISA	Maize	Maize chlorotic mottle virus	Wu et al. (2013)
		Potato	Mop top virus of potato	Arif et al. (1994)
		Citrus	Citrus psorosis virus	Martin et al. (2004)
		Cucumber	Cucumber mosaic virus	Yu et al. (2005)
		Canary	Barley yellow dwarf virus	Ilbagi et al. (2008)
		Cassava	Mosaic virus of cassava	Charoenvilaisiri et al. (2021)
	DAS-ELISA	Citrus	Citrus tristeza virus	Baral (2024)
		Maize	Chlorotic mottle virus of maize	Zhang et al. (2023)
		Tomato	Tomato spotted wilt virus	Lai et al. (2021)
		Potato	Potato virus Y (PVY) Potato virus S	Tessema et al. (2024)
		Sugarcane	Yellow leaf virus of sugarcane	Viswanathan et al. (2004)

(Continued)

Table 6.2 (Continued)

Sr. no.	Technique used	Crop	Pathogen	References
	Plate-trapped antigen ELISA	Squash	Squash mosaic virus	Nascimento et al. (2017)
		Cowpea	Cowpea severe mosaic virus	
		Cucumber	Cucumber mosaic virus	
		Cucumber	Aphid-borne mosaic virus of cowpea	
		Zucchini	Zucchini yellow mosaic virus	
		Papaya	Lethal yellowing virus of papaya	
	Dot ELISA	Potato	PVY	Faccioli and Colombarini (1996)
			Potato virus M	
		Sugarcane	Mosaic virus of sorghum	Chen et al. (2020)
3.	Aptamer-based biosensors	Soybean	Rust fungi of soybean	Krivitsky et al. (2021)
		Apple	Apple stem pitting virus	Komorowska et al. (2017)

phosphatase, β-galactosidase, and horseradish peroxidase, and the conversion of the substrate into the product is monitored. Further, resultant product must exhibit the distinct physical or chemical properties compared to the substrate, that is, the colorless chromogenic substrate that, through enzyme activity, is transformed into colored products. These colored products can be easily quantified using a spectrophotometer (Darwish, 2006). The use of ELISA offers considerable advantages, including high-throughput potential and reliable performance. Nonetheless, ELISA methodologies are labor-intensive, require expensive laboratory equipment and skilled personnel, and have limited adaptability, thereby rendering them less effective across various types of plant materials. Furthermore, ELISA is generally inadequate for early-stage pathogen detection.

6.3.1.1 Types of ELISA

6.3.1.1.1 Direct ELISA

It was initially developed by Engvall and Perlmann and V. Weemen and Schuurs in 1971, serving as the foundational format for subsequent ELISA variation. The primary detection antibody directly interacts with the target protein, thereby reducing the risk of cross-reactivity occurring with secondary antibodies and expediting the process through fewer steps compared to indirect ELISA. However, this method is constrained by lower sensitivity in comparison to other ELISA types and incurs a higher cost per reaction. The diagrammatic presentation of different ELISAs is given in Figure 6.1.

6.3.1.1.2 Indirect ELISA

This method requires the use of two antibodies: a primary antibody that specifically binds to the target protein and a secondary antibody – conjugated to an enzyme – that

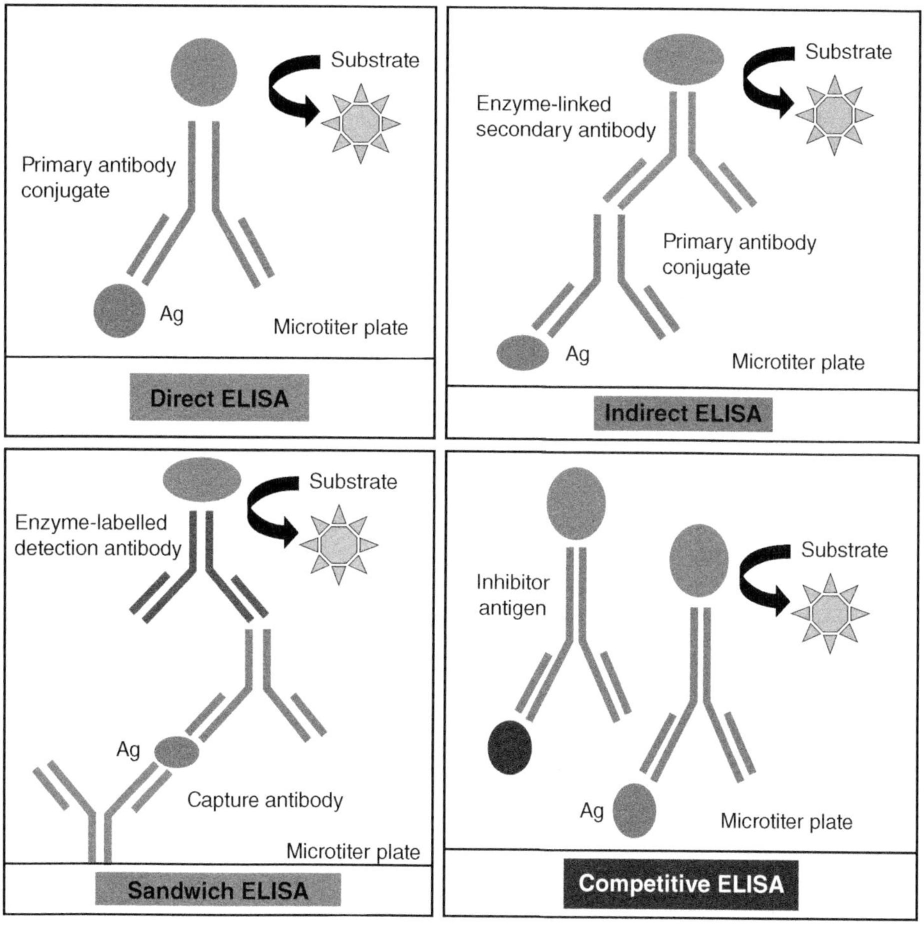

Figure 6.1 Representation of different ELISAs.

recognizes and binds to the primary antibody. In procedure, antigen detection is mediated indirectly through the secondary antibody labeled with an enzyme, thus termed "indirect ELISA." Initially, the target antigen is first immobilized onto the surface of a microtiter plate, after which blocking proteins are applied to prevent nonspecific binding. The primary antibody is then added to bind specifically to the antigen, followed by the introduction of an enzyme-linked secondary antibody. Finally, a substrate is added to initiate color development, indicating the presence of the antigen-antibody complex (Sakamoto et al., 2017). The principal drawback of indirect ELISA is the potential for cross-reactivity associated with secondary detection antibodies.

6.3.1.1.3 Sandwich ELISA

The technique is known as "sandwich" ELISA due to the antigen's placement between two antibody layers. The assay involves two antibodies: a capture antibody and a detection antibody. It begins with the immobilization of the capture antibody onto the surface of a microtiter plate. To minimize nonspecific binding, the plate is then blocked using appropriate blocking agents. Subsequently, the antigen present in the sample binds specifically to the immobilized capture antibody. Subsequently, an enzyme-labeled detection antibody binds to the antigen, forming the "sandwich" and enabling color development. The limitations of sandwich ELISA include its considerable time demands, elevated cost, and the requisite use of a precisely matched pair of antibodies compatible with the divalent or multivalent antigen.

6.3.1.1.4 Competitive ELISA

In 1973, Belanger pioneered the competitive ELISA technique to detect rat α-fetoprotein, which leveraged prior developments in indirect and sandwich ELISA methodologies (Belanger et al., 1973). Competitive ELISA is based on the principle in which the sample antigen and the inhibitor antigen compete to bind with a capture antibody. The process commences with the coating of an ELISA plate using these specific antibodies. Subsequently, a combination of unlabeled antigen from the sample (sample antigen) and labeled antigen (inhibitor antigen) is added to each well. As the concentration of the target protein increases in the sample, it competes more effectively for binding sites, resulting in reduced availability for the labeled antigen. During detection, only the enzymes bound to antibodies emit signals, resulting in an inverse correlation, where a higher concentration of the target protein produces weaker signals because fewer enzyme-labeled antigens remain attached. Advantage of this method includes minimal sample purification, the ability to quantify a broad range of antigens within a sample, applicability to small antigens, and low variability. Disadvantages include limited specificity and being unsuitable for use with highly diluted samples.

6.3.2 Lateral Flow Immunoassay

Lateral flow immunoassays (LFIAs), which utilize the immune-detection and immune-printing kits, are regarded as the earliest grower-friendly methods for pathogen surveillance. LFIA, alternatively known as the immune-chromatographic strip test method or rapid diagnostic test, operates on the principle of immune-chromatography, deploying metal nanoparticles to enable colorimetric detection of target analytes. Gold nanoparticles are particularly prevalent in creating LFIA strips, facilitating distinct and reliable visual identification (Selvarajan et al., 2020). Additional labels employed for detection purposes include latex microparticles, quantum dots, magnetic nanoparticles, and carbon nanoparticles. The fundamental principle of LFIA is based on the movement of a liquid sample containing the target analyte – conjugated to a detection label – across a test membrane via capillary action. The membrane is coated with capture molecules that interact with the detection label. Depending on whether the analyte is present or absent and the type of test (competitive or sandwich assay), the interaction produces either a positive or a negative result (Koczula & Gallotta, 2016).

LFIA tests consist of distinct components, including pads made from specialized materials. The sample is applied to a pretreated sample pad, which improves assay reproducibility and

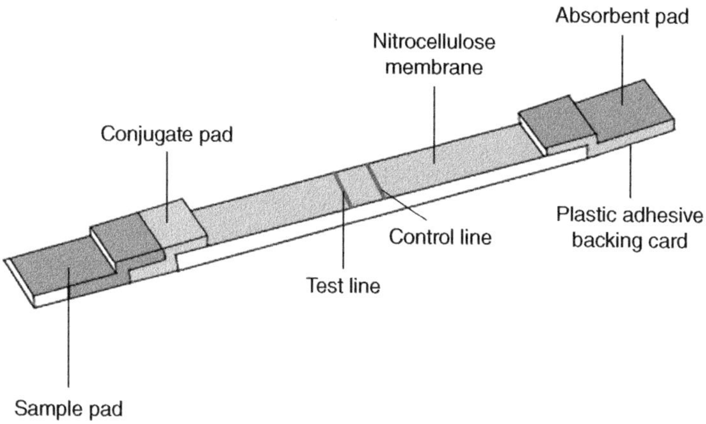

Figure 6.2 Lateral flow immunoassay.

controls the flow rate. As the sample moves along, it encounters a dried conjugate – usually immobilized on a conjugate pad made of polyester or fiberglass – where it binds to the target analyte. The mixture then advances to the reaction zone, a porous nitrocellulose membrane, where the detection of the analyte takes place. The nitrocellulose membrane is structured into two main regions: the test and control line (Figure 6.2). In the test line, specific interactions between the analyte and the reporter molecule generate the required response. The control line shows if the liquid has flowed through the strip, confirming assay functionality (Posthuma-Trumpie et al., 2009).

Nitrocellulose membranes come in a range of pore sizes, which regulate the flow rate of reactants through the device. The choice of pore size depends mainly on the analyte's size and the sample type. In commercial lateral flow assays, the capillary flow time – defined as the time required for a sample to travel approximately 4 cm across the membrane – is a key performance metric. Membranes with larger pores allow faster flow, while those with smaller pores slow down the sample movement due to increased flow time. Depending on the assay format, either an antigen or an antibody conjugated with a detection label is immobilized and dried onto the membrane. The analytical membrane is engineered to fix proteins, antibodies, or antigens in designated lines or spots, which specifically capture the analyte and conjugate through molecular interactions during the chromatographic process. Finally, an absorbent pad positioned at the end of the strip collects excess reagents, preventing background interference and sustaining consistent capillary flow throughout the test (Alhabab, 2022).

LIF is presently employed for qualitative, semiquantitative, and quantitative detection. Further, it provides an invaluable asset for the diagnostic laboratories, phytosanitary inspectors, and plant breeders, as well as extension agents. Such assays serve to enhance rather than replace established management practices (Singh & Singh, 2020). Despite their utility, LFAs exhibit certain limitations, including less potential for the multiplex detection of target proteins. Additionally, some LFA-based biosensors have shown relatively low sensitivity (Donoso & Valenzuela, 2018).

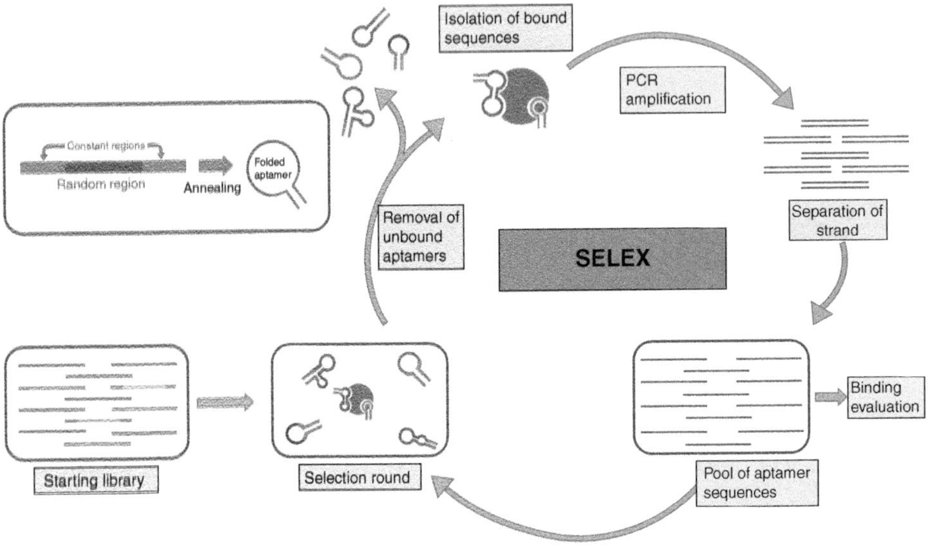

Figure 6.3 Schematic flowchart of SELEX.

6.3.3 Aptamers Binding Assay

Aptamers are short, single-stranded DNA or RNA molecules, generally less than 100 nucleotides in length, that are selected for their strong and specific binding to target molecules (Sefah et al., 2010). These ligands are typically identified through a method called systematic evolution of ligands by exponential enrichment (SELEX). Once the aptamer sequences are identified, they can be synthesized easily and modified with various functional groups such as nanomaterials or fluorescent tags (Zhao et al., 2020). The SELEX process starts by exposing a large, diverse library of nucleic acids to the target molecules to encourage binding. After each selection cycle, the sequences that bind are isolated and amplified via PCR to enrich the pool for subsequent rounds of selection (Figure 6.3). Typically, 5 to 20 rounds are necessary to enrich the pool with aptamers that strongly bind to the target (Yang et al., 2023).

6.4 Biochemical Methods

6.4.1 Flow Cytometry

Infection outbreaks have highlighted the critical need for implementing fast and economical methods to facilitate the recognition of pathogens within samples (Kennedy & Wilkinson, 2017). Viability refers to the inherent ability of an organism to survive, develop, and reproduce under ideal environmental conditions. The survival of pathogens depends on an intact plasma membrane, which serves as a protective barrier and the proper functioning of essential cellular processes such as RNA translation, DNA transcription, and enzyme activity. These processes ensure that the microorganism can carry out vital metabolic and reproductive functions. Any disruption to these components can compromise the microorganism's viability. Flow cytometry offers a fast and efficient method for assessing microbial viability and detecting plant pathogenic bacteria (Bergervoet et al., 2007). Its ability to distinguish live, damaged, and dead cells in real time makes it an indispensable methodology for microbial investigations and disease management strategies. Flow cytometry represents a refinement of fluorescence microscopy, enabling the analysis or sorting of individual cells as they pass sequentially through laser beams and detectors (Berglund et al., 1988).

6.4.2 Gas Chromatography–Mass Spectroscopy

Plants employ a variety of defense strategies to protect themselves from pathogen invasions and to defend against attacks by herbivorous insects. Among these strategies is the release of specific unstable organic compounds. Further, monitoring of these volatile organic compounds has emerged as a reliable indicator of insect pest infestation or pathogen-induced infections. The detection, quantification, and analysis of volatile organic compounds have thus become integral to modern agricultural research and formulation of advanced plant protection strategies (Makhlouf et al., 2024).

To enhance the effectiveness of compound separation and analysis, gas chromatography is frequently coupled with mass spectrometry (GC–MS) for the identification of unknown constituents in volatile samples. GC–MS has emerged as an analytical technique for profiling volatile organic compounds (Lytovchenko et al., 2009) as it utilizes a gas-phase separation and recognition system providing broad composition and chemical evidence about the specific components in volatile organic compound mixtures (Martinelli et al., 2015). Within the realm of plant pathogen analysis, GC–MS is a proven method in the identification and quantification of volatile organic compounds and secondary metabolites synthesized by fungal and bacterial pathogens (Pang et al., 2021). In GC–MS, the sample is introduced to the stream of carrier gas and it travels through the chromatographic column. During this process, the molecules of each component in the sample partition between the gas phase and the liquid stationary phase, establish a dynamic equilibrium as they constantly shift between these phases (Al-Bukhaiti et al., 2017). The separation analytes – individual components of the sample – occur due to their varying affinities for the movable phase (carrier gas) and the static phase (liquid or solid). For volatile substances, the separation may involve adsorption onto a solid stationary phase. In typical GC–MS, the stationary phase is a liquid that is confined within a narrow column (0.1–0.25 mm internal diameter) of relatively short length (10–30 m). As the sample components are separated, they exit the column sequentially, with each compound being directed to the mass spectrometer or detector.

The output from the GC is presented as a chromatogram, a graph in which the intensity of each peak corresponds to the amount of a specific compound, while the number of peaks shows the diversity of compounds present in the sample. The use of a mass spectrometer enhances the analysis by providing detailed molecular information about each separated component. Further, GC–MS has some advantages, such as it analyzes the sample quickly with high specificity at multiple stages of diseases (Stierlin et al., 2020). However, this technique has some limitations such as being time-consuming and labor-intensive for sample preparation (Fang & Ramasamy, 2015), field applicability constraints, need potential for handling errors, and the cost of the sample analysis.

6.5 Biosensors

It can be described as an analytical device that includes a biologically derived or inherently biological entity, which is integrated with a physiochemical transducer. Typically, a biosensor contains a receptor molecule serving as the molecular recognition element, which exhibits specificity toward the target analytes. The interaction between the receptor and the analyte culminates in the generation of biorecognition signals, which are subsequently transduced into measurable outputs such as electrochemical, electrical, or optical signals via a transducer (Patel et al., 2023).

A standard biosensor includes five main components: analyte, bioreceptor, transducer, electronics, and display. An analyte constitutes a substance whose components are subjected to identification or detection. The bioreceptor, a biological element, including enzymes, aptamers, or antibodies, exhibits a specificity for recognizing the analyte. This recognition process, termed biorecognition, elicits the generation of diverse signal-spanning luminescent, thermal, electrochemical, pH, or mass alterations arising from the analyte–bioreceptor interaction. The transducer is an energy-conversion device that translates biorecognition events into quantifiable signals, predominantly of an electrical nature, through a process called signalization. Transducers categorized into electrochemical, thermal, optical, electronic, and gravimetric types based on the functional principles produce signals based on the intensity or occurrence of analyte–bioreceptor interactions. The transduced signals are subsequently subjected to amplification and digitization via an electronic interface, making them suitable for further processing.

Finally, the display unit, such as a computer or printer, presents the results in a user-centric module including numerical, tabular, graphical, or visual outputs, ensuring the data is easily interpretable (Figure 6.4). In conclusion, it is an electrochemical method used to transform chemical energy produced by pathogen and biorecognition elements immobilized on electrodes into electrical energy (Banaker et al., 2022).

6.5.1 Classification of Biosensors

Biosensors are extensively utilized for detecting plant diseases. Depending on the biorecognition elements, they can be categorized into various types (Flowchart 6.1), including enzyme-based, aptamer-based, antibody-based, nanoparticle-based, and whole-cell biosensors, whereas, based on the transducer used, biosensors are classified into electrochemical biosensors, optical biosensors, thermal biosensors, gravimetric biosensors, and acoustic biosensors (Naresh & Lee, 2021).

Affinity biosensors rely on antibodies or aptamers that are highly specific to pathogens, facilitating precise and versatile detection. Nevertheless, their applicability is occasionally constrained by the potential for false-positive results. In contrast, enzymatic

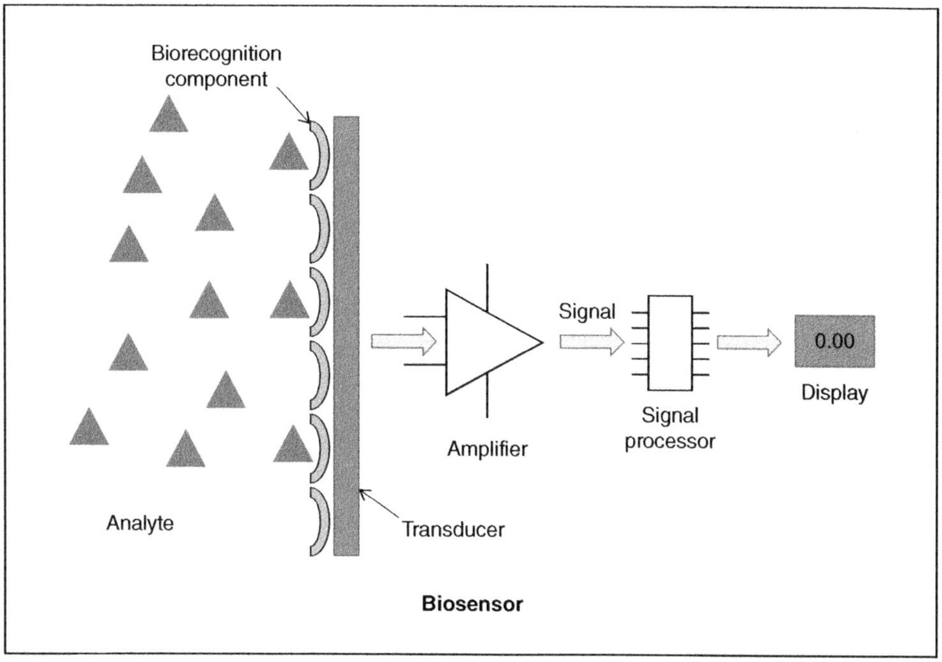

Figure 6.4 Process of biosensors.

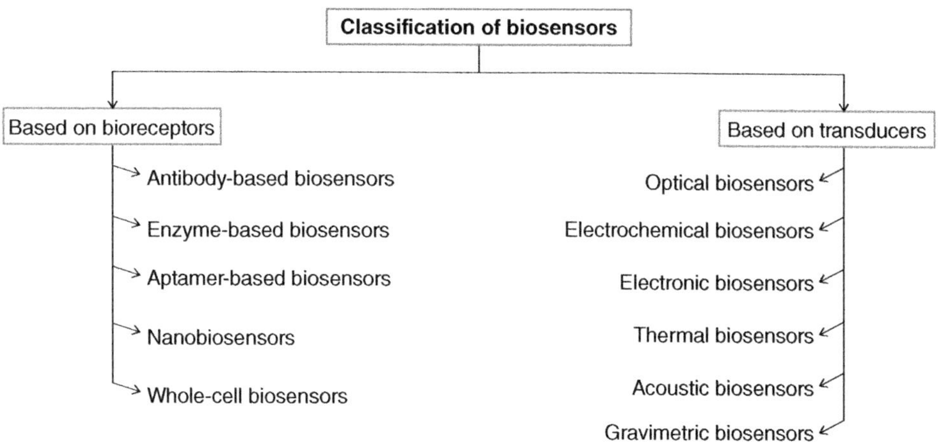

Flowchart 6.1 Biosensors based on bioreceptors and transducers.

electrochemical biosensors focus on detecting enzyme activity either formed by the effector molecules or triggered in the plant as a response to infection (Sharma et al., 2024). Among the diverse class of bioreceptors, aptamers – single-stranded oligonucleotides composed of DNA/RNA, typically ranging from 20 to 100 nucleotides – have gained prominence as advanced sensing elements (Leguillier et al., 2024). Nanomaterial-based electrochemical sensors have been developed as a promising tool for the detection of plant disease. Gold nanoparticle (AuNP)-modified cathodes enable the electrochemical detection of methyl salicylate, a volatile organic compound released by infected plants, while DNA-based approaches utilize voltammetry to analyze nucleic acid conformation and modification without requiring hydrolysis. Optical biosensors, such as colorimetric sensors, allow rapid, visual detection of pathogens through observable color changes, eliminating the need for complex instruments. Electrochemical biosensors, which rely on electron exchange during biochemical reactions, support voltammetric, amperometric, and impedimetric measurements (Patel, 2021). Additional biosensor types include gravimetric sensors, which measure mass changes; thermal sensors that detect heat release or absorption during reactions; and acoustic sensors, which correlate changes in acoustic wave properties to analyte concentration.

These advanced biosensors offer advantages such as portability, cost-effectiveness, and early disease detection capabilities (Madufor et al., 2018). However, their application is hindered by challenges such as sample preparation techniques, reduced sensitivity in complex samples, and operational complexity, necessitating further optimization for practical use. Some of the pathogens detected from various biochemical techniques are mentioned in Table 6.3.

Table 6.3 Biochemical approaches employed in detecting plant pathogens.

Sr. no.	Technique used	Crop	Pathogen	References
1.	Flow cytometry	Cucumber	Cucumber mosaic virus (CMV)	Iannelli et al. (1996)
		Cucumber, tomato, potato	CMV, Tomato mosaic virus POY	Iannelli et al. (1997)
		Brassica	*Xanthomonas campestris* pv. *campestris*	Chitarra et al. (2002)
		Potato	*Phytophthora infestans*	Day et al. (2004)
		Tomato	*Pseudomonas fluorescens*	Gamalero et al. (2005)
		Apple	*Erwinia amylovora*	Ordax et al. (2006)
		Potato	*Pectobacterium atrosepticum, Dickeya dianthicola*	Peters et al. (2007)
		Potato	*P. infestans*	Kardile et al. (2018)
2.	GC–MS	Potato	*P. infestans* *Fusarium coeruleum*	Costello et al. (2001)
		Rice	Gall midge	Agarrwal et al. (2014)
		Chilli	Pepper yellow curl virus	Agustika et al. (2021)

Sr. no.	Technique used	Crop	Pathogen	References
3.	Antibody-based biosensor	Orchids	Cymbidium mosaic potex virus Odontoglossum ring spot tobacco virus	Eun et al. (2002)
		Tobacco	Tobacco mosaic viruses	Dickert et al. (2004)
		Lettuce	Lettuce mosaic virus	Candresse et al. (2007)
		Wheat	*Puccinia striiformis urediniospores*	Skottrup et al. (2007)
		Plum	Plum pox virus	Jarocka et al. (2011)
		Potato	Potato virus x	Drygin et al. (2012)
		Watermelon	*Acidovorax avenae* subsp. *Citrulli*, Watermelon silver mottle virus, Melon yellow spot virus	Charlemroj et al. (2013)
		Maize	*Pantoea stewartia* subsp. *stewartii*	Zhang et al. (2014)
		Orchid	Cymbidium mosaic virus	Lin et al. (2014)
		Maize	*P. stewartia* subsp. *stewartii*	Feng et al. (2015)
			Pseudomonas syringae pv. *lachrymans*	Cebula et al. (2019)
			Candidatus phytoplasma	Ebrahimi et al. (2019)
		Crop of the Rosaceae family	*Eriwinia amylovora*	Razo et al. (2021)
4.	DNA hybridization-based biosensors	Banana	Banana bunchy top virus	Tang et al. (2007)
		Wheat	*Fusarium culmorum*	Zezza et al. (2006)
		Cucumber	*Botrytis cinerea*	Wang and Li (2007)
		Watermelon	*Acidovorax avenae*	Zhao et al. (2011)
			Pseudomonas syringae	Vaseghi et al. (2013)
		Banana	Bunchy top virus	Wei et al. (2014)
			Plum pox virus	Malecka et al. (2014)
			P. syringae	Wee et al. (2015)
		Potato	*P. infestans*	Zhan et al. (2018)
		Sugarcane	The sugarcane white leaf disease	Wongkaew and Poositiisak (2014)
		Cacao	*Phytophthora palmivora*	Franco et al. (2019)
5.	Gold nanorods (AuNRs)	Orchids	Orchid viruses	Lin et al. (2014)

References

Agarrwal, R., Sanmallappa, B., & Nair, S. 2014. Gas chromatography mass spectrometry based metabolic profiling reveals biomarkers involved in rice-gall midge interactions. *Journal of Integrative Plant Biology* 56(9), 837–848.

Agustika, D.K., Mercuriani, I.S., Ariyanti, N.A., Purnomo, C.W., Tryana, K., Lliscu, D.D., & Leeson, M.S. 2021. Gas chromatography-mass spectrometry analysis of compounds emitted by pepper yellow leaf curl virus-infected chili plants: a preliminary study. *Separation* 8(9), 136.

Al-Bukhaiti, W.Q., Noman, A., Qasim, A.S., & Al-Farga, A. 2017. Gas chromatography: principles, advantages and applications in food analysis. *International Journal of Agriculture Innovations and Research* 6(1), 2319–2473.

Alhabbab, R.Y. 2022. Lateral flow immunoassays for detecting viral infectious antigens and antibodies. *Micromachines* 13, 1901.

Alkhalaf, M.A., Kumari, S.G., Kasem, A.H., & Makkouk, K. 2009. Use of the two faces of nitrocellulose membrane in tissue blot immunoassay for the detection of bean yellow mosaic virus and the possibility of its mechanically transmitted from the printed membrane to the host plant. *Arab Journal of Plant Protection* 27, 91–94.

Arif, M., Torrance, L., & Reavy, B. 1994. Improved efficiency of detection of potato mop-top furovirus in potato tubers and in the roots and leaves of soil-bait plants. *Potato Research* 37, 373–381.

Atkins, S.D., & Clark, I.M. 2004. Fungal molecular diagnostics: a mini review. *Journal of Applied Genetics* 45(1), 3–15.

Badali, H., & Nabili, M. 2012. Molecular tools in medical mycology; where we are! *Jundishapur Journal of Microbiology* 6(1), 1–3.

Banakar, M., Hamidi, M., Khurshid, Z., Zafar, M.S., Sapkota, J., Azizian, R., & Rokaya, D. 2022. Electrochemical biosensors for pathogen detection: an updated review. *Biosensors* 12(11), 927.

Baral, B., 2024. DAS-ELISA-based detection and RT-PCR validation for precise and timely analysis of *Citrus tristeza virus* infection in Nepalese citrus species. *Current Research in Biotechnology* 7, 100221.

Belanger, L., Sylvestre, C., & Dufour, D. 1973. Enzyme-linked immunoassay for alpha-fetoprotein by competitive and sandwich procedures. *Clinica Chimica Acta* 48, 15–18.

Bergervoet, J.H.W., Vanderwof, J.M., & Peters, J. 2007. Detection and viability assessment of plant pathogenic microorganisms using flow cytometry. In: J. Doležel, J. Greilhuber, & J. Suda (Eds.), *Flow Cytometry with Plant Cells* (pp. 217–229). Weinheim: WILEY-VCH Verlag GmbH & Co. KGaA.

Berglund, L.D., Strobel, S., Sugawara, F., & Strobel, G.A. 1988. Flow cytometry as a method for assaying the biological activity of phytotoxins. *Plant Science* 56(2), 183–188.

Bhat, R., & Browne, G. 2010. Specific detection of *Phytophthora cactorum* in diseased strawberry plants using nested polymerase chain reaction. *Plant Pathology* 59(1), 121–129.

Black, J. 2009. *Quantitative Real-Time Polymerase Chain Reaction (qPCR) of Filamentous Fungi in Carpet*. Research Triangle Park, NC: RTI Press.

Bozorg-Ghalati, F., Mohammadpour, I., & Ranjbaran, R. 2019. Applications of fluorescence in situ hybridization in detection of disease biomarkers and personalized medicine. *Comparative Clinical Pathology* 28, 3–10.

Brown, N.A., Bass, C., Baldwin, T.K., Chen, H., Massot, F., Carion, P.W., Urban, M., Van De Meene, A.M., & Hammond-Kosack, K.E. 2011. Characterisation of the *Fusarium graminearum*-wheat floral interaction. *Journal of Pathogens* 2011, 626345. https://doi.org/10.4061/2011/626345

Bu, R., Sathiapalan, R.K., Ibrahim, M.M., Al-Mohsen, I., Almodavar, E., Gutierrez, M.I., & Bhatia, K. 2005. Monochrome Light Cycler PCR assay for detection and quantification of five common species of *Candida* and *Aspergillus*. *Journal of Medical Microbiology* 54(3), 243–248.

Bumgarner, R. 2013. Overview of DNA microarrays: types, applications, and their future. *Current Protocols in Molecular Biology* 101, 22.1.1–22.1.11.

Byzova, N.A., Vinogradova, S.V., Porotikova, E.V., Terekhova, U.D., Zherdev, A.V., & Dzantiev, B.B. 2018. Lateral flow immunoassay for rapid detection of grapevine leafroll-associated virus. *Biosensors* 8(4), 111.

Candresse, T., Lot, H., German-Retana, S., Krause-Sakate, R., Thomas, J., Souche, S., Delaunay, T., Lanneau, M., & Le Gall, O. 2007. Analysis of the serological variability of Lettuce mosaic virus using monoclonal antibodies and surface plasmon resonance technology. *Journal of General Virology* 88, 2605–2610.

Capote, N., Mara, A., Aguado, A., & Sánchez-Torres, P. 2012. Molecular tools for detection of plant pathogenic fungi and fungicide resistance. In: C.J.R. Cumagun (Ed.), *Plant Pathology* (pp. 151–202). RI, USA: InTechOpen. https://doi.org/10.5772/51021

Cebula, Z., Żołędowska, S., Dziąbowska, K., Skwarecka, M., Malinowska, N., & Białobrzeska, W. 2019. Detection of the plant pathogen *Pseudomonas syringae* pv. *lachrymans* on antibody-modified gold electrodes by electrochemical impedance spectroscopy. *Sensors* 19, 5411.

Chalupová, J., Raus, M., Sedlářová, M., & Šebela, M. 2014. Identification of fungal microorganisms by MALDI-TOF mass spectrometry. *Biotechnology Advances* 32(1), 230–241. https://doi.org/10.1016/j.biotechadv.2013.10.001

Chalupowicz, L., Dombrovsky, A., Gaba, V., Luria, N., Reuven, M., Beerman, A., Lachman, O., Dror, O., Nissan, G., & Manulis-Sasson, S. 2019. Diagnosis of plant diseases using the Nanopore sequencing platform. *Plant Pathology* 68, 229–238.

Charlermroj, R., Himananto, O., Seepiban, C., Kumpoosiri, M., Warin, N., & Oplatowska, M. 2013. Multiplex detection of plant pathogens using a microsphere immunoassay technology. *PLoS One* 8, 62344.

Charoenvilaisiri, S., Seepiban, C., Kumpoosiri, M., Rukpratanporn, S., Warin, N., & Phuangrat, B. 2021. Development of a triple antibody sandwich enzyme-linked immunosorbent assay for cassava mosaic disease detection using a monoclonal antibody to Sri Lankan cassava mosaic virus. *Virology Journal* 18, 100.

Chen, H., Ali, N., Lv, W., Shen, Y., Qing, Z., & Lin, Y. 2020. Comparison of ICRT-PCR, dot-ELISA and indirect-ELISA for the detection of sorghum mosaic virus in field grown sugarcane plants. *Sugar Tech* 22, 122–129.

Chen, J.S., Ma, E., Harrington, L.B., Da Costa, M., Tian, X., Palefsky, J.M., & Doudna, J.A. 2018. CRISPR-Cas12a target binding unleashes indiscriminate single-stranded DNase activity. *Science* 360(6387), 436–439.

Chen, X., Steed, A., Harden, C., & Nicholson, P. 2006. Characterisation of *Arabidopsis thaliana–Fusarium graminearum* interactions and identification of variation in resistance among ecotypes. *Molecular Plant Pathology* 7, 391–403.

Chitarra, L.G., Langerak, C.J., Bergervoet, J.H.W., & van den Bulk, R.W. 2002. Detection of the plant pathogenic bacterium *Xanthomonas campestris* pv. *campestris* in seed extracts of *Brassica* sp. applying fluorescent antibodies and flow cytometry. *Cytometry* 47, 118–126.

Cho, H.J., Hong, S.W., Kim, H.J., & Kwak, Y.S. 2016. Development of a multiplex PCR method to detect fungal pathogens for quarantine on exported cacti. *Plant Pathology Journal* 32(1), 53–54.

Clark, M., Panopoulou, G., Cahill, D., Büssow, K., & Lehrach, H. 1999. Construction and analysis of arrayed cDNA libraries. *Methods in Enzymology* 303, 205–233.

Compton, J. 1991. Nucleic acid sequence-based amplification. *Nature* 350(6313), 91–92. https://doi.org/10.1038/350091a0

Costello, B.P.J.D.L., Evans, P., Ewen, R.J., Gunson, E.H.E., Jones, P.R.H., Ratcliffe, N.M., & Spencer-Phillips, P.T.N. 2001. Gas chromatography–mass spectrometry analyses of volatile organic compounds from potato tubers inoculated with *Phytophthora infestans* or *Fusarium coeruleum*. *Plant Pathology* 50(4), 489–496.

Cui, C., Shu, W., & Li, P. 2016. Fluorescence in situ hybridization: cell-based genetic diagnostic and research applications. *Frontiers in Cell and Developmental Biology* 4, 89.

Darwish, I.A. 2006. Immunoassay methods and their applications in pharmaceutical analysis: basic methodology and recent advances. *International Journal of Biomedical Sciences* 2(3), 217–235.

Dasmahapatra, K., & Mallet, J. 2006. DNA barcodes: recent successes and future prospects. *Heredity* 97(4), 254–255.

Dawei, W., & Peng, Y. 2014. Assessing the impact of dominant sequencing-based gene expression profiling techniques (SGEPTs) on phytopathogenic fungi. *Chiang Mai Journal of Science* 41, 922–9444.

Day, J.P., Wattier, R.A.M., Shawl, D.S., & Sharrock, R.C. 2004. Phenotypic and genotypic diversity in phytophthora infestans on potato in Great Britain. Plant Pathology 53, 303–315.

Dickert, F.L., Hayden, O., Bindeus, R., Mann, K.J., Blaas, D., & Waigmann, E. 2004. Bioimprinted QCM sensors for virus detection – screening of plant sap. *Analytical and Bioanalytical Chemistry* 378, 1929–1934.

Donoso, A., & Valenzuela, S. 2018. In-field molecular diagnosis of plant pathogens: recent trends and future perspectives. *Plant Pathology* 67(7), 1451–1461.

Drygin, Y.F., Blintsov, A.N., Grigorenko, V.G., Andreeva, I.P., Osipov, A.P., & Varitzev, Y.A. 2012. Highly sensitive field test lateral flow immunodiagnostics of PVX infection. *Applied Microbiology and Biotechnology* 93, 179–189.

Ebrahimi, M., Norouzi, P., Safarnejad, M.R., Tabaei, O., & Haji-Hashemi, H. 2019. Fabrication of a label-free electrochemical immunosensor for direct detection of candidatus *Phytoplasma Aurantifolia*. *Journal of Electroanalytical Chemistry* 851, 113451.

Eun, A.J.C., Huang, L., Chew, F.T., Li, S.F.Y., & Wong, S.M. 2002. Detection of two orchid viruses using quartz crystal microbalance (QCM) immunosensors. *Journal of Virological Methods* 99, 71–79.

Faccioli, G., & Colombarini, A. 1996. Correlation of potato virus S and virus M contents of potato meristem tips with the percentage of virus-free plantlets produced in vitro. *Potato Research* 39, 129–140.

Fang, Y., & Ramasamy, R.P. 2015. Current and prospective methods for plant disease detection. *Biosensors* 5(3), 537–561. https://doi.org/10.3390/bios5030537.

Feng, M., Kong, D., Wang, W., Liu, L., Song, S., & Xu, C. 2015. Development of an immunochromatographic strip for rapid detection of *Pantoea stewartii* subsp. *stewartii*. *Sensors* 15, 4291–4301.

Franco, A.J.D., Merca, F.E., Rodriguez, M.S., Balidion, J.F., Migo, V.P., & Amalin, D.M. 2019. DNA-based electrochemical nanobiosensor for the detection of *Phytophthora palmivora* (Butler) Butler, causing black pod rot in cacao (*Theobroma cacao* L.) pods. *Physiological and Molecular Plant Pathology* 107, 14–20.

Frickmann, H., Zautner, A.E., Moter, A., Kikhney, J., Hagen, R.M., Stender, H., & Poppert, S. 2017. Fluorescence in situ hybridization (FISH) in the microbiological diagnostic routine laboratory: a review. *Critical Reviews in Microbiology* 43(3), 263–293.

Gamalero, E., Lingua, G., Tombolini, R., Avidano, L., Pivato, B., and Berta, G., 2005. Colonization of tomato root seedling by pseudomonas fluorescens 92rkG5: spatio-temporal dynamics, localization, organization, viability, and culture ability. *Microbial Ecology* 50, 289–297.

Gammill, L.S., & Lee, V.M. 2008. Gene discovery: macroarrays and microarrays. *Methods in Cell Biology* 87, 297–312.

Gootenberg J.S., Abudayyeh, O.O., Lee, J.W., Essletzbichler, P., Dy, A.J., Joung, J., Verdine, V., Donghia, N., Daringer, N.M., Freije, C.A., & Myhrvold, C. 2017. Nucleic acid detection with CRISPR-Cas13a/C2c2. *Science* 356(6336), 438–442.

Guigó, R. 2013. The coding and the non-coding transcriptome. In: A.J.M. Walhout, M. Vidal, & J. Dekker (Eds.), *Handbook of Systems Biology* (pp. 27–41). San Diego: Academic Press.

Hariharan, G., & Prasannath, K. 2021. Recent advances in molecular diagnostics of fungal plant pathogens: a mini review. *Frontiers in Cellular and Infection Microbiology* 10, 600234. https://doi.org/10.3389/fcimb.2020.600234.

Harrison, J., Calder, W.J., Shuman, B.N., & Buerkle, C.A. 2020. The quest for absolute abundance: the use of internal standards for DNA barcoding in microbial ecology. *Molecular Ecology Resources* 21(1), 30–43.

He, Y.H., Isono, S., Shibuya, M., Tsuji, M., Adkar Purushothama, C.R., Tanaka, K., Tanaka, K., & Sano, T. 2012. Oligo-DNA custom macroarray for monitoring major pathogenic and non-pathogenic fungi and bacteria in the phyllosphere of apple trees. *PLoS One* 7(3), e34249.

Henson, J.M., & French, R. 1993. The polymerase chain reaction and plant disease diagnosis. *Annual Review of Phytopathology* 31(1), 81–109. https://doi.org/10.1146/annurev.py.31.090193.000501

Hijri, M. 2009. The use of fluorescent in situ hybridization in plant fungal identification and genotyping. *Methods in Molecular Biology* 508, 131–145.

Hill, J., Bryant, G., & Durand, D. 1981. Detection of plant virus by using purified IgG in ELISA. *Journal of Virological Methods* 3, 27–35.

Hodgetts, J., Karamura, G., Johnson, G., Hall, J., Perkins, K., Beed, F., Nakato, V., Grant, M., Studholme, D.J., Boonham, N., & Smith, J. 2015. Development of a lateral flow device for in-field detection and evaluation of PCR-based diagnostic methods for *Xanthomonas campestris* pv. *musacearum*, the causal agent of banana *Xanthomonas* wilt. *Plant Pathology* 64(3), 559–567.

Hoshino, T., & Inagaki, F. 2017. Correction: application of stochastic labeling with random-sequence barcodes for simultaneous quantification and sequencing of environmental 16S rRNA genes. *PLoS One* 12(1), e0169431.

Hsieh, C.W., Chuang, Y.Y., Lee, M.Z., & Kirschner, R. 2020. First inventory of fungi in symptomless and symptomatic Chinese mesona indicates phytopathological threat. *Plant Disease* 104(9), 2391–2397.
Iannelli, D., D'Apice, L., Cottone, C., Viscardi, M., Scala, F., Zoina, A., Del, S.G., Spigno, P., & Capparelli, R. 1997. Simultaneous detection of cucumber mosaic virus, tomato mosaic virus and potato virus Y by flow cytometry. *Journal of Virological Methods* 69, 137–145.
Iannelli, D., D'Apice, L., Pasquini, G., Capparelli, R., Monti, L., Parrella, G., Scala, F., & Noviello, C. 1996. Cytofluorometric method for the detection of cucumber mosaic virus. *Phytopathology* 86, 959–965.
Ilbagi, F., Rabenstein, F., Habekuss, A., Ordon, F., Citir, A., Cebeci, O., & Budak, H. 2008. Molecular, serological and transmission electron microscopic analysis of the barley yellow dwarf virus- *PAV* and the cereal yellow dwarf virus-RPV in canary seed (*Phalaris canariensis* L.). *Cereal Research Communications* 36(2), 225–234.
Ishii, H., & Holloman, D.W. 2015. *Fungicide Resistance in Plant Pathogens*. Tokyo: Springer.
Jarocka, U., Wąsowicz, M., Radecka, H., Malinowski, T., Michalczuk, L., & Radecki, J. 2011. Impedimetric immunosensor for detection of plum pox virus in plant extracts. *Electroanalysis* 23, 2197–2204.
Johannes, W., Stefan, D.K., Philipp, A.M., Christin, R., Matthias H., Gerald A.U. 2020. Digital DNA microarray generation on glass substrates. *Scientific Reports* 10, 5770.
Kang, H., Peng, Y., Hua, K., Deng, Y., Bellizzi, M., Gupta, D.R., Mahmud, N.U., Urashima, A.S., Paul, S.K., Peterson, G., & Zhou, Y. 2021. Rapid detection of wheat blast pathogen *Magnaporthe oryzae* Triticum pathotype using genome-specific primers and Cas12a-mediated technology. *Engineering* 7(9), 1326–1335.
Kardile, H.B., Sharma, N.K., Sharma, A., Sharma, S., Bhardwaj, V., & Chakrabarti, S.K. 2018. Development of detection assay for early infection of late blight in potato using flow cytometry. *Potato Journal* 45(1), 50–54.
Kennedy, D., & Wilkinson, M.G. 2017. Application of flow cytometry to the detection of pathogenic bacteria. *Current Issues in Molecular Biology* 23, 21–38.
Khater, M., Escosura-Muniz, A.D., & Merkoci, A. 2017. Biosensors for plant pathogen detection. *Biosensors and Bioelectronics* 93, 72–86.
Kim, N.K., Lee, H.J., Kim, S.M., & Jeong, R.D. 2022. Rapid and visual detection of barley yellow dwarf virus by reverse transcription recombinase polymerase amplification with lateral flow strips. *The Plant Pathology Journal* 38(2), 159–166.
Koczula, K.M., & Gallotta, A. 2016. Lateral flow assays. *Essays Biochem* 60(1), 111–20.
Kokoskova, B., & Janse, J.D. 2009. Enzyme-linked immunosorbent assay for the detection and identification of plant pathogenic bacteria (in particular for *Erwinia amylovora* and *Clavibacter michiganesis* subsp. *sepedonicus*). *Methods in Molecular Biology, Plant Pathology* 508, 75–87.
Komorowska, B., Hasiow-Jaroszewska, B., & Minicka, J. 2017. Application of nucleic acid aptamers for detection of apple stem pitting virus isolates. *Molecular and Cellular Probes* 36, 62–65.
Krivitsky, V., Granot, E, Avidor, Y., Borberg, E., Voegele, R.T., & Patolsky, F. 2021. Rapid collection and aptamer-based sensitive electrochemical detection of soybean rust fungi airborne urediniospores. *ACS Sensors* 6(3), 1187–1198.

Kukurba, K.R., & Montgomery, S.B. 2015. RNA sequencing and analysis. *Cold Spring Harbor Protocols* 2015(11), 951–969.

Lai, P.C., Abney, M.R., Chen, Y.J., Bag, S., & Srinivasan, R. 2021. Discrepancies in serology-based and nucleic acid-based detection and quantitation of tomato spotted wilt orthotospovirus in leaf and root tissues from symptomatic and asymptomatic peanut plants. *Pathogens* 10, 1476.

Leguillier, V., Heddi, B., & Vidic, J. 2024. Recent advances in aptamer-based biosensors for bacterial detection. *Biosensors* 14(5), 210.

Lei, R., Kuang, R., Peng, X., Jiao, Z., Zhao, Z., Cong, H., Fan, Z., & Zhang, Y. 2023. Portable rapid detection of maize chlorotic mottle virus using RT-RAA/CRISPR-Cas12a based lateral flow assay. *Frontiers Plant Science* 14, 1088544.

Li, S.Y., Cheng, Q.X., Wang, J.M., Li, X.Y., Zhang, Z.L., Gao, S., Cao, R.B., Zhao, G.P., & Wang, J. 2018. CRISPR-Cas12a-assisted nucleic acid detection. *Cell Discovery* 4(1), 20.

Li, Y., Li, S., Wang, J., & Liu, G. 2019. CRISPR/Cas systems towards next-generation biosensing. *Trends in Biotechnology* 37, 730–743.

Liebe, S., Daniela, S., Christ, D.S., Ehricht, R., & Varrelmann, M. 2016. Development of a DNA microarray-based assay for the detection of sugar beet root rot pathogens. *Phytopathology* 106(1), 76–86.

Lin, H.Y., Huang, C.H., Lu, S.H., Kuo, I.T., & Chau, L.K. 2014. Direct detection of orchid viruses using nanorod-based fiber optic particle plasmon resonance immunosensor. *Biosensors and Bioelectronics* 51, 371–378.

Lo′pez-Soriano, P., Noguera, P., Gorris, M.T., Puchades, R., Maquieira, A., & Marco-Noales. 2017. Lateral flow immunoassay for on-site detection of *Xanthomonas arboricola* pv. *pruni* in symptomatic field samples. *PLoS ONE* 12(4), e0176201.

Luchi, N., Ghelardini, L., Belbahri, L., Quartier, M., & Santini, A. 2013. Rapid detection of *Ceratocystis platani* inoculum by quantitative real-time PCR assay. *Applied and Environmental Microbiology* 79(17), 5394–5404. https://doi.org/10.1128/AEM.01155-13

Lukumbuzya, M., Schmid, M., Pjevac, P., & Daims, H. 2019. A multicolor fluorescence in situ hybridization approach using an extended set of fluorophores to visualize microorganisms. *Frontiers in Microbiology* 10, 1383.

Lytovchenko, A., Beleggia, R., & Schauer, N. 2009. Application of GC-MS for the detection of lipophilic compounds in diverse plant tissues. *Plant Methods* 5, 4.

Mackay, I.M. 2004. Real-time PCR in the microbiology laboratory. *Clinical Microbiology and Infection* 10(3), 190–212.

Madufor, N.J.K., Perold, W.J., & Opara, U.L. 2018. Detection of plant diseases using biosensors: a review. *Acta Horticulturae* 1201, 83–90.

Makhlouf, L., El Fakhouri, K., Kemal, S.A., Aasfar, A., Meftah Kadmiri, I., & El Bouhssini, M. 2024. Advances in analytical techniques for assessing volatile organic compounds in pulse crops: a comprehensive review. *Frontiers in Horticulture* 3, 1394041.

Malapi-Wight, M., Salgado-Salazar, C., Demers, J.E., Clement, D.L., Rane, K.K., & Crouch, J.A. 2016. *Sarcococca* blight: use of whole-genome sequencing for fungal plant disease diagnosis. *Plant Disease* 100, 1093–1100.

Malecka, K., Michalczuk, L., Radecka, H., & Radecki, J. 2014. Ion-channel genosensor for the detection of specific DNA sequences derived from plum pox virus in plant extracts. *Sensors* 14, 18611–18624.

Martın, S., Alioto, D., Milne, R.G., Garnsey, S. M., Laura Garcia, M., & Grau, O. 2004. Detection of citrus psorosis virus by ELISA, molecular hybridization, RT-PCR and immunosorbent electron microscopy and its association with citrus psorosis disease. *European Journal of Plant Pathology* 110, 747–757.

Martinelli, F., Scalenghe, R., Davino, S., Panno, S., Ruisi, P., Villa, P., Stroppiana, D., Boschetti, M., Goulart, L.R., Davis, C.E., & Dandekar, A.M. 2015. Advanced methods of plant disease detection. A review. *Agronomy for Sustainable Development* 35, 1–25.

Masanga, J.O., Matheka, J.M., Omer, R.A., Ommeh, S.C., Monda, E.O., & Alakonya, A.E. 2015. Downregulation of transcription factor aflR in *Aspergillus flavus* confers reduction to aflatoxin accumulation in transgenic maize with alteration of host plant architecture. *Plant Cell Reports* 34, 1379–1387. https://doi.org/10.1007/s00299-015-1794-9

McCartney, H.A., Foster, S.J., Fraaije, B.A., & Ward, E. 2003. Molecular diagnostics for fungal plant pathogens. *Pest Management Science* 59(2), 129–142.

Midorikawa, G.E.O., Miller, R.N.G., & Bittencourt, D.M. 2018. Molecular identification and detection of foodborne and feedborne mycotoxigenic fungi. In: A.F. El Sheikha, R. Levin, & J. Xu (Eds.), *Molecular Techniques in Food Biology* (pp. 385–407). USA: John Wiley & Sons Ltd. https://doi.org/10.1002/9781119374623.ch17

Milner, H., Ji, P., Sabula, M., & Wu, T. 2019. Quantitative polymerase chain reaction (Q-PCR) and fluorescent in situ hybridization (FISH) detection of soilborne pathogen *Sclerotium rolfsii*. *Applied Soil Ecology* 136, 86–92.

Moter, A., & Göbel, U.B. 2000. Fluorescence in situ hybridization (FISH) for direct visualization of microorganisms. *Journal of Microbiological Methods* 41, 85–112.

Musil, M., & Gallo, J. 1993. Determination of broad bean stain virus serotypes by enzyme-linked immunosorbent assay. *Acta Virologica* 37(4), 265–70.

Nakayashiki, T., Kurtzman, C.P., Edskes, H.K., & Wickner, R.B. 2005. Yeast prions [URE3] and [PSI+] are diseases. *Proceedings of the National Academy of Sciences* 102(30), 10575–10580.

Naresh, V., & Lee, N. 2021. A review on biosensors and recent development of nanostructured materials-enabled biosensors. *Sensors* 21, 1109.

Nascimento, A.K.Q., Lima, J.A.A., & Barbosa, G.D.S. 2017. A simple kit of plate-trapped antigen enzyme-linked immunosorbent assay for identification of plant viruses1. *Revista Ciencia Agronmica* 48, 216–220.

Nikitin, M., Deych, K., Grevtseva, I., Girsova, N., Kuznetsova, M., & Pridannikov, M. 2018. Preserved microarrays for simultaneous detection and identification of six fungal potato pathogens with the use of real-time PCR in matrix format. *Biosensors* 8, 129.

Nilsson, R.H., Abarenkov, K., Larsson, K.-H., & Kõljalg, U. 2011. Molecular identification of fungi: rationale, philosophical concerns, and the UNITE database. *Open Applied Informatics Journal* 5, 81–86. https://doi.org/10.2174/1874136301105010081

Ordax, M., Marco-Noales, E., Lopez, M.M., & Biosca, E.G. 2006. Survival strategy of *Erwinia amylovora* against copper: induction of the viable-but-nonculturable state. *Applied and Environmental Microbiology* 72(5), 3482–3488.

Panferov, V.G., Safenkova, I.V., Byzova, N.A., Varitsev, Y.A., Zherdev, A.V., & Dzantiev, B.B. 2017. Silver-enhanced lateral flow immunoassay for highly-sensitive detection of potato leafroll virus. *Food and Agricultural Immunology* 29, 1.

Pang, Z., Chen, J., Wang, T., Gao, C., Li, Z., Guo, L., Xu, J., & Cheng, Y. 2021. Linking plant secondary metabolites and plant microbiomes: a review. *Frontiers in Plant Science* 12, 1–22.

Panwar, V., McCallum, B., & Bakkeren, G. 2012. Endogenous silencing of *Puccinia triticina* pathogenicity genes through *in planta*-expressed sequences leads to the suppression of rust diseases on wheat. *Plant Journal* 73, 521–532.

Patel, P. 2021. A review on plant disease diagnosis through biosensor. *International Journal of Biosensors & Bioelectronics* 7(2), 50–52.

Patel, R., Mitra, B., Vinchurkar, M., Adami, A., Patkar, R., Giacomozzi, F., & Baghini, M.S. 2023. Plant pathogenicity and associated/related detection systems. A review. *Talanta* 251, 123808.

Peters, J., Sledz, W., Bergervoet, J.H.W., & Van der Wolf, J.M. 2007. An enrichment microsphere immunoassay for the detection of *Pectobacterium atrosepticum* and *Dickeya dianthicola* in potato tuber extracts. *European Journal of Plant Pathology* 117, 97–107.

Posthuma-Trumpie, A., Korf, J., & Ameronen, A.V. 2009. Lateral flow (immuno) assay: its strengths, weaknesses, opportunities and threats. a literature surve. *Analytical and Bioanalytical Chemistry* 393, 569–582.

Potgieter, L., Feurtey, A., Dutheil, J.Y., & Stukenbrock, E.H. 2020. On variant discovery in genomes of fungal plant pathogens. *Frontiers in Microbiology* 11, 626.

Qi, M., & Yang, Y. 2002. Quantification of Magnaporthe grisea during infection of rice plants using real-time polymerase chain reaction and northern blot/phosphoimaging analyses. *Phytopathology* 92(8), 870–876.

Qin, D. 2019. Next-generation sequencing and its clinical application. *Cancer Biology and Medicine* 16(1), 4–10.

Rajesh, T., & Jaya, M. 2017. Next-generation sequencing methods. In: P. Gunasekaran, S. Noronha, & A. Pandey (Eds.), *Current Developments in Biotechnology and Bioengineering* (pp. 143–158). Amsterdam: Elsevier.

Razo, S.C., Safenkova, Drenova, N.V., Kharchenko, A.A., Tsymbal, Y.S., Varitsev, Y.A., Zherdev, A.V., Pakina, E.N., & Dzantiev, B.B. 2021. New lateral flow immunoassay for on-site detection of *Erwinia amylovora* and its application on various organs of infected plants. *Physiological and Molecular Plant Pathology* 114, 101637.

Richards, J., Kania, S., Wilson, A., Kent, E., & Gerhold, R. 2023. Novel methods of immunogenic antigen selection for serological diagnosis of Parelaphostrongylus tenuis infection. *Scientifc Reports* 13, 10989.

Roe, A.D., Rice, A.V., Bromilow, S.E., Cooke, J.E., & Sperling, F.A. 2010. Multilocus species identification and fungal DNA barcoding: insights from blue stain fungal symbionts of the mountain pine beetle. *Molecular Ecology Resources* 10(6), 946–959.

Rollins, L., Coats, K., Elliott, M., & Chastagner, G. 2016. Comparison of five detection and quantification methods for *Phytophthora ramorum* in stream and irrigation water. *Plant Disease* 100(5), 1202–1211. https://doi.org/10.1094/PDIS-08-15-0903-RE

Safenkova, I.V., Zaitsev, I.A., Varitsev, Y.U.A., Zherdev, A.V., & Dzantiev, B.B. 2015. Lateral flow immunoassay for rapid diagnosis of potato blackleg caused by *Pectobacterium atrosepticum*. *Biosciences Biotechnology Research Asia* 12(3), 1937–1945.

Sakamoto, S., Putalun, W., Vimolmangkang, S., Phoolcharoen, W., Shoyama, Y., Tanaka, H., & Morimoto, S. 2017. Enzyme-linked immunosorbent assay for the quantitative/qualitative analysis of plant secondary metabolites. *Journal of Natural Medicines* 72(1), 32–42.

Sato, T., Takayanagi, A., Nagao, K., Tomatsu, N., Fukui, T., Kawaguchi, M., Kudoh, J., Amagai, M., Yamamoto, N., & Shimizu, N. 2010. Simple PCRbased DNA microarray system to identify human pathogenic fungi in skin. *Journal of Clinical Microbiology* 48(7), 2357–2364.

Savolainen, V., Cowan, R.S., Vogler, A.P., Roderick, G.K., & Lane, R. 2005. Towards writing the encyclopaedia of life: an introduction to DNA barcoding. *Philosophical Transactions of the Royal Society B* 360(1462), 1805–1811.

Sefah, K., Shangguan, D., Xiong, X, Donoghue, M.B., & Tan, W. 2010. Development of DNA aptamers using Cell-SELEX. *Nature Protocols* 5, 1169–1185.

Selvarajan, R., Selvam K.P., Balasubramanian, V., & Sundaram, S.S. 2020. A rapid and sensitive lateral flow immunoassay (LFIA) test for the on-site detection of banana bract mosaic virus in banana plants. *Journal of Virological Methods* 284, 113929.

Shahraki, A.H., Chaganti, S.R., & Heath, D. 2019. Assessing high-throughput environmental DNA extraction methods for meta-barcode characterization of aquatic microbial communities. *Journal of Water Health* 17(1), 37–49.

Shakoori, A.R. 2017. Fluorescence in situ hybridization (FISH) and its applications. In: T.A. Bhat, & A.A. Wani (Eds.), *Chromosome Structure and Aberrations* (pp. 343–367), India: Springer.

Sharma, G., Dwibedi, V., Seth, C.S., Singh, S., Ramamurthy, P.C., Bhadrecha, P., & Singh, J. 2024. Direct and indirect technical guide for the early detection and management of fungal plant diseases. *Current Research in Microbial Sciences* 7, 100276.

Singh, V., & Singh, U. 2020. Lateral flow device for diagnosis of plant pathogens. *Biotech Today* 2, 23–25.

Skottrup, P., Frokiaer, H., Hearty, S., O'Kennedy, R., Hejgaard, J., Nicolaisen, M., & Justesen, A.F. 2007. Monoclonal antibodies for the detection of *Puccinia striiformis* urediniospores. *Mycological Research* 111, 332–338.

Stierlin, E., Michel, T., & Fernandez, X. 2020. Field analyses of lavender volatile organic compounds: performance evaluation of a portable gas chromatography-mass spectrometry device. *Phytochemical Analysis* 31, 778–785.

Tang, Y.B., Xing, D., Zhu, D.B., & Liu, J.F. 2007. An improved electrochemiluminescence polymerase chain reaction method for highly sensitive detection of plant viruses. *Analytica Chimica Acta* 582, 275–280.

Tessema, L., Kakuhenzire, R., Seid, E., Tafesse, S., Tadesse, Y., & Negash, K. 2024. Detection of six potato viruses using double antibody sandwich ELISA from in vitro, screen house and field grown potato crops in Ethiopia. *Discover Applied Sciences* 6, 79.

Thangavelu, R.M., Kadirvel, N., Balasubramaniam, P., & Viswanathan, R. 2022. Ultrasensitive nano-gold labelled, duplex lateral flow immunochromatographic assay for early detection of sugarcane mosaic viruses. *Scientific Reports* 12, 4144.

Tsai, H.L., Huang, L.C., Ann, P.J., & Liou, R.F. 2006. Detection of orchid *Phytophthora* disease by nested PCR. *Botanical Studies* 47(4), 379–387.

Tsuda, S., Kameya-Iwaki, M., Hanada, K., Kouda, Y., Hikdtd, M., & Tomaru, K. 1992. A novel detection and identification technique for plant viruses: rapid immunofilter paper assay (RIPA). *Plant Disease* 76, 466–469.

Úrbez-Torres, J.R., Haag, P., Bowen, P., Lowery, T., & O'Gorman, D.T. 2015. Development of a DNA macroarray for the detection and identification of fungal pathogens causing decline of young grapevines. *Phytopathology* 105(10), 1373–1388.

Vaseghi, A., Safaie, N., Bakhshinejad, B., Mohsenifar, A., & Sadeghizadeh, M. 2013. Detection of *Pseudomonas syringae* pathovars by thiol-linked DNA Gold nanoparticle probes. *Sensors and Actuators B: Chemical* 181, 644–651.

Velculescu, V.E., Zhang, L., Vogelstein, B., & Kinzler, K.W. 1995. Serial analysis of gene expression. *Science* 270, 484–487.

Viswanathan, R., Karuppaiah, R., & Malathi, P. 2004. Detection of sugarcane yellow leaf virus: the causal agent of yellow leaf syndrome in sugarcane by DAS-ELISA. *Archives of Phytopathology and Plant Protection* 37, 169–176.

Wakeham, A.J., Keane, G., & Kennedy, R. 2016. Field evaluation of a competitive lateral-flow assay for detection of *Alternaria brassicae* in vegetable brassica crops. *Plant Disease* 100, 1831–1839.

Wang, L., & Li, P.C.H. 2007. Flexible microarray construction and fast DNA hybridization conducted on a microfluidic chip for greenhouse plant fungal pathogen detection. *Journal of Agricultural and Food Chemistry* 55, 10509–10516.

Wee, E.J.H., Lau, H.Y., Botella, J.R., & Trau, M. 2015. Re-purposing bridging flocculation for on-site, rapid, qualitative DNA detection in resource-poor settings. *Chemical Communications* 51, 5828–5831.

Wei, J., Liu, H., Liu, F., Zhu, M., Zhou, X., & Xing, D. 2014. Miniaturized paper-based gene sensor for rapid and sensitive identification of contagious plant virus. *ACS Applied Materials & Interfaces* 6, 22577–22584.

Werres, S., & Steffens, C. 1994. Immunological techniques used with fungal plant pathogens: aspects of antigens, antibodies and assays for diagnosis. *Annals of Applied Biology* 125(3), 615–643.

Wongkaew, P., & Poosittisak, S. 2014. Diagnosis of sugarcane white leaf disease using the highly sensitive DNA based voltammetric electrochemical determination. *American Journal of Plant Sciences* 05, 2256.

Wu, J., Wg, Q., Liu, H., Qian, Y., Zie, Y., & Zhou, X. 2013. Monoclonal antibody-based serological methods for maize chlorotic mottle virus detection in China. *Journal of Zhejiang University Science* 14(7), 555–562.

Yang, C., Shi, Y., Zhang, Y., He, J., Li, M., Huang, W., Yuan, R., & Xu, W. 2023. Modular DNA tetrahedron nanomachine-guided dual-responsive hybridization chain reactions for discernible bivariate assay and cell imaging. *Analytical Chemistry* 95, 10337–10345.

Yeo, S.F., & Wong, B. 2002. Current status of nonculture methods for diagnosis of invasive fungal infections. *Clinical Microbiology Reviews* 35(3), 465–484. https://doi.org/10.1128/CMR.15.3.465-484. 2002

Yu, C., Wu, J., & Zhou, X. 2005. Detection and subgrouping of cucumber mosaic virus isolates by tas-elisa and immunocapture RT-PCR. *Journal of Virological Methods* 123, 155–161.

Zezza, F., Pascale, M., Mule, G., & Visconti, A. 2006. Detection of *Fusarium culmorum* in wheat by a surface plasmon resonance-based DNA sensor. *Journal of Microbiological Methods* 66, 529–537.

Zhan, F., Wang, T., Iradukunda, L., & Zhan, J. 2018. A gold nanoparticle-based lateral flow biosensor for sensitive visual detection of the potato late blight pathogen, *Phytophthora infestans*. *Analytica Chimica Acta* 1036, 153–161.

Zhang, C., Guo, M., Dong, J., Liu, L., Zhou, X., & Wu, J. 2023. Visual and super-sensitive detection of maize chlorotic mottle virus by dot-ELISA and Au nanoparticle-based immunochromatographic test strip. *Viruses* 15(7), 1607.

Zhang, F., Zou, M., Chen, Y., Li, J., Wang, Y., & Qi, X. 2014. Lanthanide-labeled immunochromatographic strips for the rapid detection of *Pantoea stewartii* subsp. *stewartii*. *Biosensors and Bioelectronics* 51, 29–35.

Zhang, N., McCarthy, M.C., & Smart, C.D. 2008. A macroarray system for the detection of fungal and oomycete pathogens of solanaceous crops. *Plant Disease* 92(6), 953–960.

Zhang, Y., Zhang, Y., & Xie, K. 2020. Evaluation of CRISPR/Cas12a-based DNA detection for fast pathogen diagnosis and GMO test in rice. *Molecular Breeding* 40, 11.

Zhao, Q., Tao, J., Feng, W., Uppal, J.S., Peng, H., & Le, X.C. 2020. Aptamer binding assays and molecular interaction studies using fluorescence anisotropy—a review. *Analytica Chimica Acta* **1125**, 267–278.

Zhao, W., Lu, J., Ma, W., Xu, C., Kuang, H., & Zhu, S. 2011. Rapid on-site detection of *Acidovorax avenae* subsp. *citrulli* by gold-labeled DNA strip sensor. *Biosensors and Bioelectronics* 26, 4241–4244.

Zhao, X., Zhai, L., Chen, J., Zhou, Y., Gao, J., Xu, W., Li, X., Zhong, T., Xiao, Y., & Yu, X. 2024. The use and development of immunological methods have attracted great interest due to their high selectivity and sensitivity. *Journal of Agricultural and Food Chemistry* 72(28), 15201–15415.

7

Microbes: A New Approach Towards Sustainable Environment

Abha Sharma[1]*, Abhay Sharma*[2*]*, Agrim Rattan*[3]*, Guneshori Maisnam*[4]*, and Devashish*[5]

[1] *Amity Institute of Organic Agriculture, Amity University Noida, Noida, Uttar Pradesh, India*
[2] *Regional Horticultural Research & Training Station, Kullu, Himachal Pradesh, India*
[3] *Atmospheric Sciences, IIT Delhi, Delhi, India*
[4] *Department of Rural Development and Planning, Nagaland University, Medziphema Campus, Nagaland, India*
[5] *Ramakrishna Mission Vivekananda Educational and Research Institute, Kolkata, India*

[*] Corresponding author: abhayevs@gmail.com

7.1 Introduction

Microorganisms are the predominant form of life on Earth due to their remarkable adaptability and versatile nutritional behaviors. Their omnipresence in the biosphere allows them to profoundly influence the environment in which they thrive. A wide variety of microbial forms, including bacteria, fungi, actinomycetes, viruses, mycoplasma, and protozoans, mediate these direct and indirect impacts, which are crucial for the development of contemporary human society, spanning food production, pharmaceuticals, textiles, agriculture, and environmental sustainability (Chandra & Sobti, 2020). Beneficial microorganisms have demonstrated efficacy in improving crop growth and yield. Microbial technologies provide an economically and environmentally sound means of improving agricultural productivity and establishing sustainable soil health (Chahal et al., 2022). Contemporary agricultural practices increasingly utilize microbial consortia, combining two or more compatible microbial species, rather than single cultures, to optimize plant growth and development. Microbial consortia – communities composed of multiple interacting microbial species – closely resemble the complexity of natural microbial ecosystems. In nature, microorganisms rarely function in isolation when associated with plant or animal hosts. Instead, their influence often arises from interactions developed in concert with other microbes over time. Due to their evolutionary co-adaptation, these consortia interact with host plants both biotically and abiotically, often in symbiotic relationships, and are able to carry out crucial tasks in agroecosystems (Nunes et al., 2024). Through synergistic interactions, microbial consortia can improve the colonization success and

Microbial Consortia in Plant Science and Sustainable Agriculture, First Edition. Edited by Ajay Tomar, et al.

functional efficiency of beneficial microbial strains. The application of such consortia has been shown to support plant development and enhance resilience to environmental stresses (Rodriguez et al., 2025).

The enzyme systems of microorganisms play one of their most important functions, which facilitates the degradation of various organic materials. This function is not only pivotal for nutrient recycling but also underscores their essential role in biogeochemical cycles, including carbon, oxygen, and nitrogen recycling (Gupta et al., 2016). Microorganisms have undergone significant diversification, enabling their participation in numerous green processes and cleaner technologies. By leveraging microbial technologies, sustainable development goals can be advanced. For example, microbes contribute to economic growth, job creation, enhanced food and pharmaceutical production, and eco-friendly chemical manufacturing. Cleaner production practices, the use of renewable resources, and advancements in healthcare are supported by these positions (Akinsemolu, 2018).

Ecosystems contribute to human well-being through a range of benefits that arise from the interplay of living and non-living components. There are four main categories into which they fall (Figure 7.1).

1. **Regulating services** maintaining air quality, mitigating the climate, preventing erosion, controlling infections, and purifying water are examples of regulating services.
2. **Supporting services** encompass primary production, generation of oxygen, and formation of soil, which underpin all other ecosystem services.
3. **Provisioning services** supply resources such as food, fiber, fresh water, fuel, and genetic materials.
4. **Cultural services** offer spiritual enrichment, recreational opportunities, aesthetic experiences, and cognitive development (Li et al., 2014).

These services are interdependent and contribute to biodiversity conservation, resource management, and environmental preservation (Polasky et al., 2011). Soil, as the physical foundation for most human activities, harbors a significant proportion of Earth's biodiversity.

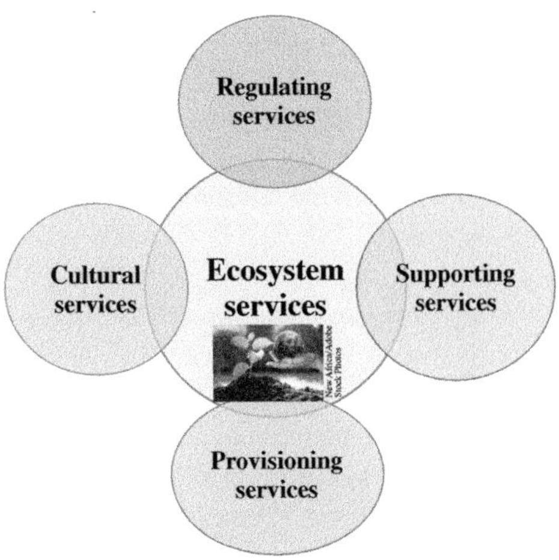

Figure 7.1 Categorization of ecosystem services.

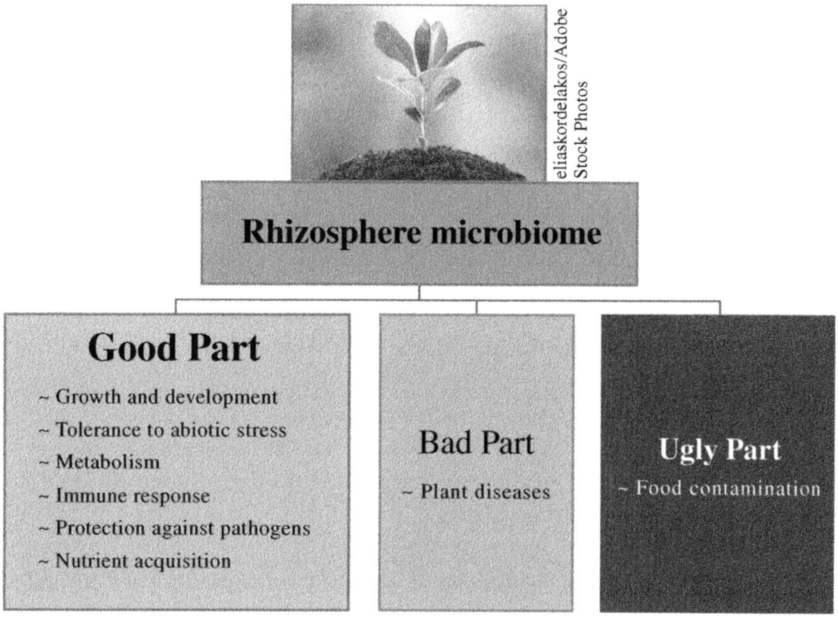

Figure 7.2 Interactions and functions of the rhizomicrobiome.

While extensive studies have examined soil's chemical and physical properties, its biological diversity and functionality remain less understood. Soil organisms provide an extensive spectrum of ecosystem services, which are important for the sustainability of managed and natural ecosystems (Adhikari & Hartemink, 2016).

Microbial communities, primarily composed of archaea, bacteria, and microfungi, are fundamental to the healthy functioning of soil ecosystems. Their sheer abundance, cumulative mass, and metabolic activities drive numerous aboveground processes, influencing biodiversity and productivity in terrestrial ecosystems. Through organic matter degradation and nutrient cycling, microbes contribute to the fertility of soil and ecosystem resilience. Moreover, microbial communities regulate soil homeostasis by removing contaminants and providing crucial ecosystem services such as enhancement of fertility, stress resistance, and ecosystem stability (Sacca et al., 2017). Understanding these microbial processes and their contributions to ecosystem functioning is imperative for achieving long-term environmental sustainability.

Microbial consortia, introduced as synergistic mixtures, perform multiple roles within the rhizosphere and are among the most promising solutions for promoting sustainable and environmentally friendly agricultural practices (Figure 7.2).

7.1.1 Microbial Diversity and Ecosystem Functioning

Microbial diversity refers to the wide variety of microorganisms, such as archaea, bacteria, viruses, and fungi, found in different environments. This diversity is not just about classifying species but also highlights the essential roles these microorganisms play in ecosystems. The complex relationships within microbial communities, as well as their interactions with the environment, are fundamental to many critical ecological processes. Microorganisms inhabit a vast range of environments, from extreme ecosystems like acidic hot springs and hydrothermal vents to more recognizable environments like soil, water, and the human body. Their remarkable capacity to survive and flourish across these varied habitats stems from their highly adaptable metabolic systems, which enable them to utilize diverse energy sources. For example, certain bacterial species derive energy from inorganic substances, while others rely on light through photosynthetic processes (Gadd, 2010).

Encouraging the nutrient cycle in ecosystems requires microbial diversity. Essential elements like carbon, nitrogen, phosphorus, and sulfur are transformed and transported more readily by these microorganisms (Marschner, 2007). According to Kuypers et al. (2018), bacteria break down organic molecules in the carbon cycle, generating methane in anaerobic environments and releasing carbon dioxide through respiration. Within the nitrogen cycle, certain bacterial species participate in processes such as denitrification, which reduces nitrate to nitrogen gas that is then released into the atmosphere, and nitrification, which converts ammonia to nitrate (Falkowski et al., 2008).

Microorganisms found in soil, such as bacteria, fungi, protozoa, and archaea, constitute the greatest source of terrestrial biodiversity and contribute significantly to Earth's biomass. The productivity and variety of aboveground ecosystems are influenced by them, and they are crucial for nutrient cycling, organic matter decomposition, and biogeochemical processes (Fierer, 2017).

As the world faces challenges like overexploitation of natural resources, rising food demands, and the need for sustainable agriculture, the potential of microbial communities to address these issues is immense. Despite significant progress, our knowledge of microbial community dynamics remains incomplete, particularly regarding how diversity influences function, the extent of functional redundancy, and how these communities respond to environmental disturbances (Verstraeten et al., 2012).

Microbial diversity and biomass in soil are critical for maintaining ecosystem sustainability and soil health, particularly through their roles in carbon and nutrient cycling. Ecosystems with higher microbial diversity and biomass are better equipped to sustain ecological processes, even under environmental stress. Research shows that factors like plant litter quality and quantity, along with soil microbial and faunal diversity, are key to ensuring long-term ecosystem functioning (Bodelier, 2011).

Despite the established recognition of microbial importance in supporting ecosystem services, the intricate processes at play necessitate further investigation for complete elucidation. Future studies should focus on comprehensively evaluating soil microorganisms and their diverse contributions to maintaining and improving ecosystem services.

The purpose of this chapter is to highlight the critical function that soil microbial communities perform in providing ecosystem services as well as their growing importance as markers of ecosystem health. Gaining deeper insights into microbial behavior and their interactions – especially those taking place in the rhizosphere alongside plant roots – is vital for forecasting how ecosystems may react to global environmental shifts and for optimizing ecological performance.

Turbé et al. (2010) and Sacca et al. (2017) have identified three main functional groups into which soil biota can be divided: chemical engineers, biological regulators, and ecosystem engineers. To maintain soil environments that are balanced, productive, and biodiverse, each of these groups has a distinct function.

a. **Chemical engineers**, mostly made up of bacteria and fungi: chemical engineers are essential to the cycling of nutrients and the breakdown of organic matter, both of which are essential for plant nutrition. These microbes are responsible for over 90% of the energy dynamics in the soil and are essential to procedures like bioremediation, which uses biochemical transformations to help neutralize environmental contaminants.
b. **Biological regulators** are responsible for managing the population levels of various organisms within the soil ecosystem. They achieve this regulation through interactions like grazing, predation, and parasitism. This group comprises small soil-dwelling invertebrates such as nematodes, enchytraeids (commonly known as pot worms), Collembola (springtails), and Acari (mites).
c. **Ecosystem engineers** actively reshape the physical structure of the soil, impacting the distribution of resources and the microhabitats available for other soil organisms. They enhance soil porosity, form biological structures, and help in the aggregation and movement of soil particles. These activities lead to zones of heightened microbial activity and influence key processes like water infiltration and retention. Notable representatives of this group include earthworms, termites, ants, and some small burrowing mammals.

Microbial diversity plays a pivotal role in aquatic ecosystems, where microbes are key drivers of biogeochemical processes that control nutrient dynamics and help maintain water quality. In both freshwater and marine environments, microorganisms contribute significantly to ecosystem productivity. Photosynthetic microbes, such as cyanobacteria, are essential to primary production; they serve as the foundation for aquatic food chains by converting carbon dioxide into organic molecules (Azam & Malfatti, 2007). Meanwhile, heterotrophic bacteria are essential in breaking down organic materials, recycling nutrients, and supporting the secondary levels of production in these systems.

Beyond nutrient turnover, aquatic microbes engage in various ecological relationships that range from mutualism to pathogenicity. On legume roots, for instance, symbiotic nodules of nitrogen-fixing bacteria convert atmospheric nitrogen into forms that plants can use and absorb energy-dense carbohydrates (Zengler & Palsson, 2012). In mutualistic relationships, both species benefit. Such interactions are crucial for sustaining productivity and stability across ecosystems.

Because it offers functional redundancy, microbial diversity also increases ecosystem resilience. In a variety of microbial communities, various species can perform comparable tasks, which allow ecosystem functions to persist even when environmental conditions fluctuate, or specific species are negatively impacted. For instance, in soil systems, continuous organic matter decomposition and nutrient cycling are ensured by a diverse range of decomposers, despite changes in temperature, moisture, or pH.

Recent developments in technology have greatly improved our capacity to investigate the ecological roles of microbial diversity. Metagenomics and high-throughput sequencing are two examples of techniques that have revolutionized our ability to describe microbial communities and comprehend their metabolic behavior. These approaches allow scientists to detect and analyze microbial populations across various habitats, providing detailed insights into their ecological roles. Additionally, bioinformatics has become indispensable for managing and interpreting the large datasets generated, offering a deeper understanding of microbial dynamics and interactions (Tiedje et al., 1999).

Despite these scientific advances, our understanding of microbial ecosystems remains incomplete. Future studies should aim to untangle the complex relationships between microbes and their environments, assess how global changes affect microbial structure and function, and explore how microbial communities can be applied to enhance ecosystem management and restoration efforts. By combining microbial ecology with other fields like soil ecology, hydrology, and climate science, more comprehensive models of ecosystem function and resilience can be created (Janssen, 2006).

A central focus in environmental microbiology is the investigation of microbial roles in nutrient cycles and in cleaning up environmental pollutants. This includes understanding how microbes facilitate nutrient recycling, degrade harmful substances, and help alleviate the impacts of contamination. Gaining insight into these microbial processes is essential for creating effective and sustainable environmental management strategies to address pressing global ecological challenges.

7.1.1.1 Microbial Fertilizers

Microbial fertilizers, a subset of biofertilizers, consist of beneficial microorganisms that carry out specific roles to support plant health (Goyal et al., 2023). When applied to soil, these fertilizers help establish diverse microbial populations that enhance nutrient uptake for plants. Research indicates that microbial fertilizers enhanced with plant growth-promoting rhizobacteria (PGPR) substantially improve soil fertility and encourage plant growth (Table 7.1). Additionally, they strengthen plant defenses against both biological threats, such as pests and diseases, and environmental challenges like drought and high salinity. As a sustainable and eco-friendly solution, microbial fertilizers are increasingly seen as vital tools in advancing environmentally responsible farming. This review explores the mechanisms by which these fertilizers enhance plant growth and stress resilience, while also highlighting their potential role in soil restoration (Bamdad et al., 2022).

Based on their primary purpose, microbial fertilizers can be divided into three groups: microbial fungicides, compound microbial fertilizers, and bioorganic fertilizers.

Table 7.1 Beneficial growth-promoting bacterial strains of different types of microbial fertilizers on different crops.

Sr. no.	Types of microbial fertilizers	Crop	Plant growth-promoting rhizobacteria	References
1	Bioorganic fertilizer	Lettuce	Actinobacteria, Proteobacteria, Chloroflexi, Acidobacteria, Gemmatimonadota, Ascomycota, and Basidiomycota	Jin et al. (2022)
2	Bioorganic fertilizer	Tobacco	Actinobacteria, Chloroflexi, Proteobacteria, Acidobacteria, Firmicutes, Gemmatimonadota, Streptomyces, Bacillus, Arthrobacter, and Paenibacillus	Ma et al. (2018a)

Sr. no.	Types of microbial fertilizers	Crop	Plant growth-promoting rhizobacteria	References
3	Bioorganic fertilizer	Beet, potato, winter, wheat	Actinobacteria, Proteobacteria, Acidobacteria, Arthrobacter, and Paenibacillus	Jakienė et al. (2015), Li et al. (2021)
4	Bioorganic fertilizer	Cauliflower	Proteobacteria, Actinobacteria, Acidobacteria, Gemmatimonadetes, Bacteroidetes, and Chloroflexi	Xiao et al. (2022)
5	Bioorganic fertilizer	Tomato	Proteobacteria, Actinobacteriota, Bacteroidota, Firmicutes, and Verrucomicrobiota	Wang et al. (2021)
6	Microbial inoculants	Watermelon	*Pseudomonas flavobacterium, Aspergillus, Myceliophthora, Trichoderma, Humicola,* and *Neocosmospora*	Zhao et al. (2018)
7	Microbial inoculants	Radish	Proteobacteria, Bacterioidetes, Acidobacteria, Actinobacteria, and Planctomycetes	Jatav et al. (2012)
8	Microbial inoculants	Rice	Proteobacteria, Acidobacteria, Bacteroidetes, Gemmatimonadetes, Actinobacteria, Planctomycetes, Ascomycota, and Chytridiomycota	Xie et al. (2021)
9	Microbial inoculants	Prunusdavidana	Proteobacteria, Bacteroidetes, Acidobacteria, Gemmatimonadetes, Actinobacteria, Patescibacteria, Chloroflexi, Verrucomicrobia, Nitrospirae, Latescibacteria, and Rokubacteria	Shi et al. (2022)
10	Microbial inoculants	Cucumber	Betaproteobacteria, Gammaproteobacteria, Deltaproteobacteria, Alphaproteobacteria, Actinobacteria, Acidobacteria, Gemmatimonadetes, Bacteroidetes, Chloroflexi, Planctomycetes, Firmicutes, Nitrospirae, Verrucomicrobia, Armatimonadetes, Cyanobacteria, TM7, Fibrobacteres, and Chlorobi	Simranjit et al. (2019)
11	Compound microbial fertilizer	Soybean	Phosphorus-solubilizing bacteria, nitrogen-fixing bacteria	Fu et al. (2023)
12	Compound microbial fertilizer	Sugarcane	*Gluconacetobacter diazotrophicus, Trichoderma harzianum,* and *Pseudomonas fluorescens*	Shukla et al. (2020)

7.1.1.2 Bioorganic Fertilizer

Bioorganic fertilizers primarily originate from processed organic materials like animal manure, crop residues, municipal waste, and other biodegradable sources. These fertilizers are enriched with selected beneficial microorganisms, offering both organic nutrients and microbial functions. By integrating the advantages of conventional organic fertilizers with those of microbial agents, bioorganic fertilizers not only supply high levels of organic matter but also introduce specialized microbes that perform targeted biological functions (Yap & Al-Mutairi, 2023). The increasing amounts of organic waste and pollutants in recent years have gradually reduced farming quality, negatively impacting soil health. Using appropriate fertilizer application strategies, returning crop residues to the field, and similar practices can maximize the benefits of organic fertilizers and reduce the environmental risks associated with excessive chemical fertilizer use, thus promoting the development of bioorganic fertilizers (Kiruba & Saeid, 2022). Compared to chemical fertilizers, bioorganic fertilizers offer the advantages of traditional organic fertilizers, combined with the specific functions of beneficial microorganisms, maximizing overall fertilizer efficiency. Research indicates that bioorganic fertilizers improve soil quality by acting as beneficial soil conditioners, promoting soil aggregate formation, and increasing nutrient availability for crops (Mitter et al., 2021). Furthermore, microbial fertilizers have been shown to be more effective than chemical fertilizers in improving soil microbiota and promoting the cycling of carbon and nitrogen in the soil. Crucially, bioorganic fertilizers introduce beneficial microorganisms that colonize the rhizosphere (the area where microorganisms, soil, and plant roots interact), facilitating improved communication and beneficial interactions between these components.

7.1.1.3 Microbial Agents

Microbial agents, often referred to as microbial inoculants, are a type of biofertilizer composed of live, beneficial microorganisms (O'Callaghan et al., 2022). These inoculants are developed by isolating functional microbes – such as bacteria, fungi, actinomycetes, and algae – that offer specific agricultural benefits. Among these, PGPRs have gained significant attention due to their crucial role in supporting plant health. Common PGPR genera include *Klebsiella, Azotobacter, Azospirillum*, and *Bacillus*, which naturally inhabit the rhizosphere. Notably, *Bacillus* and *Pseudomonas* species, frequently found in the root zone, have been extensively utilized in formulating microbial inoculants, taking advantage of the diverse microbial populations present in soil (Comeau et al., 2021).

Research shows that these bioinoculants can directly stimulate plant growth by producing phytohormones and enhancing plant resilience to both biological threats (such as pathogens and pests) and environmental stressors (like drought and salinity). Importantly, these microbial agents are safe for plants, environmentally friendly, and contribute to improved plant health, growth, and development. In addition to promoting nutrient uptake, they help balance soil nutrients, generate plant growth regulators, and assist in detoxifying contaminated soils by breaking down heavy metals, pesticides, and fungicides. They also play a role in soil restoration through biological control mechanisms. Early investigations into microbial inoculants typically focused on single microbial strains with limited impact.

However, more recent studies emphasize the development of multi-strain, multifunctional inoculants, highlighting the growing importance and complexity of this research field (Baez-Rogelio et al., 2016).

7.1.1.4 Complex Microbial Fertilizer

To improve soil fertility and plant growth, several advantageous bacteria, including *Bacillus subtilis, Bacillus licheniformis, Azospirillum brasilense*, and *Streptomyces*, are combined in composite microbial fertilizers, which are novel agricultural inputs. These fertilizers are designed to optimize functions like phosphorus and potassium solubilization, nitrogen fixation, and organic matter decomposition, creating a more efficient and sustainable approach to crop nutrition (Yang et al., 2020). By leveraging the complementary metabolic activities of these microbes, composite microbial fertilizers improve nutrient availability and uptake, leading to healthier and more productive crops.

Compound microbial fertilizers, on the other hand, are formulated using agricultural residues – such as poultry manure and straw – as their primary components, combined with beneficial microorganisms. This type of fertilizer not only recycles organic waste but also introduces beneficial microbes that enhance soil health and nutrient cycling. The result is a sustainable fertilizer that improves soil structure, increases organic matter content, and supports long-term agricultural productivity.

Advanced microbial fertilizers, often referred to as complex or composite microbial fertilizers, combine the strengths of chemical fertilizers, organic matter, and beneficial microorganisms in a single formulation. This integrated solution delivers the quick nutrient availability typical of chemical fertilizers, the sustained soil health benefits of organic materials, and the biological functions of living microbes. When these elements work together, they boost crop yield, soil quality, and plant growth while lowering the environmental damage caused by traditional fertilization techniques (Chojnacka et al., 2020). This balanced nutrient delivery system supports plants throughout their life cycle, fostering healthier crops and promoting environmentally sustainable farming.

The beneficial microbes included in these fertilizers significantly contribute to plant development through their biochemical processes. These microbial fertilizers dissolve soil minerals and release secondary metabolites that increase the availability of important nutrients like phosphorus and potassium. These microorganisms thrive in the rhizosphere – the soil zone adjacent to plant roots – where they help form a dynamic and beneficial microbial community. In interaction with root exudates, the microbes can stimulate the plant's production of secondary metabolites, which enhance stress resilience, improve defense responses, and regulate key growth mechanisms.

These microbial fertilizers also contain organisms that generate gibberellins, auxins, and cytokinins, which are plant hormones. These growth regulators increase the effectiveness of nutrient intake, encourage root growth, and improve overall plant health. The close relationship formed between plants and these microbes leads to a self-sustaining, mutually beneficial system that supports robust plant development while decreasing the dependence on synthetic agricultural inputs.

In conclusion, composite and complex microbial fertilizers mark a significant leap forward in sustainable agriculture. By utilizing beneficial microbes alongside organic and inorganic components, these fertilizers not only improve soil fertility and nutrient accessibility but also promote healthier plant growth. These microbial fertilizers are instrumental in lessening the reliance on chemical fertilizers and contribute to ecological balance by recycling organic agricultural waste.

7.2 Regulating Services

Essential ecosystem processes like the breakdown of organic pollutants, the control of diseases and pests, water purification and its regulation, and climate regulation are all greatly affected by microorganisms living beneath the soil surface. With significant ramifications for international ecological and economic systems, these activities are crucial for preserving ecosystem stability and functionality.

7.2.1 Climate Regulation

Soil microorganisms play a crucial role in regulating the climate, especially through their impact on the exchange of carbon (C) between the land and the atmosphere. Bacteria and fungi are central to this process, as they govern soil carbon storage and emissions, critical determinants of atmospheric greenhouse gas concentrations. The capacity of soil to regulate climate is highly contingent on land use practices and intensification, which have been shown to alter soil carbon sequestration potential and emission rates significantly. These changes carry substantial economic and ecological consequences (Macías & Camps Arbestain, 2010).

Climate change can trigger feedback loops that significantly affect microbial processes in the soil. For example, increasing global temperatures can enhance the rate of microbial respiration by heterotrophic organisms, which may lead to greater carbon emissions and contribute further to atmospheric carbon levels (Bardgett et al., 2008). According to the Intergovernmental Panel on Climate Change (IPCC), persistently high levels of greenhouse gas emissions are probably going to have a detrimental impact on ecosystem services and functioning (IPCC, 2014). Furthermore, it has been demonstrated that increasing atmospheric CO_2 concentrations change the makeup of microbial populations more than fungi or nematodes, particularly bacteria. Numerous factors, such as the kinds of plants present and the soil's nitrogen level, affect these microbial alterations (Drigo et al., 2007).

7.2.2 Microbial Communities and the Carbon Cycle

Microbial communities play a vital role in the global carbon cycle by facilitating processes such as carbon sequestration, promoting plant growth, and breaking down organic materials. Significant carbon reserves are found in ecosystems like permafrost regions, grasslands, and tropical forests. The activity of soil microorganisms greatly influences the stability and long-term storage of this carbon, as well as its potential release into the atmosphere as greenhouse gases. By regulating carbon flow through ecosystems, microbes directly impact the pace of climate change and support essential ecological processes, including nutrient cycling, which relies heavily on microbial function.

Furthermore, bacteria play a key role in the decomposition and conversion of decomposing organic matter into nutritional forms that other species may absorb and utilize. Microbial enzymes are therefore frequently acknowledged as essential regulators of Earth's biogeochemical cycles. The interaction of respiration and photosynthesis is essential to maintaining the equilibrium of the terrestrial carbon cycle. By converting atmospheric CO_2 into organic matter, autotrophs – such as green plants and photo- or chemoautotrophic microbes – introduce carbon into the soil. Microorganisms are crucial to the cycling and control of atmospheric carbon dioxide because they use this carbon for energy and metabolism.

7.2.3 Water Regulation and Purification

Soil acts as both a natural filter, purifying water as it permeates its layers, and a reservoir, storing it for plant uptake (Gavrilescu, 2021). In Europe, the ecosystem service of water purification provided by soil has been valued at approximately €16 billion annually (La Notte et al., 2017). Microorganisms are key contributors to this function, as they help break down pollutants found in both soil and groundwater. For example, *Rhodococcus wratislaviensis*, a bacterium known for its ability to degrade herbicides, has been identified in soil and groundwater samples from aquifers contaminated with terbuthylazine, demonstrating its potential for bioremediation in polluted environments (Turbé et al., 2010).

Soil microbes affect water movement via altering the amount and content of soil organic matter, in addition to directly decomposing pollutants. Better water infiltration, storage, and filtration can result from these microbially driven alterations that improve soil structure and porosity. These combined roles highlight how crucial microbial communities are to maintaining water availability and quality in terrestrial environments.

7.2.4 Disease and Pest Regulation

Microorganisms play a critical role in controlling plant diseases by employing a variety of direct and indirect strategies. These include inhibiting pathogens through antagonistic activity, competing for resources, interfering with pathogen communication systems, and triggering the plant's own defense responses. Such disease-suppressing abilities have been documented in numerous soil-dwelling bacteria, such as *Streptomyces, Pseudomonas*, and *Bacillus*, as well as in fungi like *Trichoderma* and *Fusarium* (Ciancio et al., 2016).

Resistance to the establishment of non-native species is another crucial ecosystem service provided by diverse microbial communities. Complex terrestrial ecosystems, characterized by high microbial diversity, exhibit greater resilience to biological invasions compared to less diverse systems (van Elsas et al., 2008). Utilizing beneficial microorganisms offers a sustainable alternative to synthetic pesticides in agriculture. However, the efficacy of biological control agents can be inconsistent under field conditions, particularly in managing diseases of woody plants (Cazorla & Mercado-Blanco, 2016).

Incorporating organic materials into the soil can significantly improve ecosystem services, particularly in the breakdown of organic matter and the control of soilborne plant diseases. For example, composts that can suppress pathogens have effectively managed plant diseases by altering the soil's microbial makeup and boosting its inherent ability to resist disease, a phenomenon known as soil suppressiveness (Blaya et al., 2016). Certain bacterial groups, including γ- and β-Proteobacteria (such as Pseudomonadaceae, Burkholderiaceae, and Xanthomonadales) as well as members of the Firmicutes phylum, are closely linked to this suppressive effect. Among fungi, species within the Ascomycota phylum are especially active in this role (Mendes et al., 2011). Notably, a study investigating farmers' views on ecosystem services in arable systems found that the natural suppression of plant diseases was regarded as the most valuable service – surpassing other critical functions like water and nutrient retention, climate regulation, and environmental detoxification (Rutgers et al., 2012).

7.2.5 Biodegradation of Organic Waste and Xenobiotics

Soil contamination poses a significant threat to the functioning of soils and the stability of entire ecosystems (Rillig et al., 2019). Major contributors to this pollution include improper waste disposal, industrial and mining activities, agricultural runoff, and incidents involving military or radioactive materials. Natural attenuation of pollutants relies on a mix of physical, chemical, and biological mechanisms – such as dilution, dispersion, adsorption, volatilization, transformation, mineralization, and stabilization – that, under favorable conditions, can naturally reduce the toxicity, concentration, mobility, or volume of pollutants without human intervention (Mulligan & Yong, 2004). Applying efficient microbial consortia has been shown to enhance the composting process of agricultural by-products, promoting better waste utilization and supporting sustainable farming (Mali et al., 2024).

By converting pollutants through metabolic processes like co-metabolism, in which deterioration happens inadvertently alongside regular microbial metabolism, or by using them as sources of energy, carbon, nitrogen, or nutrition, microorganisms play a crucial part in biodegradation. The presence of pollutants often acts as a selective force, encouraging the proliferation of microbial species that are more tolerant or capable of breaking down contaminants (Boivin et al., 2002). This microbial adaptability highlights the resilience of soil ecosystems. While controlled experiments have identified specific bacterial strains capable of degrading various pollutants (Franchi et al., 2016), complete remediation often requires the synergistic activity of diverse microbial communities, each contributing to different stages of the degradation pathway (Pino & Peñuela, 2011).

The challenge of soil pollution is particularly evident in the case of complex contaminants like polychlorinated biphenyls (PCBs), which require multiple microbial species and a variety of metabolic routes functioning under both aerobic and anaerobic conditions for effective breakdown (Ancona et al., 2016). Microbes support soil remediation through both general soil biodegradation and rhizoremediation, where root-associated microbial activity plays a central role. Enhancing these processes may involve adding nutrients or soil amendments that stimulate microbial activity or target specific degradative functions in the rhizosphere (Paul et al., 2006; Luo et al., 2008).

Key microbial species and functional genes involved in the degradation of synthetic chemicals can be identified, thanks to emerging genomic methods, offering powerful tools for monitoring bioremediation through bioindicators (organisms) and biomarkers (genes) (Eyers et al., 2004).

The service of natural pollutant attenuation provided by microbes is especially valuable because of its broader benefits, including improved plant productivity and cleaner water resources, underscoring its critical role in maintaining ecosystem health.

7.2.6 Metal Detoxification

Microorganisms exhibit exceptional physiological adaptations that enable them to withstand metal-induced stress, utilizing a variety of mechanisms to manage metal toxicity. A key aspect of this resistance involves the regulation and activity of metal uptake proteins, with different bacterial species employing unique regulatory systems. In some cases, bacteria control transporter function through specialized regulatory proteins that can bind metal ions with extremely high affinity – down to the femtomolar range. In response to metal exposure, microbes may also increase the synthesis of extracellular substances such as polymers or siderophores. These compounds contain functional groups that can bind metal ions, either limiting their entry into cells or promoting their expulsion (Carrondo, 2003).

Additional microbial detoxification strategies include the conversion of metals into less soluble forms, such as oxides, sulfides, or crystalline aggregates, which are often deposited near or on cell membranes. Within the cytoplasm, proteins like bacterioferritin and metallothioneins can capture and store metal ions, reducing their toxicity. Some microorganisms also carry out redox transformations, using oxidative or reductive processes to convert harmful metal species into more stable, less toxic forms.

Specific microbes have developed detoxification mechanisms in response to heavy metals such as copper, mercury, lead, zinc, and cadmium. For instance, cadmium accumulation in New Zealand's agricultural soils, resulting from the widespread use of superphosphate fertilizers, highlights the need for microbial detoxification strategies (Loganathan et al., 2003). Common microbial defense mechanisms include metal sequestration, detoxification, and ion efflux. Metals may be sequestered by binding to cell membranes, cell walls, or extracellular polysaccharides. Detoxification can also occur through enzymatic reduction; for example, mercury reductase converts Hg^{2+} to elemental mercury (Hg^0), which evaporates due to its low boiling point (Nies, 1999). Gram-negative bacteria, such as *Cupriavidus metallidurans* (formerly *Alcaligenes eutrophus*), expel toxic metals from the cytoplasm using cation/proton antiporters located in the membrane of the cell (Silver & Phung, 1996).

The capacity of microorganisms to modify heavy metals is increasingly being applied in bioremediation, offering a sustainable method for reducing metal contamination in soils.

7.3 Supporting Services

Despite the fact that people do not directly use supporting services, essential ecological processes keep other ecosystem functions going. These services, such as soil formation, nitrogen and water cycling, primary production, and the provision of habitats for biodiversity, are all significantly influenced by soil microbial communities. Microbial activity has a major impact on soil fertility, which is defined as the soil's ability to support plant growth by guaranteeing sufficient nutrient availability and organic matter breakdown (Tang, 2024). Particularly, bacteria and fungi function as "biological catalysts," encouraging the mineralization, degradation, and transformation of carbon (C) and other essential nutrients. Stoichiometric homeostasis, which allows microbes to maintain a constant ratio of biomass elements, governs these activities (Spohn, 2016).

7.3.1 Microbial Interactions in Soil Formation

Soil formation is a complex process shaped by the interactions among plants, micro- and mesofauna, and microorganisms, particularly through their role in mineral weathering (Sokolova, 2011). Microorganisms play a crucial role in mineral dissolution, a process that is as significant as the absorption of carbon dioxide by plants and the nitrogen fixation carried out by soil microbes (Schulz et al., 2013). Soil biota are also essential for breaking down and decomposing organic matter, which leads to the formation of humus. Microorganisms decompose organic materials, such as fallen leaves and plant litter, utilizing them as energy sources, thereby increasing microbial populations in the soil. This decomposition process involves the breakdown of readily decomposable materials, while more resistant compounds persist, eventually forming humus. Humus binds soil particles like clay, silt, and sand into aggregates, helping to develop soil structure and create distinct soil horizons (Martin & Haider, 1971).

Microorganisms are fundamental to soil development due to their widespread presence, rapid reproduction, diverse metabolic capabilities, and capacity to inhabit and adjust to challenging environments. Their activity drives the transformation of plant litter's chemical makeup and regulates the cycling of carbon and nitrogen in the soil, and is intimately linked to the decomposition of plant litter (Berg & McClaugherty, 2014). The breakdown of litter and the creation of soil depend heavily on microbial enzymatic activity (Helfrich et al., 2015). Nutrient immobilization is impacted by the chemical composition of soil organic matter, which is influenced by the growth, biomass, and necromass of microbes (Cotrufo et al., 2013). Furthermore, the conversion of microbial biomass and necromass supports soil development, fertility, and stability (Fontaine et al., 2011). These processes can be tracked by measuring amino acids and amino sugars, which are more abundant in microbial biomass than in plant litter and are essential for nitrogen immobilization and carbon storage (Liang et al., 2007).

7.3.2 Nutrient Cycling

The geological process of microbial weathering occurs at the Earth's surface and is fueled by the growth and reproduction of microorganisms. Microbial metabolites, including enzymes, are key to the breakdown of minerals and rocks, leading to alterations in their composition and the release of elements like silicon (Si), iron (Fe), manganese (Mn), and aluminum (Al) from silicates, oxides, and carbonates (Bin et al., 2008). This process impacts the composition and quantity of soil minerals, influencing both soil formation and fertility.

While the breakdown of siliceous rocks is mostly caused by freeze–thaw cycles, the weathering of calcareous rocks usually includes the acidic decomposition of calcium carbonate. Calcium rocks offer fewer elements, which may restrict microbial growth, whereas siliceous rocks release vital nutrients for microbial life. On the other hand, poisonous elements like nickel (Ni) and cadmium (Cd) that are released by serpentinite rocks may prevent microbial activity and plant development (Bratteler et al., 2006).

Both direct and indirect processes are used by microorganisms to contribute to mineral weathering. Direct effects occur when microbial cells adhere to mineral surfaces, often facilitated by the binding of previously absorbed compounds. Indirect effects involve microbial metabolites, such as chelating agents, acids, and bases, which accelerate mineral breakdown. Microbial activity is particularly important in the release of phosphorus (P) and sulfur (S) from bedrock, as these elements are essential for life but are absent in the initial soil composition. Conversely, carbon and nitrogen are limited in newly formed soils and must be introduced through biological processes (Kobierska et al., 2011).

Soil aggregation, a crucial factor in determining soil structure and function, is significantly influenced by microbial activity. The early stages of soil formation and nutrient input rely on the biological weathering of bedrock and the creation of interfaces for nutrient cycling in areas lacking vegetation. Consequently, microorganisms are vital to the fundamental processes of soil formation and ecosystem development.

7.3.3 Organic Matter Transformation

According to Pries et al. (2017), soil organic carbon (SOC) is the biggest organic carbon store on land and is essential for controlling the global carbon cycle. Environmental variables like soil temperature, moisture content, and the quantity and caliber of organic matter inputs have a major impact on SOC turnover rates (Ghee et al., 2013). Climate change unavoidably impacts SOC dynamics by changing soil temperature and water availability (Liang et al., 2017). SOC can be categorized into different pools based on their responsiveness to environmental changes. Particulate organic carbon (POC) represents the SOC fraction that is not mineral-protected and is generally more labile. In contrast, mineral-associated organic carbon (MAOC) is more resistant to environmental changes due to its interaction with minerals (Lavallee et al., 2020). Dissolved organic matter (DOM) serves as a key connection between the MAOC and POC pools (Ma et al., 2018b). The labile forms of DOM are closely tied to microbial activity and provide a rapid response to climate change (Kramer & Chadwick, 2018). The solubilization of DOM is primarily driven by both physicochemical and microbial processes, with factors like aridity and clay content influencing the molecular composition of DOM. Soil microorganisms are crucial for terrestrial carbon cycling, facilitating both the decomposition of catabolic carbon and the stabilization of anabolic carbon (Ding et al., 2023).

Soils provide a wide range of essential ecosystem services, including supporting food production, regulating atmospheric greenhouse gas levels, controlling erosion, managing water quality and availability, and maintaining biodiversity of plants, animals, and microorganisms (Díaz et al., 2006).

SOC dynamics are fundamental to the functioning of these ecosystem services (Lorenz & Lal, 2014). The benefits of SOC include its role as a source of vital nutrients for biological activity (such as nitrogen, phosphorus, and sulfur, collectively known as soil organic matter or SOM), its contribution to cation exchange capacity and water retention, its enhancement of soil structure and its ability to reduce erosion, and its function as a carbon sink.

Converting natural vegetation to agricultural land has been shown to lead to significant SOC losses, typically ranging from 25 to 50% (Lal, 2008). Ongoing depletion of SOM poses a considerable risk to future food security, atmospheric greenhouse gas concentrations, and biodiversity. Over the past two decades, numerous reviews have thoroughly examined the physical and chemical mechanisms involved in SOM stabilization and turnover (Dungait et al., 2012).

7.3.4 Role of Microbes in Plant Growth Promotion

The interactions between plants and bacteria in the rhizosphere – the area of soil that comes into direct contact with plant roots – determine the fertility of the soil and the health of the plants. Soil bacteria are also vital for biogeochemical cycling, having long been important to agricultural productivity (Shoebitz et al., 2009). Free-living bacteria known as plant PGPR, plant health-promoting rhizobacteria, or nodule-promoting rhizobacteria increase plant growth by colonizing root surfaces. The rhizosphere's ecological balance depends on these interactions. Nitrogen-fixing bacteria, including *Rhizobium, Bradyrhizobium, Azorhizobium*, and others, engage in mutually beneficial relationships with plants by converting atmospheric nitrogen into a usable form. Furthermore, free-living nitrogen fixers such as *Azospirillum, Enterobacter, Klebsiella*, and *Pseudomonas* are also effective at colonizing plant roots (Prasad et al., 2015). PGPRs offer considerable promise for enhancing sustainable agriculture, primarily through three mechanisms: synthesizing compounds that benefit plants, enhancing nutrient uptake, and suppressing plant diseases.

The promotion of plant growth by PGPR occurs through both direct and indirect mechanisms. Indirect benefits arise from mitigating the negative effects of harmful pathogens (Podile & Kishore, 2006). This includes the production of siderophores, which are molecules that bind to metals like iron, preventing pathogens from accessing them. Additionally, PGPRs can produce antibiotics and biological agents to control soilborne plant diseases, as well as enzymes that degrade fungal cell walls, such as chitinase and β-1,3-glucanase. Direct growth promotion occurs through the secretion of plant hormones, including auxins, cytokinins, gibberellins, and ethylene. One example is the production of indole-3-acetic acid, an auxin that promotes root growth. Certain PGPR also produce ACC deaminase, an enzyme that breaks down ACC, a precursor to ethylene, thus reducing ethylene levels and stimulating root growth (Li et al., 2000). Beyond producing hormones, Additionally, PGPR stabilizes soil aggregates, increases plant resilience to stress, improves soil structure and organic matter content, and helps make mineral phosphates and other essential nutrients soluble. PGPRs can enhance nutrient availability and reduce dependency on chemical fertilizers by retaining soil nitrogen and other nutrients within the plant–soil system (Hayat et al., 2010).

7.3.5 Soil Microorganisms as Indicators

Numerous biological indicators have been suggested for assessing soil health. For an indicator to be considered effective, it should be easy to measure, cost-efficient, responsive to environmental and management changes, relevant for policy decisions, and suited to the intended evaluation purpose. Among these, microbial indicators stand out for their reliability (Floch et al., 2011). Soil microbes fulfill many essential criteria: they react sensitively to both land use practices and climate shifts, are closely linked with critical soil and ecosystem functions, reveal how land management affects plant productivity and health, are relatively straightforward for practitioners to understand, and can often be assessed using affordable and accessible methods (Bennett et al., 2010).

Soil quality can also be gauged through its capacity to support key ecosystem services, including food and fiber production, environmental regulation, habitat support, and biodiversity. A tiered strategy is often applied, utilizing genotypic, phenotypic, or functional indicators. This may involve evaluating soil microbial biomass, community composition, respiration rates, carbon processing, and various biochemical pathways (Ritz et al., 2009). Evidence from this approach indicates that microbial functional genes are especially valuable as indicators of soil processes related to water dynamics, nutrient cycling, and carbon storage.

Advances in soil microbial ecology have made it possible to study microbial diversity and functions through a broad array of biochemical and molecular tools. These range from community-level fingerprinting techniques to precise species-level identification and functional gene analysis. However, researchers have frequently emphasized the need to standardize methodologies across studies to ensure data comparability (Fierer et al., 2009; Philippot et al., 2012). The most comprehensive insights are achieved by combining conventional culturing techniques with modern biochemical and molecular methods (Nardo et al., 2005; Pulleman et al., 2012). While "omics" technologies offer deep insights into microbial communities, they are most effective when used alongside classical, culture-based approaches to fully understand microbial behavior and ecosystem roles (Marx, 2017).

7.4 Provisioning Services

Ecosystems provide us with tangible benefits known as provisioning services, which include food, water, fuel, fiber, genetic resources, and medicines. The rhizosphere is a rich environment for microorganisms that produce various bioactive compounds, such as antibiotics, enzymes, biosurfactants, and osmoprotectants. Microbial antibiotics, in particular, have the potential to serve as new treatments to combat the increasing issue of multidrug-resistant pathogens (Compant et al., 2005). Soil ecosystems provide the necessary conditions for food production through cultivation, harvesting, and hunting. These provisioning services constitute tangible goods directly derived from soil ecosystems and utilized by humans (Quijas et al., 2010). Soil ecological services provide essential resources, including biochemical compounds, genomic resources, food, fodder, fiber, energy, freshwater, and medicines. These tangible resources, originating from soil ecosystems, are finite and capable of regeneration (Pandey, 2020). Soil is fundamental to all life, providing sustenance, hydration, shelter, and structural materials for both animals and plants (De Groot et al., 2002). As plants are a primary food source for many animals, soil indirectly supports animal populations. Plants help purify the air, which is necessary for animals to survive, and soil supplies vital nutrients for human consumption, which makes it possible to produce nutrient-rich animal feed (Murphy & Allen, 2003). Soil is also vital for aquaculture, providing suitable aquatic environments for a significant portion of global fish consumption. Soil components, such as humic acids, fulvic acids, and carbon particles, protect aquatic organisms, while humates facilitate mineral breakdown, improving root zone health and water penetration. Enhanced development of root increases the soil's nutrient-holding capacity, promoting economic diversification, supporting healthy ecosystems, and providing resources for vulnerable populations (Razafindratsima et al., 2021). Medicinal plants, sourced from soil ecosystems, treat various ailments, offer drought resistance, and generate income for poorer populations through the sale and direct consumption of wild fruits, firewood, and charcoal (Tieminie et al., 2021). Aquifers, wetlands, mangroves, rivers, and other freshwater ecosystems are essential for reducing the effects of floods and droughts. Water flows are controlled and water storage is provided by soil ecosystems, which also provide, clean, and safeguard freshwater supplies. Water withdrawals from irrigation make up 60% of total water use, and cropping strategies can increase water retention. According to Lapointe et al. (2014), aquaculture development and sustainable fisheries management can improve the freshwater supply provided by aquatic ecosystems. According to Schoenholtz et al. (2000), soils sustain forests and offer physical stability, whereas fisheries and fish management techniques must be integrated for sustainable freshwater management. Soils sustain forests, which maintain healthy aquatic ecologies and ensure a reliable supply of contaminant-free freshwater. Beyond water filtration and purification, forests also reduce reservoir sedimentation, prevent soil erosion, and lessen the risk of floods, mudslides, and other hazards that could impact downstream water supplies (Thiele-Bruhn, 2021).

7.4.1 Importance of Microorganisms in Food Processing

Mycelium-based food products, which provide a sustainable substitute for conventional food sources, are made possible in large part by microorganisms in the production and processing of food (Acevedo-Rocha et al., 2019). Leading the way in the production of oyster mushrooms and food ingredients produced from mycelium through submerged fermentation processes are companies like Kernel MycoFood, Leep Foods, and Meati. Prominent companies such as Quorn Foods manufacture mycelium-based goods all around the world. Additionally, the fermentation of tempeh depends heavily on microorganisms; Nordic Tempeh is produced by firms like Contemporary Foods and Tempty Foods using solid-state fermentation on a variety of grains and legumes. Additionally, microbes are being used more and more to recycle nutrients from food industry waste. For example, Mush Labs uses fungi to cut waste and make money from food production waste (El-Ramady et al., 2022). In summary, microorganisms have a vital role in creating sustainable food substitutes and tackling environmental issues associated with waste disposal, demonstrating their enormous potential to support sustainable growth in the food sector.

7.4.2 Importance of Microorganisms in Food Production

Food production relies heavily on microorganisms, which also present a number of prospects for sustainable development. Enhancing soil fertility and plant growth is one of their main benefits. To promote more sustainable agricultural methods and lessen the demand for chemical fertilizers, rhizosphere microorganisms help in nutrient exchange. This improves crop yields and maintains the general well-being of agroecosystems (Moo-Young et al., 1993). The diverse

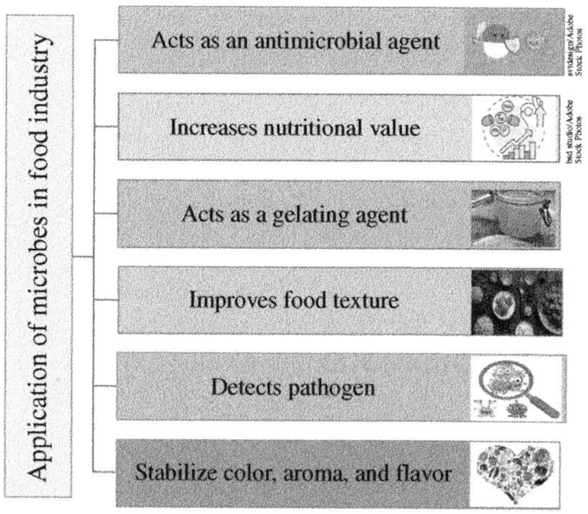

Figure 7.3 Role of microorganisms in the food industry.

roles of microorganisms in the food industry are illustrated in Figure 7.3. Microbes play a crucial role in the precision fermentation process used to produce food components and additives, in addition to maintaining the health of the soil. Utilizing modified microorganisms, this method maximizes the synthesis of target chemicals, including vitamins, sweeteners, fragrance compounds, nutraceuticals, and flavor enhancers (Caplice & Fitzgerald, 1999).

Compared to conventional chemical synthesis or plant-based extraction techniques, the microbial generation of these compounds provides a more sustainable option (Mazhar et al., 2022). Because of their high protein content and availability of omega-3 fatty acids, microalgae are also showing promise as a sustainable food source. This makes them a desirable substitute for traditional animal-based proteins.

Furthermore, microorganisms contribute significantly to food preservation and processing. Their ability to thrive with minimal resource inputs enhances their potential for sustainable food production. Yeasts, for example, are widely used in fermentation processes like bread and beer production, helping preserve flavor, extend shelf life, reduce food waste, and improve resource efficiency, all of which support environmental sustainability (Hannon et al., 2010).

7.4.3 Use of Microorganisms in Pharmaceuticals

Microorganisms are essential to the pharmaceutical industry, particularly in the manufacturing of medications and dietary supplements. In medical history, the discovery of antibiotics – which come from a variety of microbiological origins and are a class of synthetic compounds – is a significant accomplishment. Microorganisms, including bacteria, actinomycetes, and fungi, have been crucial in the development of a broad spectrum of antibiotics effective against pathogenic bacteria (Barzee et al., 2021). Microorganisms are also employed as vectors in vaccines, triggering immune responses and conferring protection against future infections. Probiotics, such as *Streptococcus* bacteria and *Lactobacillus*, commonly found in fermented dairy and other food products, offer documented benefits for gastrointestinal health and overall well-being. Furthermore, the pharmaceutical industry utilizes microorganisms derived from algae and fungi to produce dietary supplements (Baker, 2018). The COVID-19 pandemic highlighted the crucial role microorganisms play in vaccine development, especially for vaccines that utilize bacterial and viral vector systems. Human health depends on the development of vaccinations, drugs, and supplements using microbial technologies. But it also brings up environmental issues, especially when it comes to how medical waste is disposed of. In an effort to resolve these challenges, ongoing studies are examining how microorganisms can be employed in bioremediation processes to treat waste materials and remove harmful substances from contaminated water and soil (Sharma et al., 2022). To conclude, microorganisms are vital in the creation of pharmaceuticals, vaccines, and antibiotics, thus playing a crucial role in human well-being. Nonetheless, sustainable methods must be employed to address the environmental issues that arise from their application (Kaushik & Kapila, 2009).

7.5 Cultural Services

Soil, a diverse and vital component of terrestrial ecosystems, provides numerous nonmaterial services and benefits, collectively known as cultural services (Motiejūnaitė et al., 2019). Soil biodiversity contributes to a range of ecological and cultural services, including opportunities for leisure, social interaction, a sense of belonging, intellectual development, artistic expression, and introspection. In addition to activities like gathering wild herbs and watching wildlife, these services highlight the inherent relevance of nature in human civilizations and well-being. Participating in outdoor activities promotes the expanding nature tourism sector and has several health advantages. While the valuation and assessment of these cultural and environmental benefits can be challenging, scholars recognize their essential role in ecosystem preservation (Fish et al., 2016). A focus on social indicators, such as artistic endeavors, nature tourism, clothing, design, and outdoor recreation for health benefits, can contribute to making these often-intangible concepts more readily appreciated (Satterfield et al., 2013). Soil ecosystems play a vital role in safeguarding artifacts and built structures from past human activities, acting as both cultural and natural archives. Soils themselves can be viewed as natural monuments, showcasing distinct structures and offering protection against human disturbances or natural destructive forces (Bouma, 1989). Healthy soils consistently perform a range of critical ecological functions. The preservation of historical artifacts emphasizes the need for maintaining a record of both cultural and natural heritage, including paleosols (ancient soils), which offer valuable contributions to earth sciences, ecology, soil classification, and pedological development, as well as to historical and cultural contexts and economic and social value. Paleosols, in particular, provide crucial insights into past environments, soil processes, and earth sciences, making them an essential resource. Such inventories should also include preservation plans for key pedosites (important soil locations).

A variety of soil properties, including depth, texture, acidity, moisture content, permeability, and nutrient levels, all contribute to the preservation of cultural property by preventing its deterioration and mineralization. These soil features play a significant role in preserving historical legacies, serving as invaluable archives for both cultural and natural history. However, land use changes present a significant risk to these cultural services, often resulting in soil degradation and oxidation. The knowledge derived from studying these preserved artifacts, along with the aesthetic appreciation of cultural heritage, provides indirect benefits to humanity (Costantini, 2023).

7.6 Conclusion

This chapter has highlighted the diverse and essential benefits that soil ecosystem services provide to both human populations and the environment, emphasizing their crucial role in fostering sustainable bio- and circular economies. The value of many soil-derived ecosystem services is immeasurable, as they are foundational to life itself. However, the contributions of the soil's biotic components often go underappreciated due to their perceived disconnect from market systems and the challenges associated with quantifying them. Although 60% of the value of ecosystem services worldwide is provided by terrestrial ecosystems, the application of efficient soil management techniques is frequently hampered by a lack of thorough knowledge about the complex relationships between soil health, management techniques, and the advantages they offer. Despite the difficulties in assigning precise value to these services, the undeniable social and economic repercussions of global soil degradation are clear. Ensuring long-term human well-being requires active conservation and improvement of the Earth's soil resources is essential to shift agricultural, forestry, and grazing systems toward more sustainable methods. This transition will not only protect the crucial ecosystem services that soil provides but also reduce the strain on other vital ecological functions. Embracing a biomaterial-based circular economy presents a promising solution for addressing climate change. Therefore, to achieve effective soil protection, it is essential that environmental risk assessments fully account for the broad spectrum of ecosystem services offered by microbial communities.

References

Acevedo-Rocha, C.G., Gronenberg, L.S., Mack, M., Commichau, F.M., & Genee, H.J. 2019. Microbial cell factories for the sustainable manufacturing of B vitamins. *Current Opinion in Biotechnology* 56, 18–29.

Adhikari, K., & Hartemink, A.E. 2016. Linking soils to ecosystem services – A global review. *Geoderma* 262, 101–111

Akinsemolu, A.A. 2018. The role of microorganisms in achieving the sustainable development goals. *Journal of cleaner production* 182, 139–155.

Ancona, V., Caracciolo, A.B., Grenni, P., Di Lenola, M., Campanale, C., Calabrese, A., Uricchio, VF., Mascolo, G., & Massacci, A. 2017. Plant-assisted bioremediation of a historically PCB and heavy metal-contaminated area in Southern Italy. *New Biotechnology* 38, 65–73.

Azam, F., & Malfatti, F. 2007. Microbial structuring of marine ecosystems. *Nature Reviews Microbiology* 5(10), 782–791.

Baez-Rogelio. A., Morales-García, Y.E., Quintero-Hernandez, V., & Munoz-Rojas, J. 2017. Next generation of microbial inoculants for agriculture and bioremediation. *Microbial Biotechnology* 10(1), 19–21.

Baker, S.E. 2018. Protein hyper production in fungi by design. *Applied Microbiology and Biotechnology* 102(20): 8621–8628.

Bamdad, H., Papari, S., Lazarovits, G., & Berruti, F. 2022. Soil amendments for sustainable agriculture: microbial organic fertilizers. *Soil Use and Management* 38(1), 94–120.

Bardgett, R.D., Freeman, C., & Ostle, N.J. 2008 Microbial contributions to climate change through carbon cycle feedbacks. *The ISME Journal* 2(8), 805–814.

Barzee, T.J., Cao, L., Pan, Z., & Zhang, R., 2021. Fungi for future foods. *Journal of Future Foods* 1(1), 25–37.

Bennett, L.T., Mele, P.M., Annett, S., & Kasel, S. 2010. Examining links between soil management, soil health, and public benefits in agricultural landscapes: an Australian perspective. *Agriculture, Ecosystems and Environment* 139(1-2), 1–2.

Berg, B., & McClaugherty, C. 2020. Decomposition as a process – some main features. In: *Plant litter: Decomposition, Humus Formation, Carbon Sequestration* (pp. 13–43). Springer Verlag.

Bin, L.I., Ye, C.H., Lijun, Z.H., & Ruidong, Y.A., 2008. Effect of microbial weathering on carbonate rocks. *Earth Science Frontiers* 15(6), 90–99.

Blaya, J., Marhuenda, F.C., Pascual, J.A., & Ros, M. 2016. Microbiota characterization of compost using omics approaches opens new perspectives for Phytophthora root rot control. *PloS One* 11(8), e0158048.

Bodelier, P.L. 2011. Toward understanding, managing, and protecting microbial ecosystems. *Frontiers in microbiology* 2, 80.

Boivin, M.E., Breure, A.M., Posthuma, L., & Rutgers, M. 2002. Determination of field effects of contaminants – significance of pollution-induced community tolerance. *Human and Ecological Risk Assessment* 8(5), 1035–1055.

Bouma, J. 1989. Using soil survey data for quantitative land evaluation. *Advances in Soil Science* 9, 177–213.

Bratteler, M., Lexer, C., & Widmer, A. 2006. Genetic architecture of traits associated with serpentine adaptation of Silene vulgaris. *Journal of Evolutionary Biology* 19, 1149–1156.

Caplice, E., & Fitzgerald, G.F. 1999. Food fermentations: role of microorganisms in food production and preservation. *International Journal of Food Microbiology* 50, 131–149.

Carrondo, M.A. 2003. Ferritins, iron uptake and storage from the bacterioferritin viewpoint. *The EMBO Journal* 22, 1959–1968

Cazorla, F.M., & Mercado-Blanco, J. 2016. Biological control of tree and woody plant diseases: an impossible task? *BioControl* 61, 233–242.

Chahal, G.K., Kaur, A., & Ghai, N. 2022. Mitigation of salt stress with *Azospirillium* and *Azotobacter* inoculation in maize (*Zea mays* L.). *Cereal Research Communications* 50, 915–927.

Chandra, R., & Sobti, R. (Eds.). 2020. *Microbes for Sustainable Development and Bioremediation* (1st ed.). CRC Press. https://doi.org/10.1201/9780429275876

Chojnacka, K., Moustakas, K., & Witek-Krowiak, A. 2020. Bio-based fertilizers: a practical approach towards circular economy. *Bioresource Technology* 295, 122223.

Ciancio, A., Pieterse, C.M., & Mercado-Blanco, J. 2016. Harnessing useful rhizosphere microorganisms for pathogen and pest biocontrol. *Frontiers in Microbiology* 10, 1935.

Comeau, D., Balthazar, C., Novinscak, A., Bouhamdani, N., Joly, D.L., & Filion, M. 2021. Interactions between *Bacillus* spp., *Pseudomonas* spp. and *Cannabis sativa* promote plant growth. *Frontiers in Microbiology* 12, 715758.

Compant, S., Duffy, B., Nowak, J., Clément, C., & Barka, E.A. 2005. Use of plant growth-promoting bacteria for biocontrol of plant diseases: principles, mechanisms of action, and future prospects. *Applied and Environmental Microbiology* 71(9), 4951–4959.

Costantini, E.A. 2023. Possible policies and actions to protect the soil cultural and natural heritage of Europe. *Geoderma Regional* 32, e00599.

Cotrufo, M.F., Wallenstein, M.D., Boot, C.M., Denef, K., & Paul, E. 2013. The Microbial efficiency-matrix stabilization (MEMS) framework integrates plant litter decomposition with soil organic matter stabilization: do labile plant inputs form stable soil organic matter? *Global Change Biology* 19(4), 988–995.

De Groot, R.S., Wilson, M.A., & Boumans, R.M. 2002. A typology for the classification, description and valuation of ecosystem functions, goods and services. *Ecological Economics* 41(3), 393–408.

Díaz, S., Fargione, J., Chapin III, F.S., & Tilman, D. 2006. Biodiversity loss threatens human well-being. *Plos Biology* 4(8), e277.

Ding, W., Chang, N., Zhang, G., Kang, J., Yi, X., Zhang, J., Zhang, J., Wang, L., & Li, H. 2023. Soil organic carbon changes in China's croplands: a newly estimation based on DNDC model. *Science of The Total Environment* 905, 167107.

Drigo, B., Kowalchuk, G.A., Yergeau, E., Bezemer, T.M., Boschker, H.T., & Van Veen, J.A., 2007. Impact of elevated carbon dioxide on the rhizosphere communities of *Carex arenaria* and *Festuca rubra*. *Global Change Biology* 13(11), 2396–2410.

Dungait, J.A., Hopkins, D.W., Gregory, A.S., & Whitmore, A.P. 2012. Soil organic matter turnover is governed by accessibility not recalcitrance. *Global Change Biology* 18(6), 1781–1796.

El-Ramady, H., Abdalla, N., Badgar, K., Llanaj, X., Törős, G., Hajdú, P., Eid, Y., & Prokisch, J. 2022. Edible mushrooms for sustainable and healthy human food: nutritional and medicinal attributes. *Sustainability* 14(9), 4941.

Eyers, L., George, I., Schuler, L., Stenuit, B., Agathos, S.N., & El Fantroussi, S. 2004. Environmental genomics: exploring the unmined richness of microbes to degrade xenobiotics. *Applied Microbiology and Biotechnology* 66: 123–130.

Falkowski, P.G., Fenchel, T., & Delong, E.F. 2008. The microbial engines that drive Earth's biogeochemical cycles. *Science* 320, 1034–1039.

Fierer, N., Grandy, A.S., Six, J., & Paul, E.A. 2009. Searching for unifying principles in soil ecology. *Soil Biology and Biochemistry* 41(11), 2249–2256.

Fierer, N. 2017. Embracing the unknown: disentangling the complexities of the soil microbiome. *Nature Reviews Microbiology* 15(10), 579–590.

Fish, R., Church, A., & Winter, M. 2016. Conceptualising cultural ecosystem services: a novel framework for research and critical engagement. *Ecosystem Services* 21, 208–217.

Floch, C., Chevremont, A.C., Joanico, K., Capowiez, Y., & Criquet, S. 2011. Indicators of pesticide contamination: soil enzyme compared to functional diversity of bacterial communities via Biolog® Ecoplates. *European Journal of Soil Biology* 47(4), 256–263.

Fontaine, S., Henault, C., Aamor, A., Bdioui, N., Bloor, J.M., Maire, V., Mary, B., Revaillot, S., & Maron, P.A. (2011). Fungi mediate long term sequestration of carbon and nitrogen in soil through their priming effect. *Soil Biology and Biochemistry* 43, 86–96.

Franchi, E., Agazzi, G., Rolli, E., Borin, S., Marasco, R., Chiaberge, S., Conte, A., Filtri, P., Pedron, F., Rosellini, I., & Barbafieri, M. 2016. Exploiting hydrocarbon-degrading indigenous bacteria for bioremediation and phytoremediation of a multicontaminated soil. *Chemical Engineering and Technology* 39(9), 1676–1684.

Fu, C., Ma, W., Qiang, B., Jin, X., Zhang, Y., & Wang, M. 2023. Effect of chemical fertilizer with compound microbial fertilizer on soil physical properties and soybean yield. *Agronomy* 13(10), 2488.

Gadd, G.M. 2010. Metals, minerals and microbes: geomicrobiology and bioremediation. *Microbiology* 156(3), 609–643.

Gavrilescu, M. 2021. Water, soil, and plants interactions in a threatened environment. *Water* 13(19), 2746.

Ghee, C., Neilson, R., Hallett, P.D., Robinson, D., & Paterson, E. 2013. Priming of soil organic matter mineralisation is intrinsically insensitive to temperature. *Soil Biology and Biochemistry* 66, 20–28.

Goyal, S., Dhanker, R., Hussain, T., Ferreira, A., Gouveia, L., Kumar, K., & Mohamed HI. 2023. Modern advancement in biotechnological applications for wastewater treatment through microalgae: a review. *Water, Air and Soil Pollution* 234(7), 417.

Gupta, A., Gupta, R., & Singh, R.L. 2016. Microbes and environment. *Principles and Applications of Environmental Biotechnology for a Sustainable Future* 15, 43–84.

Hannon, M., Gimpel, J., Tran, M., Rasala, B., & Mayfield, S. 2010. Biofuels from algae: challenges and potential. *Biofuels* 1, 763–784

Hayat, R., Ali, S., Amara, U., Khalid, R., & Ahmed, I. 2010. Soil beneficial bacteria and their role in plant growth promotion: a review. *Annals of Microbiology* 60, 579–598.

Helfrich, M., Ludwig, B., Thoms, C., Gleixner, G., & Flessa, H. 2015. The role of soil fungi and bacteria in plant litter decomposition and macroaggregate formation determined using phospholipid fatty acids. *Applied Soil Ecology* 96, 261–264.

IPCC. 2014. Climate change 2014 synthesis report. In: R.K. Pachauri, & L.A. Meyer (Eds.), *Contribution of Working Groups I, II and III to the Fifth Assessment Report of the Intergovernmental Panel on Climate Change* (pp. 1–151). Geneva, Switzerland.

Jakienė, E., Spruogis, V., Romaneckas, K., Dautarte, A., & Avizienyte, D. 2015. The bio-organic nano fertilizer improves sugar beet photosynthesis process and productivity. *Zemdirbyste-Agriculture* 102(2), 141–146.

Janssen, P.H. 2006. Identifying the dominant soil bacterial taxa in libraries of 16S rRNA and 16S rRNA genes. *Applied and Environmental Microbiology* 72(3), 1719–1728.

Jatav, M.K., Kumar, M., Trehan, S.P., Dua, V.K., Lal, S.S., & Sharma, R.P. 2012. Influence of microorganisms inoculation on nutrient economy in Potatoradish crop sequence in North Western Himalayas. *Vegetable Science* 39(1), 21–25.

Jin, N., Jin, L., Wang, S., Li, J., Liu, F., Liu, Z., Luo, S., Wu, Y., Lyu, J., & Yu, J. 2022. Reduced chemical fertilizer combined with bio-organic fertilizer affects the soil microbial community and yield and quality of lettuce. *Frontiers in Microbiology* 13, 863325.

Kaushik, K., & Kapila, K. 2009. Women in medical microbiology: reflections on contributions. *Indian Journal of Medical Microbiology* 27(4), 285.

Kiruba, N., & Saeid, A. 2022. An insight into microbial inoculants for bioconversion of waste biomass into sustainable "bio-organic" fertilizers: a bibliometric analysis and systematic literature review. *International Journal of Molecular Sciences* 23: 13049.

Kobierska, F., Jonas, T., Magnusson, J., Zappa, M., Bavay, M., Bosshard, T., Paul, F., & Bernasconi, S.M. 2011. Climate change effects on snow melt and discharge of a partly glacierized watershed in Central Switzerland (SoilTrec Critical Zone Observatory). *Applied Geochemistry* 26, S60–S62.

Kuypers, M.M., Marchant, H.K., & Kartal, B. 2018. The microbial nitrogen-cycling network. *Nature Reviews Microbiology* 16(5), 263–276.

Kramer, M.G., & Chadwick, O.A. 2018. Climate-driven thresholds in reactive mineral retention of soil carbon at the global scale. *Nature Climate Change* 8(12): 1104–1108.

La Notte, A., Maes, J., Dalmazzone, S., Crossman, N.D., Grizzetti, B., & Bidoglio, G. 2017. Physical and monetary ecosystem service accounts for Europe: a case study for in-stream nitrogen retention. *Ecosystem Services* 23, 18–29.

Lal, R. 2008. Carbon sequestration. Philosophical Transactions of the Royal Society B: Biological Sciences 363(1492), 815–830.

Lapointe, N.W., Cooke, S.J., Imhof, J.G., Boisclair, D., Casselman, J.M., Curry, R.A., Langer, O.E., McLaughlin, R.L., Minns, C.K., Post, J.R., & Power, M. 2014. Principles for ensuring healthy and productive freshwater ecosystems that support sustainable fisheries. *Environmental Reviews* 22(2), 110–134.

Lavallee, J.M., Soong, J.L., & Cotrufo, M.F. 2020. Conceptualizing soil organic matter into particulate and mineral-associated forms to address global change in the 21st century. *Global Change Biology* 26(1), 261–273.

Li, X., Li, Y., Zhang, Z., & Wang, J. 2021. Distribution of soil bacterial communities—Actinobacteria, Proteobacteria, Acidobacteria, and genera *Arthrobacter* and *Paenibacillus*. *Journal of Environmental Microbiology* 15(4), 123–134.

Li, F., Wang, R., Hu, D., Ye, Y., Yang, W., & Liu, H. 2014. Measurement methods and applications for beneficial and detrimental effects of ecological services. *Ecological Indicators* 47, 102–111.

Li, J., Ovakim, D.H., Charles, T.C., & Glick, B.R. 2000. An ACC deaminase minus mutant of Enterobacter cloacae UW4No longer promotes root elongation. *Current Microbiology* 41, 101–105.

Li, W., Zhang, F., Cui, G., Wang, Y., Yang, J., Cheng, H., Liu, H., & Zhang, L. 2021. Effects of bio-organic fertilizer on soil fertility, microbial community composition, and potato growth. *Science Asia* 47, 347.

Liang, C., Schimel, J.P., & Jastrow, J.D. 2017. The importance of anabolism in microbial control over soil carbon storage. *Nature Microbiology* 2(8), 1–6.

Liang, C., Zhang, X., & Balser, T.C. 2007. Net microbial amino sugar accumulation process in soil as influenced by different plant material inputs. *Biology and Fertility of Soils* 44, 1–7.

Loganathan, P., Hedley, M.J., Grace, N.D., Lee, J., Cronin, S.J., Bolan, N.S., & Zanders, J.M. 2003. Fertiliser contaminants in New Zealand grazed pasture with special reference to cadmium and fluorine-a review. *Soil Research* 41(3), 501–532.

Lorenz, K., & Lal, R. 2014. Soil organic carbon sequestration in agroforestry systems. a review. *Agronomy for Sustainable Development* 34, 443–454.

Luo, W., D'Angelo, E.M., & Coyne, M.S. 2008. Organic carbon effects on aerobic polychlorinated biphenyl removal and bacterial community composition in soils and sediments. *Chemosphere* 70(3), 364–373.

Ma, L., Zhang, H.Y., Zhou, X.K., Yang, C.G., Zheng, S.C., Duo, J.L., & Mo, M.H. 2018a. Biological control tobacco bacterial wilt and black shank and root colonization by bio-organic fertilizer containing bacterium *Pseudomonas aeruginosa* NXHG29. *Applied Soil Ecology* 129, 136–144.

Ma, T., Zhu, S., Wang, Z., Chen, D., Dai, G., Feng, B., Su, X., Hu, H., Li, K., Han, W., & Liang, C. 2018b. Divergent accumulation of microbial necromass and plant lignin components in grassland soils. *Nature Communications* 9(1), 3480.

Macías, F., & Camps Arbestain, M. 2010. Soil carbon sequestration in a changing global environment. *Mitigation and Adaptation Strategies for Global Change* 15(6), 511–529.

Mali, R.H., Ghante, P.H., Magar, S.J., Patil, N.H., & Kuthumbare, R.S. 2024. Evaluation of different microbial consortia for decomposing different wastes: an In vitro and In vivo study approach. *Journal of Advances in Biology & Biotechnology* 27(11), 1140–1147.

Marschner, P. 2007. Plant microbe interactions in the rhizosphere and nutrient cycling. In P. Marschner, & Z. Rengel (Eds.), *Nutrient cycling in terrestrial ecosystems* (Soil biology series, pp. 159–182). Heidelberg: Springer

Martin, J.P., & Haider, K. 1971. Microbial activity in relation to soil humus formation. *Soil Science* 111(1), 54–63.

Marx, V. 2017. Microbiology: the return of culture. *Nature Methods* 14(1), 37–40.

Mazhar, S., Yasmeen, R., Chaudhry, A., Summia, K., Ibrar, M., Amjad, S., & Ali, E. 2022. Role of Microbes in modern food industry. *International Journal of Innovations in Science and Technology* 4(1), 65–77.

Mendes, R., Kruijt, M., De Bruijn, I., Dekkers, E., Van Der Voort, M., Schneider, J.H., Piceno, Y.M., DeSantis, T.Z., Andersen, G.L., Bakker, P.A., & Raaijmakers, J.M. 2011. Deciphering the rhizosphere microbiome for disease-suppressive bacteria. *Science* 332(6033), 1097–1100.

Mitter, E.K., Tosi, M., Obregón, D., Dunfield, K.E., & Germida J.J. 2021. Rethinking crop nutrition in times of modern microbiology: innovative biofertilizer technologies. *Frontiers in Sustainable Food Systems* 5, 606815.

Moo-Young, M., Chisti, Y., & Vlach, D. 1993. Fermentation of cellulosic materials to mycoprotein foods. *Biotechnology Advances* 11(3), 469–479.

Motiejūnaitė, J., Børja, I., Ostonen, I., Bakker, M.R., Bjarnadottir, B., Brunner, I., Irsėnaite, R., Mrak, T., Oddsdottir, E.S., & Lehto, T. 2019. Cultural ecosystem services provided by the biodiversity of forest soils: a European review. *Geoderma* 343, 19–30.

Mulligan, C.N., & Yong, R.N. 2004. Natural attenuation of contaminated soils. *Environment International* 30(4), 587–601.

Murphy, S.P., & Allen, L.H. 2003. Nutritional importance of animal source foods. *The Journal of Nutrition* 133(11), 3932S–3935S.

Nies, D. H. 1999. Microbial heavy-metal resistance. *Applied Microbiology and Biotechnology*, 51(6), 730–750.

Nunes, P.S.O., Lacerda-Junior, G.V., Mascarin, G.M., Guimaraes, R.A., Medeiros, F.H.V., Arthurs, S., & Bettiol, W. 2024. Microbial consortia of biological products: do they have a future? *Biological Control* 188, 105439.

O'Callaghan, M., Ballard, R.A., & Wright, D. 2022. Soil microbial inoculants for sustainable agriculture: limitations and opportunities. *Soil Use and Management* 38(3), 1340–1369.

Pandey, D. 2020. Agricultural sustainability and climate change nexus. In: P. Singh, R. Singh, & V. Srivastava (Eds.), *Contemporary Environmental Issues and Challenges in Era of Climate Change* (pp. 77–97). Springer Nature.

Paul, D., Pandey, G., Meier, C., Roelof van der Meer, J., & Jain, R.K. 2006. Bacterial community structure of a pesticide-contaminated site and assessment of changes induced in community structure during bioremediation. *FEMS Microbiology Ecology* 57(1), 116–127.

Philippot, L., Ritz, K., Pandard, P., Hallin, S., & Martin-Laurent, F. 2012. Standardisation of methods in soil microbiology: progress and challenges. *FEMS Microbiology Ecology* 82(1), 1–10.

Pino, N., & Penuela, G. 2011. Simultaneous degradation of the pesticides methyl parathion and chlorpyrifos by an isolated bacterial consortium from a contaminated site. *International Biodeterioration and Biodegradation* 65(6), 827–831.

Podile, A., & Kishore, G. 2006. Plant growth-promoting rhizobacteria. In: *Plant-Associated Bacteria* (pp.195–230). Springer Nature.

Polasky, S., Nelson, E., Pennington, D., & Johnson, K.A. 2011. The impact of land-use change on ecosystem services, biodiversity and returns to landowners: a case study in the state of Minnesota. *Environmental and Resource Economics* 48, 219–242.

Prasad, R., Kumar, M., & Varma, A. 2015. Role of PGPR in soil fertility and plant health. In: D. Egamberdieva, S. Shrivastava, & A. Varma (Eds.), *Plant-Growth-Promoting Rhizobacteria (PGPR) and Medicinal Plants* (pp. 247–260). Cham: Springer.

Pries, C.E., Castanha, C., Porras, R.C., & Torn, M.S. 2017. The whole-soil carbon flux in response to warming. *Science* 355, 1420–1423.

Pulleman, M., Creamer, R., Hamer, U., Helder, J., Pelosi, C., Peres, G., & Rutgers, M. 2012. Soil biodiversity, biological indicators and soil ecosystem services – an overview of European approaches. *Current Opinion in Environmental Sustainability* 4(5), 529–538.

Quijas, S., Schmid, B., & Balvanera, P. 2010. Plant diversity enhances provision of ecosystem services: a new synthesis. *Basic and Applied Ecology* 11(7), 582–593.

Razafindratsima, O.H., Kamoto, J.F., Sills, E.O., Mutta, D.N., Song, C., Kabwe, G., Castle, S.E., Kristjanson, P.M., Ryan, C.M., Brockhaus, M., & Sunderland, T. 2021. Reviewing the evidence on the roles of forests and tree-based systems in poverty dynamics. *Forest Policy and Economics* 131, 102576.

Rillig, M.C., Ryo, M., Lehmann, A., Aguilar-Trigueros, C.A., Buchert, S., Wulf, A., Iwasaki, A., Roy, J., & Yang, G. 2019. The role of multiple global change factors in driving soil functions and microbial biodiversity. *Science* 366(6467), 886–890.

Ritz, K., Black, H.I., Campbell, C.D., Harris, J.A., & Wood, C. 2009. Selecting biological indicators for monitoring soils: a framework for balancing scientific and technical opinion to assist policy development. *Ecological Indicators* 9(6), 1212–1221.

Rodriguez, A.M.D., Cota, F.I.P., Chavez, L.A.C., Ortega, L.F.G., Alvarado, M.I.E., Santoyo, G., & Villalobos, S.S. 2025. Microbial inoculants in sustainable agriculture: advancements, challenges, and future directions. *Plants* 14(2), 191.

Rutgers, M., Van Wijnen, H.J., Schouten, A.J., Mulder, C., Kuiten, A.M., Brussaard, L., & Breure, A.M. 2012. A method to assess ecosystem services developed from soil attributes with stakeholders and data of four arable farms. *Science of the Total Environment* 415, 39–48.

Sacca, M.L., Barra Caracciolo, A., Di Lenola, M., & Grenni, P. 2017. Ecosystem services provided by soil microorganisms. In: M. Lukac, P. Grenni, & M. Gamboni (Eds.), *Soil Biological Communities and Ecosystem Resilience* (pp. 9–24). Springer International Publishing.

Satterfield, T., Gregory, R., Klain, S., Roberts, M., & Chan, K.M. 2013. Culture, intangibles and metrics in environmental management. *Journal of Environmental Management* 117, 103–114.

Schoenholtz, S.H., Van Miegroet, H., & Burger, J.A. 2000. A review of chemical and physical properties as indicators of forest soil quality: challenges and opportunities. *Forest Ecology and Management* 138(1-3), 335–356.

Sharma, P., Bano, A., Singh, S.P., Dubey, N.K., Chandra, R., & Iqbal, H.M. 2022. Recent advancements in microbial-assisted remediation strategies for toxic contaminants. *Cleaner Chemical Engineering* 2, 100020.

Shi, H., Lu, L., Ye, J., & Shi, L. 2022. Effects of two *Bacillus velezensis* microbial inoculants on the growth and rhizosphere soil environment of *Prunus davidiana*. *International Journal of Molecular Sciences* 23(21) 13639.

Shoebitz, M., Ribaudo, C.M., Pardo, M.A., Cantore, M.L., Ciampi, L., & Cura, J.A. 2009. Plant growth promoting properties of a strain of *Enterobacter ludwigii* isolated from *Lolium perenne* rhizosphere. *Soil Biology and Biochemistry* 41(9), 1768–1774.

Shukla, S.K., Sharma, L., Jaiswal, V.P., Pathak, A.D., Tiwari, R., Awasthi, S.K., & Gaur, A. 2020. Soil quality parameters vis-a-vis growth & yield attributes of sugarcane as influenced by integration of microbial consortium with NPK fertilizers. *Scientific Reports* 10(1), 19180.

Silver, S., & Phung, L.T. 1996. Bacterial heavy metal resistance: new surprises. *Annual Review of Microbiology* 50, 753–789.

Simranjit, K., Kanchan, A., Prasanna, R., Ranjan, K., Ramakrishnan B, Singh, A.K., & Shivay, Y.S. 2016. Microbial inoculants as plant growth stimulating and soil nutrient availability enhancing options for cucumber under protected cultivation. *World Journal of Microbiology and Biotechnology* 35, 1–4.

Sokolova, T.A. 2011. The role of soil biota in the weathering of minerals: a review of literature. *Eurasian Soil Science* 44, 56–72.

Spohn, M. 2016. Element cycling as driven by stoichiometric homeostasis of soil microorganisms. *Basic and Applied Ecology* 17(6), 471–478.

Tang, L. 2024. Soil fertility, plant nutrition and nutrient management. *Plants* 14(1), 34.

Thiele-Bruhn, S. 2021. The role of soils in provision of genetic, medicinal and biochemical resources. *Philosophical Transactions of the Royal Society B* 376, 20200183.

Tiedje, J.M., Asuming-Brempong, S., Nüsslein, K., Marsh, T.L., & Flynn, S.J. 1999. Opening the black box of soil microbial diversity. *Applied Soil Ecology* 13(2), 109–122.

Tieminie, R.N., Loh, C.E., Tieguhong, J.C., Nghobuoche, M.F., Mandiefe, P.S., & Tieguhong, M.R. 2021. Non-timber forest products and climate change adaptation among forest dependent communities in Bamboko forest reserve, southwest region of Cameroon. *Environmental Systems Research* 10, 1–3.

Turbé, A., De Toni, A., Benito, P., Lavelle, P.A., Lavelle, P.E., Ruiz, N., Van der Putten W.H., Labouze, E., & Mudgal, S. 2010. *Soil biodiversity: Functions, Threats and Tools for Policy Makers* (pp. 1–250). Bio Intelligence Service.

Verstraeten, A., Neirynck, J., Genouw, G., Cools, N., Roskams, P., & Hens, M. 2012. Impact of declining atmospheric deposition on forest soil solution chemistry in Flanders, Belgium. *Atmospheric Environment* 62, 50–63.

Wang, X., Yang, Y., Gao, B., Wan, Y., Li, Y.C., Xie, J., & Tang, Y. 2021. Slow-released bio-organic–chemical fertilizer improved tomato growth: synthesis and pot evaluations. *Journal of Soils and Sediments*. 21, 319–327.

Xiao, X., Li, J., Lyu, J., Feng, Z., Zhang, G., Yang, H., Gao, C., Jin, L., & Yu, J. 2022. Chemical fertilizer reduction combined with bio-organic fertilizers increases cauliflower yield via regulation of soil biochemical properties and bacterial communities in Northwest China. *Frontiers in Microbiology* 27, 13: 922149.

Xie, H., Wu, K., Iqbal, A., Ali, I., He, L., Ullah, S., Wei, S., Zhao, Q., Wu, X., Huang, Q., & Jiang, L. 2021. Synthetic nitrogen coupled with seaweed extract and microbial inoculants improves rice (*Oryza sativa* L.) production under a dual cropping system. *Italian Journal of Agronomy* 16, 1800.

Yang, W., Gong, T., Wang, J., Li, G., Liu, Y., Zhen, J., Ning, M., Yue, D., Du, Z., & Chen, G. 2020. Effects of compound microbial fertilizer on soil characteristics and yield of wheat (*Triticum aestivum* L.). *Journal of Soil Science and Plant Nutrition* 20, 2740–2748.

Yap, C.K., & Al-Mutairi, K.A. 2023. Effective microorganisms as halal-based sources for biofertilizer production and some socio-economic insights: a review. *Foods* 12(8), 1702.

Zengler, K., & Palsson, B.O. 2012. A road map for the development of community systems (CoSy) biology. *Nature Reviews Microbiology* 10(5), 366–372.

Zhao, J., Wang, Y., Liang, H., Huang, J., Chen, Z., & Nie, Y. 2018. The rhizosphere microbial community response to a bio-organic fertilizer: finding the mechanisms behind the suppression of watermelon *Fusarium wilt* disease. *Acta Physiologiae Plantarum* 40, 1–14.

8

Identification of Plant-derived Microbial Consortium Resistant to Biotic Stress

Ashwini Ananda[1], Aditya Kumar[2], Abha Sharma[2], Victor Thingujam[3], and Aashita Shree[2]*

[1] *AICRP-Vegetable Crops, School of Agricultural Sciences, Nagaland University, Medziphema, Nagaland, India*
[2] *Amity Institute of Organic Agriculture, Amity University Noida, Noida, Uttar Pradesh, India*
[3] *ICAR Research Complex for Eastern Region-Farming System Research Centre for Hill and Plateau Region, Ranchi, Jharkhand, India*

* Corresponding author: Aditkumar302@gmail.com

8.1 Introduction

Biotic stress in plants, caused by pathogens, pests, nematodes, and weeds, poses significant challenges to agricultural productivity and plant health. These stressors can lead to increased reactive oxygen species, affecting the physiological and molecular functions of plants, ultimately reducing crop yields (Chaudhary et al., 2022). Plants have evolved complex interactions with their surrounding environment, including the modulation of root exudates to communicate with and alter the rhizospheric microbial community, which can enhance stress tolerance (Sharma et al., 2023). The plant-associated microbiota plays a crucial role in stress resistance. These microbial communities, including bacteria, fungi, and other microorganisms, assist in nutrient uptake, modulate plant metabolism, and provide protection against both biotic and abiotic stresses (Omae & Tsuda, 2022; Singh et al., 2023). The presence of beneficial microbes can lead to improved plant growth and health by enhancing nutrient acquisition, regulating phytohormones, and inducing systemic resistance (Khanna et al., 2022).

The concept of a microbial consortium, which involves the use of multiple microbial strains, offers advantages over single-strain inoculants. These consortia can provide synergistic effects, enhancing plant growth and stress resistance more effectively than individual strains. For instance, mycorrhizal fungi and plant growth-promoting rhizobacteria (PGPRs) can interact synergistically to improve plant health and stress tolerance (Santoyo et al., 2021a). This approach allows for a more comprehensive and robust response to environmental challenges. Plant-derived microbial consortia are particularly relevant in sustainable

Microbial Consortia in Plant Science and Sustainable Agriculture, First Edition. Edited by Ajay Tomar, et al.

agriculture. They provide an environmentally sustainable alternative for chemical fertilizers and pesticides, enhancing plant health and yield while preserving soil fertility and mitigating environmental pollution (Ali et al., 2023; Chaudhary et al., 2022). Utilizing the inherent interactions between plants and their microbiota, sustainable agricultural methods may be formulated to improve crop resilience and yield amid escalating environmental pressures.

8.2 Microbial Diversity in Plant Systems

8.2.1 Types of Beneficial Plant-associated Microbes

8.2.1.1 Rhizospheric Bacteria (e.g. PGPRs)

Rhizospheric bacteria, specifically PGPRs, reside in the soil adjacent to the plant roots. These bacteria promote plant development by transforming unavailable nutrients into easily absorbed forms, hence enhancing nutrient uptake and boosting plant vitality (Das et al., 2022; Tian et al., 2020). PGPRs contribute to the induction of systemic resistance in plants, hence enhancing their capacity to withstand infections and environmental challenges (Yu et al., 2022; Almario et al., 2022). The relationship between plants and rhizospheric bacteria is essential for sustainable agriculture, since it reduces the reliance on chemical fertilizers and pesticides (Das et al., 2022).

8.2.1.2 Endophytic Microbes (Bacteria and Fungi)

Endophytic microorganisms, comprising both bacteria and fungi, inhabit plant tissues without causing damage. These microorganisms promote plant development and stress resilience by regulating physiological processes and improving nutrient absorption (Tian et al., 2020; Rivera-Vega et al., 2022). Endophytic fungi, for example, can shift plant resource allocation toward growth rather than resistance, thereby increasing tolerance to herbivory and other stresses (Rivera-Vega et al., 2022). The relationships between plants and endophytic microbes are mutualistic, characterized by nutrient exchange and signaling pathways that provide advantages to both organisms (Almario et al., 2022).

8.2.1.3 Phyllospheric Microbes (Leaf-associated Microbial Communities)

Phyllospheric microbes inhabit the aerial parts of plants, such as leaves, and form diverse microbial communities. These microbes can enhance plant immunity and stress tolerance by interacting with plant signaling pathways, such as those involving salicylic acid. The phyllosphere is a dynamic environment where beneficial microbes can be recruited by plants to enhance local and systemic resistance against pathogens (Sommer et al., 2024). Environmental factors and the plant's immune responses significantly influence interactions within the phyllosphere, which highlights the complexity of plant–microbe interactions in this particular region (Santoyo et al., 2021b).

8.2.2 Common Microbial Genera Associated with Plant Protection

Microbial genera are essential in plant protection, functioning as biocontrol agents against a range of plant pathogens. Genera such as *Bacillus, Pseudomonas, Streptomyces*, and *Rhizobium* are extensively documented for their plant-protective properties among bacteria.

Bacillus species are recognized for their capacity to produce secondary metabolites such as lipopeptides and polyketides, which contribute to the suppression of plant pathogens and the induction of systemic resistance in plants (Andrić et al., 2020; Dimkić et al., 2021). *Pseudomonas* species serve as effective biocontrol agents by producing compounds like phenazines and cyclic lipopeptides, which inhibit fungal pathogens and enhance plant health (Dimkić et al., 2021). *Streptomyces* and *Rhizobium* enhance plant protection via analogous mechanisms, such as antibiosis and nutrient competition (Pfeiffer et al., 2021).

Fungi, including *Trichoderma, Beauveria*, and mycorrhizal fungi, also play significant roles in plant protection. *Trichoderma* species are renowned for their mycoparasitic abilities, producing enzymes and secondary metabolites that degrade pathogen cell walls and enhance plant defense mechanisms (Poveda & Eugui, 2022; Anckaert et al., 2024; Guzmán-Guzmán et al., 2023). *Beauveria*, although primarily known as an entomopathogenic fungus, can also contribute to plant health by controlling insect pests that may otherwise facilitate pathogen entry (Guzmán-Guzmán et al., 2023). Mycorrhizal fungi, particularly from the arbuscular mycorrhizal group, establish symbiotic associations with plant roots, facilitating nutrient absorption and offering protection against soilborne pathogens via induced systemic resistance (ISR) (Anckaert et al., 2024).

8.2.3 Role of Microbial Consortia in Plant–Microbe Interactions

Microbial consortia play a vital role in plant–microbe interactions, significantly improving plant health and growth via multiple mechanisms. Consortia, consisting of various interacting microorganisms, provide additive or synergistic advantages that exceed those of individual microbial species. Microbial consortia in the rhizosphere, such as PGPB, and fungi like *Trichoderma* and Mycorrhizae, enhance nutrient availability, modulate phytohormones, and improve biocontrol and stress tolerance in plants (Santoyo et al., 2021a). The interactions within these consortia are intricate, involving mechanisms like siderophore-mediated iron scavenging, which mitigates pathogen infection and promotes plant health by restricting pathogen access to vital nutrients (Gu et al., 2020). Microbial consortia can induce systemic resistance in plants by activating defense pathways that inhibit pathogen growth and development (Negi et al., 2024). The development of effective microbial consortia necessitates the meticulous selection of compatible strains capable of synergistic interaction. The combination of *Streptomyces microflavus* and *Trichoderma harzianum* induces stronger resistance responses in plants than single inoculants, despite initial antagonism observed *in vitro* (Prigigallo et al., 2023). This underscores the significance of *in vivo* experiments for comprehending the complete potential of microbial consortia in enhancing plant health.

Additionally, microbial consortia play a significant role in sustainable agriculture by decreasing the reliance on chemical pesticides and fertilizers. They improve soil health through the promotion of biogeochemical cycling and the bioremediation of soil pollutants, thereby supporting sustainable agricultural practices (Wu et al., 2023). The establishment of stable and functional microbial consortia, particularly those sourced from the *Brachypodium* rhizosphere, highlights their potential in sustainable agriculture. This provides a basis for the assembly of effective consortia grounded in microbial interactions and colocalization (Chen et al., 2024).

8.3 Mechanisms of Biotic Stress Resistance Mediated by Microbial Consortia

8.3.1 Direct Mechanisms

The direct mechanisms of biocontrol agents in combating plant pathogens are multifaceted and involve several strategies. One of the primary methods is the production of antibiotics and antimicrobial compounds. Species such as *Bacillus* and *Pseudomonas* are recognized for their capacity to synthesize various lipopeptide compounds, including iturin, surfactin, and fengycin, which demonstrate efficacy against numerous plant pathogens (Dimkić et al., 2021). These compounds inhibit pathogen growth and induce systemic resistance in plants, thereby enhancing their defense mechanisms (Dimkić et al., 2021; Hashem et al., 2019).

Another significant mechanism is the secretion of hydrolytic enzymes, including chitinases, glucanases, and proteases. These enzymes play a crucial role in degrading the cell walls of fungal pathogens, thereby inhibiting their growth and spread. *Trichoderma* species, for example, are well-documented for their ability to produce such enzymes, which contribute to their effectiveness as biocontrol agents (Manzar et al., 2022; Pellan et al., 2021). The presence of fungal pathogen biomass often induces the production of these enzymes, highlighting their role in direct pathogen suppression (Dimkić et al., 2021).

Hyperparasitism and competition against pathogens are also vital direct mechanisms employed by biocontrol agents. Hyperparasitism involves the biocontrol agent directly attacking and parasitizing the pathogen, as seen in some *Trichoderma* species that exhibit mycoparasitism by synthesizing antifungal compounds and directly attacking pathogen structures (Pellan et al., 2021). Additionally, competition for nutrients and space is a common strategy where biocontrol agents outcompete pathogens for essential resources, thereby limiting pathogen proliferation. This competitive exclusion is a key feature of many biocontrol agents, including certain yeasts and bacteria, which can effectively suppress pathogen growth by occupying ecological niches and utilizing available nutrients more efficiently (Wu et al., 2024; Dimkić et al., 2021).

8.3.2 Indirect Mechanisms

ISR and the priming of plant defenses serve as crucial indirect mechanisms through which plants improve their immunity against pathogens. ISR takes place through beneficial microbes, including PGPRs, which engage with plant roots and trigger defense responses across the plant. This process encompasses intricate signaling pathways, primarily reliant on jasmonic acid and ethylene. Recent studies, however, have also underscored the role of salicylic acid and small RNAs in the regulation of ISR (Yu et al., 2022; Vlot et al., 2020; Meena et al., 2020). Priming entails the use of natural or synthetic compounds that enhance plants' ability to respond swiftly and effectively to pathogen attacks. The primed state may endure over time and influence plant parts that were not directly exposed to the priming agent, thereby enhancing systemic resistance (Hönig et al., 2023; Pršić & Ongena, 2020).

Microbial signaling molecules significantly modulate plant immune responses. Beneficial microbes synthesize various molecules, including quorum-sensing molecules, lipids, and proteins, that function as elicitors of ISR. These molecules initiate immune responses through interactions with plant receptors, resulting in the activation of defense mechanisms that limit pathogen growth (Nishad et al., 2020; Orozco-Mosqueda et al., 2023). The interaction between plants and beneficial microbes enhances plant immunity and affects the plant microbiome, which may result in additional alterations in plant defense and plant–microbe interactions (Vlot et al., 2020; Rabari et al., 2022).

Beneficial microbes utilize siderophore production as an indirect mechanism to bolster plant defenses. Siderophores are compounds that chelate iron, synthesized by microbes to sequester iron from the environment, thereby reducing its availability to pathogenic microbes. Competition for iron is essential, as iron serves as a critical nutrient for microbial growth. Siderophores suppress pathogen proliferation by limiting iron availability, thereby indirectly enhancing plant health and resistance (Meena et al., 2020; Romera et al., 2019). The integration of ISR, priming, microbial signaling, and siderophore production illustrates the advanced strategies employed by plants and their associated microbes to bolster immunity and secure survival against various biotic stresses as mentioned in Figure 8.1.

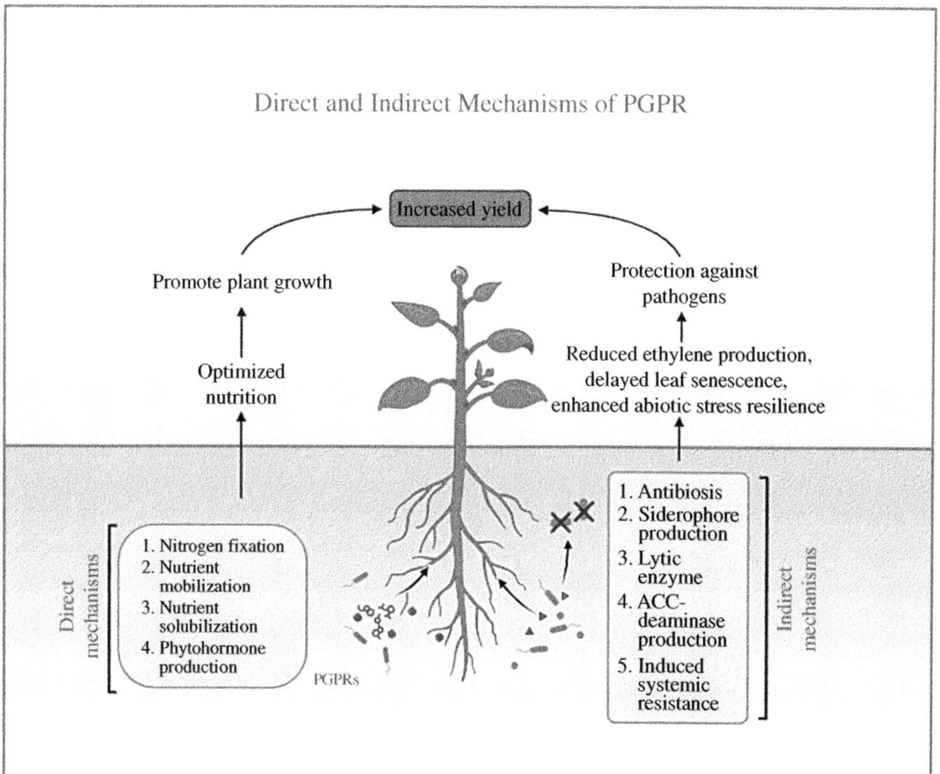

Figure 8.1 The direct and indirect mechanisms by which PGPRs enhance plant growth, yield, and stress resilience.

8.4 Methods for Identification and Characterization of Microbial Consortia

8.4.1 Culture-dependent Approaches

Culture-dependent approaches are essential for isolating and studying microorganisms from various plant-associated environments such as the rhizosphere, endosphere, and phyllosphere. These environments host diverse microbial communities that play crucial roles in plant health and growth. The rhizosphere, a narrow zone of soil influenced by root secretions, is particularly rich in beneficial bacteria that enhance plant growth and stress tolerance. However, many of these bacteria are difficult to culture using traditional methods, necessitating the development of improved cultivation techniques. For instance, replacing agar with gellan gums and modifying media preparation can significantly increase the recovery of cultivable bacteria from the rhizosphere, as demonstrated in studies on wheat (Youseif et al., 2021). In addition to isolation, dual culture assays are employed to assess the antagonistic activity of isolated bacteria against plant pathogens. This method involves coculturing the bacteria with a target pathogen to observe inhibitory effects.

8.4.2 Culture-independent Molecular Tools

Culture-independent molecular techniques have transformed microbial community profiling, providing intricate insights into the complex ecosystems of microorganisms. A principal technique utilized is 16S rRNA sequencing, which focuses on the 16S ribosomal RNA gene to identify and categorize bacteria within a population. This approach is prevalent because of its cost-efficiency and the accessibility of curated reference datasets, while it may experience amplification bias and restricted genomic breadth (Beaudry et al., 2021; Muthappa et al., 2022). Recent innovations, including nanopore sequencing for comprehensive 16S rRNA gene analysis, have enhanced taxonomic resolution, facilitating species-level categorization (Lee et al., 2023; Johnson et al., 2019). Nonetheless, obstacles persist, such as the necessity for standardized protocols and the impact of bioinformatic techniques on community composition assessments (Beaudry et al., 2021; Kapustina et al., 2021).

In addition to 16S rRNA sequencing, metagenomics and functional genomics have emerged as powerful tools for novel gene identification and understanding microbial functions. Shotgun metagenomic sequencing, for instance, samples a broader range of genomic regions, providing a more comprehensive view of microbial diversity and function (Han et al., 2020; Beaudry et al., 2021). This approach has been enhanced by platforms like bio Bakery 3, which integrates taxonomic, functional, and strain-level profiling, thereby increasing the accuracy of microbial community studies (Beghini et al., 2020). Metagenomics not only identifies novel genes but also elucidates the functional potential of microbial communities, offering insights into their roles in various environments (Lin & Liu, 2023; Benjamino et al., 2021).

Together, these culture-independent molecular tools have significantly advanced our understanding of microbial communities. They provide a detailed picture of microbial diversity and function, which is crucial for applications ranging from environmental monitoring to human health research. Despite the challenges, ongoing improvements in sequencing technologies and bioinformatics are continually enhancing the resolution and accuracy of microbial community profiling, paving the way for new discoveries in microbiology.

8.4.3 Biochemical and Physiological Screening

Biochemical and physiological screening methods are crucial in the discovery and development of new antimicrobial agents, particularly in the face of rising antibiotic resistance. Enzymatic assays play a pivotal role in this process by enabling the identification of compounds that exhibit antimicrobial activity. These assays often involve target-based screening, where specific enzymes or proteins are isolated and tested for interactions with potential antimicrobial compounds. This approach allows researchers to identify molecules that can inhibit the function of essential bacterial enzymes, thereby halting bacterial growth or killing the bacteria outright. However, one limitation of enzymatic assays is that they may not always predict the compound's effectiveness in a whole-cell context due to issues like poor permeability or efflux (Rütten et al., 2022; Landeta & Mejia-Santana, 2021).

In addition to enzymatic assays, screening for secondary metabolite production is another vital aspect of antimicrobial discovery. Secondary metabolites, often produced by microorganisms such as actinomycetes, are a rich source of bioactive compounds, including antibiotics. Techniques such as genome mining and *in situ* Raman hyperspectral analysis have been developed to enhance the detection and characterization of these metabolites. Genome mining, for instance, involves the use of bioinformatics tools like anti-SMASH to identify biosynthetic gene clusters responsible for the production of secondary metabolites (Blin et al., 2021, 2023). Meanwhile, Raman spectroscopy allows for the rapid and efficient screening of microbial colonies for secondary metabolite production, providing insights into the spatial distribution and heterogeneity of these compounds within colonies (Suwa et al., 2024).

Both enzymatic assays and secondary metabolite screening are integral to the development of new antimicrobials. While enzymatic assays provide a targeted approach to identifying potential inhibitors, secondary metabolite screening offers a broader perspective by uncovering novel compounds with antimicrobial properties. Together, these methods contribute to a comprehensive strategy for combating antibiotic-resistant pathogens and advancing the field of antimicrobial drug discovery (Yang et al., 2024; Stirk & Van Staden, 2022; Stokes et al., 2019).

8.5 Plant-associated Factors Influencing Microbial Consortia Selection

Plant-associated factors play a crucial role in shaping microbial consortia selection, with plant genotype, root exudates, developmental stages, and environmental factors being key determinants. The genotype of a plant significantly influences the composition of root exudates, which in turn affects microbial recruitment in the rhizosphere. For instance, studies on carrot and barley genotypes have shown that different genotypes secrete distinct root exudates, leading to the recruitment of specific microbial communities. In carrots, novel genotypes with unique breeding histories were found to modify soil microbial diversity and composition, particularly affecting bacterial recruitment (Anderson et al., 2024). Similarly, in barley, the composition of root exudates varied between landrace and modern cultivars, influencing the abundance and phenotype of rhizobacterial populations such as Pseudomonas (Pacheco-Moreno et al., 2024).

The developmental stage of a plant significantly affects the makeup of microbial communities. Research on maize indicates that the concentration of specific root exudates, including phytohormones such as indole-3-acetic acid and abscisic acid, varies across developmental stages, influencing rhizobacterial communities differently during vegetative and reproductive phases (Lopes et al., 2023). The differentiation in ecological roles of microbial communities is more pronounced at various developmental stages, with bacterial communities exerting greater influence during early stages and fungal communities becoming more significant in later stages (Xiong et al., 2021). Environmental factors such as soil type, climate, and nutrient availability influence the assembly and function of microbial communities. Soil properties and root exudation patterns are essential factors influencing the structure of microbial communities along the soil–root continuum. The composition of the microbiome in various root compartments of *Melilotus officinalis* is influenced by soil properties and root exudate metabolism, with notable shifts occurring across different anatomical boundaries (Zhou et al., 2022). Furthermore, climate factors like temperature interact with root exudates, affecting microbial diversity and community composition, thereby underscoring the intricate relationship between environmental conditions and plant–microbe interactions (Fang et al., 2024). The selection of microbial consortia in the rhizosphere involves various factors, including plant genotype, root exudates, developmental stages, and environmental conditions. Comprehending these interactions is essential for enhancing plant health and productivity, as well as for formulating strategies to manage plant-associated microbiomes in sustainable agriculture.

8.6 Application of Microbial Consortia in Biotic Stress Management

8.6.1 Development of Bioinoculants and Biopesticides

The application of microbial consortia in biotic stress management is a promising approach in modern agriculture, particularly in the development of bioinoculants and biopesticides. Microbial consortia, which involve the combination of multiple microbial species, offer a multifaceted strategy to enhance plant health and resilience against biotic stresses such as diseases and pests. These consortia leverage the diverse ecological functions of different microorganisms, promoting a biological equilibrium that can effectively manage plant diseases and pests while also stimulating plant growth (Nunes et al., 2024; Minchev et al., 2021; Singh et al., 2024). The application of microbial consortia as biopesticides is increasingly recognized for offering a more dependable and sustainable alternative to chemical pesticides. These consortia utilize various mechanisms, including direct microbial antagonism and ISR, to manage a broad spectrum of pathogens, providing enhanced functionality over single-strain applications (Minchev et al., 2021; Negi et al., 2024). The exploration of bioinoculants developed from microbial consortia aims to enhance plant growth and productivity. Consortia enhance nutrient availability, modulate phytohormones, and increase tolerance to biotic and abiotic stresses, thus supporting sustainable agricultural practices (Wang et al., 2024; Santoyo et al., 2021b). Microbial consortia serve as a versatile and effective approach in managing biotic stress, facilitating the development of environmentally sustainable agricultural practices.

8.6.2 Field Application Strategies (Seed Treatment, Foliar Spray, Soil Drenching) and Compatibility with Integrated Pest and Disease Management Programs

The use of microbial consortia for managing biotic stress is increasingly recognized as a sustainable alternative to chemical pesticides in agriculture. Consortia consisting of various beneficial microorganisms provide multiple mechanisms to improve plant health and increase resilience to pests and diseases. Strategies for field application, including seed treatment, foliar spray, and soil drenching, are frequently utilized to effectively deliver these microbial consortia. Seed treatment consists of applying microbial consortia to seeds, which can improve seedling vigor and offer initial protection against soilborne pathogens (Singh et al., 2024; Minchev et al., 2021). Foliar sprays facilitate direct application to plant leaves, providing a swift response to foliar pathogens and pests (Minchev et al., 2021). Soil drenching specifically addresses root-associated pathogens by applying the consortia directly to the soil, thereby enhancing root health and nutrient uptake (Wu et al., 2023).

The compatibility of microbial consortia with integrated pest and disease management (IPM) programs is a significant advantage. IPM strategies aim to reduce reliance on chemical inputs by integrating biological, cultural, and mechanical control methods. Microbial consortia fit well within this framework as they can be used alongside other IPM components to enhance overall pest and disease control efficacy (Nunes et al., 2024; Negi et al., 2024). For instance, microbial consortia can be combined with resistant crop varieties or cultural practices such as crop rotation to manage soilborne diseases more effectively (Zhang et al., 2022). Additionally, the use of microbial consortia can reduce the need for chemical pesticides, aligning with the goals of IPM to minimize environmental impact and promote sustainable agriculture (Bhatt et al., 2021; Dwisandi et al., 2024). Microbial consortia offer a versatile and environmentally friendly approach to managing biotic stress in crops. Their application through various field strategies and compatibility with IPM programs make them a promising tool for enhancing crop resilience and productivity while reducing chemical pesticide use. This approach not only supports sustainable agriculture but also contributes to the long-term health of agricultural ecosystems.

8.6.3 Case Studies of Successful Microbial Consortia in Disease Suppression

The application of microbial consortia in biotic stress management has emerged as a promising strategy in sustainable agriculture, offering a natural and effective means to enhance plant resilience against pathogens. This approach leverages the synergistic interactions among different microbial species to suppress diseases and promote plant growth, thereby reducing the reliance on chemical pesticides and fertilizers.

One notable case study involves the use of rhizobacterial consortia in rice cultivation to combat the necrotrophic fungus *Rhizoctonia solani*. In this study, a consortium of PGPRs was applied to rice plants, resulting in enhanced resistance to the pathogen. The treatment not only promoted plant growth but also triggered antioxidative defense mechanisms, such as the activation of enzymes like ascorbate peroxidase and catalase, which help mitigate oxidative stress in plants (Bhattacharyya et al., 2020). This demonstrates the potential of microbial consortia to enhance plant immunity and stress tolerance under field conditions.

Another successful application of microbial consortia is observed in the organic paddy fields of Sikkim, India. Here, a consortium comprising strains like *Pseudomonas kribbensis* and *Bacillus megaterium* was developed to improve rice growth and yield. The consortia exhibited biocontrol potential against the phytopathogen *Colletotrichum gloeosporioides*, which affects large cardamom. Field studies showed that the consortia significantly increased agronomic performance, including root length and grain yield, highlighting their effectiveness in organic farming systems (Sherpa et al., 2021).

In the context of groundnut cultivation, a PGPR consortium was developed to control early-stage diseases caused by *Aspergillus* and *Fusarium* species. The consortium, which included strains like *Acinetobacter baumannii* and *Bacillus subtilis*, demonstrated broad antagonism against these pathogens. The application of this consortium not only improved germination rates and root induction but also enhanced overall plant growth, showcasing its potential in disease management and crop productivity (Syed et al., 2020).

The use of *Bacillus* spp. consortia for soil bioaugmentation has also shown promising results in improving soil health and plant growth. A study conducted in West Bengal, India, introduced a multi-strain *Bacillus* consortium into overexploited agricultural soil, resulting in significant improvements in plant growth and soil microbial diversity. This approach aligns with sustainable agricultural practices by enhancing soil biological health and supporting the cultivation of resilient crops like soybean (Mukhopadhyay et al., 2023). In the case of arecanut cultivation, microbial consortia have been evaluated for their efficacy in managing fruit rot disease. Although chemical treatments like *Bordeaux* mixture showed higher disease control efficiency, microbial consortia were found to be effective in intermediate efficacy groups. This suggests that microbial consortia can be integrated with other treatments to enhance disease management and yield sustainability in arecanut farming (Patil et al., 2022).

Furthermore, the use of arsenic-tolerant bacterial consortia in phytoremediation has been explored to enhance arsenic uptake in plants. In a study involving *Adiantum capillus-veneris*, the inoculation of arsenic-tolerant strains like *Paracoccus versutus* significantly increased arsenic accumulation and improved plant growth. This highlights the potential of microbial consortia in environmental remediation and stress management (Marwa et al., 2020). In brinjal cultivation, a consortial formulation of entomopathogenic fungi and biocontrol agents was developed to manage diseases like *Phomopsis* leaf blight and *Fusarium* wilt. The application of this consortium in an IPM module resulted in reduced disease incidence and increased crop yield. This approach not only improved plant health but also enhanced soil nutrient status, demonstrating its utility in organic farming systems (Upamanya et al., 2020).

In summary, the application of microbial consortia in biotic stress management offers a sustainable and effective approach to enhance plant resilience against diseases.

These consortia leverage the synergistic interactions among microbial species to suppress pathogens, promote plant growth, and improve soil health. As demonstrated in various case studies, microbial consortia can be tailored to specific crops and environmental conditions, providing a versatile tool for sustainable agriculture.

8.7 Conclusion

The integration of microbial consortia into agricultural systems offers a promising and sustainable strategy for addressing biotic stress in plants. Beneficial microbes, such as rhizospheric bacteria, endophytes, and phyllospheric microorganisms, give multiple advantages in plant protection by directly preventing pathogens and indirectly regulating immune responses. The development and utilization of microbial consortia promote plant growth, increase disease resistance, and reduce dependence on chemical pesticides and fertilizers. Approaches for field application, including seed treatment, foliar spraying, and soil drenching, enable the efficient use of microbial consortia across various cropping systems. In addition, alignment with IPM programs enhances their viability in sustainable agriculture. Successful case studies illustrate how microbial consortia can improve crop resilience and yield, all while preserving environmental health. Future investigations should concentrate on enhancing microbial formulations, increasing field applicability, and gaining an increased awareness of plant–microbe interactions. Utilizing microbial consortia can lead agriculture toward a more resilient and environmentally sustainable future, therefore ensuring the availability of food amid rising environmental challenges.

References

Ali, S., Tyagi, A., Mir, R., Rather, I., Anwar, Y., & Mahmoudi, H. 2023. Plant beneficial microbiome a boon for improving multiple stress tolerance in plants. *Frontiers in Plant Science* 14, 1266182. https://doi.org/10.3389/fpls.2023.1266182

Almario, J., Fabiańska, I., Saridis, G., & Bucher, M. 2022. Unearthing the plant-microbe quid pro quo in root associations with beneficial fungi. *The New Phytologist* 234, 1967–1976. https://doi.org/10.1111/nph.18061

Anckaert, A., Declerck, S., Poussart, L., Lambert, S., Helmus, C., Boubsi, F., Steels, S., Argüelles-Arias, A., Calonne-Salmon, M., & Ongena, M. 2024. The biology and chemistry of a mutualism between a soil bacterium and a mycorrhizal fungus. *Current Biology* 34, 4934–4950.e8. https://doi.org/10.1016/j.cub.2024.09.019

Anderson, H., Cagle, G., Majumder, E., Silva, E., Dawson, J., Simon, P., & Freedman, Z. 2024. Root exudation and rhizosphere microbial assembly are influenced by novel plant trait diversity in carrot genotypes. *Soil Biology and Biochemistry* 197, 109516. https://doi.org/10.1016/j.soilbio.2024.109516

Andrić, S., Meyer, T., & Ongena, M. 2020. *Bacillus* responses to plant-associated fungal and bacterial communities. *Frontiers in Microbiology* 11, 1350. https://doi.org/10.3389/fmicb.2020.01350

Beaudry, M., Wang, J., Kieran, T., Thomas, J., Bayona-Vásquez, N., Gao, B., Devault, A., Brunelle, B., Lu, K., Wang, J., Rhodes, O., & Glenn, T. 2021. Improved microbial community characterization of 16S rRNA via metagenome hybridization capture enrichment. *Frontiers in Microbiology* 12, 644662. https://doi.org/10.3389/fmicb.2021.644662

Beghini, F., McIver, L., Blanco-Míguez, A., Dubois, L., Asnicar, F., Maharjan, S., Mailyan, A., Thomas, A., Manghi, P., Valles-Colomer, M., Weingart, G., Zhang, Y., Zolfo, M., Huttenhower, C., Franzosa, E., & Segata, N. 2020. Integrating taxonomic, functional, and strain-level profiling of diverse microbial communities with bioBakery 3. *eLife* 10, e65088. https://doi.org/10.1101/2020.11.19.388223

Benjamino, J., Leopold, B., Phillips, D., & Adams, M. 2021. Genome-based targeted sequencing as a reproducible microbial community profiling assay. *mSphere* 6, e01325-20. https://doi.org/10.1128/mSphere.01325-20

Bhatt, P., Bhatt, K., Sharma, A., Zhang, W., Mishra, S., & Chen, S. 2021. Biotechnological basis of microbial consortia for the removal of pesticides from the environment. *Critical Reviews in Biotechnology* 41, 317–338. https://doi.org/10.1080/07388551.2020.1853032

Bhattacharyya, C., Banerjee, S., Acharya, U., Mitra, A., Mallick, I., Haldar, A., Haldar, S., Ghosh, A., & Ghosh, A. 2020. Evaluation of plant growth promotion properties and induction of antioxidative defense mechanism by tea rhizobacteria of Darjeeling, India. *Scientific Reports* 10, 15536. https://doi.org/10.1038/s41598-020-72439-z

Blin, K., Shaw, S., Augustijn, H.E., Reitz, Z.L., Biermann, F., Alanjary, M., Fetter, A., Terlouw, B.R., Metcalf, W.W., Helfrich, E.J.N., Wezel, G.P.V., Medema, M.H., & Weber, T. 2023. antiSMASH 7.0: new and improved predictions for detection, regulation, chemical structures and visualisation. *Nucleic Acids Research* 51, W46–W50. https://doi.org/10.1093/nar/gkad344

Blin, K., Shaw, S., Kloosterman, A., Charlop-Powers, Z., Wezel, G., Medema, M., & Weber, T. 2021. antiSMASH 6.0: improving cluster detection and comparison capabilities. *Nucleic Acids Research* 49, W29–W35. https://doi.org/10.1093/nar/gkab335

Chaudhary, P., Agri, U., Chaudhary, A., Kumar, A., & Kumar, G. 2022. Endophytes and their potential in biotic stress management and crop production. *Frontiers in Microbiology* 13, 933017. https://doi.org/10.3389/fmicb.2022.933017

Chen, M., Acharya, S., Yee, M., Cabugao, K., & Chakraborty, R. 2024. Developing stable, simplified, functional consortia from *Brachypodium rhizosphere* for microbial application in sustainable agriculture. *Frontiers in Microbiology* 15, 1401794. https://doi.org/10.3389/fmicb.2024.1401794

Das, P., Singh, K., Nagpure, G., Mansoori, A., Singh, R., Ghazi, I., Kumar, A., & Singh, J. 2022. Plant-soil-microbes: a tripartite interaction for nutrient acquisition and better plant growth for sustainable agricultural practices. *Environmental Research* 214, 113821. https://doi.org/10.1016/j.envres.2022.113821

Dimkić, I., Janakiev, T., Petrović, M., Degrassi, G., & Fira, D. 2021. Plant-associated *Bacillus* and *Pseudomonas* antimicrobial activities in plant disease suppression via biological control mechanisms – a review. *Physiological and Molecular Plant Pathology* 117, 101754. https://doi.org/10.1016/j.pmpp.2021.101754

Dwisandi, R., Miranti, M., Widiastuti, A., Prismantoro, D., Awal, M., Mispan, M., Joshi, R., & Doni, F. 2024. Microbial secondary metabolites for modulating plant biotic stress resistance: bridging the lab-field gap. *Plant Stress* 15, 100720. https://doi.org/10.1016/j.stress.2024.100720

Fang, K., Kou, Y., Tang, N., Liu, J., Zhang, X., He, H., Xia, R., Zhao, W., Li, D., & Liu, Q. 2024. Differential responses of soil bacteria, fungi and protists to root exudates and temperature. *Microbiological research* 286, 127829. https://doi.org/10.1016/j.micres.2024.127829

Gu, S., Yang, T., Shao, Z., Wang, T., Cao, K., Jousset, A., Friman, V., Mallon, C., Mei, X., Wei, Z., Xu, Y., Shen, Q., & Pommier, T. 2020. Siderophore-mediated interactions determine the disease suppressiveness of microbial consortia. *mSystems* 5, e00811-19. https://doi.org/10.1128/mSystems.00811-19

Guzmán-Guzmán, P., Kumar, A., De Los Santos-Villalobos, S., Parra-Cota, F., Orozco-Mosqueda, M., Fadiji, A., Hyder, S., Babalola, O., & Santoyo, G. 2023. *Trichoderma* species: our best fungal allies in the biocontrol of plant diseases – A review. *Plants* 12, 432. https://doi.org/10.3390/plants12030432

Han, D., Gao, P., Li, R., Tan, P., Xie, J., Zhang, R., & Li, J. 2020. Multicenter assessment of microbial community profiling using 16S rRNA gene sequencing and shotgun metagenomic sequencing. *Journal of Advanced Research* 26, 111–121. https://doi.org/10.1016/j.jare.2020.07.010

Hashem, A., Tabassum, B., & Abd Allah, E. 2019. *Bacillus subtilis*: a plant-growth promoting rhizobacterium that also impacts biotic stress. *Saudi Journal of Biological Sciences* 26, 1291–1297. https://doi.org/10.1016/j.sjbs.2019.05.004

Hönig, M., Roeber, V., Schmülling, T., & Cortleven, A. 2023. Chemical priming of plant defense responses to pathogen attacks. *Frontiers in Plant Science* 14, 1146577. https://doi.org/10.3389/fpls.2023.1146577

Johnson, J., Spakowicz, D., Hong, B., Petersen, L., Demkowicz, P., Chen, L., Leopold, S., Hanson, B., Agresta, H., Gerstein, M., Sodergren, E., & Weinstock, G. 2019. Evaluation of 16S rRNA gene sequencing for species and strain-level microbiome analysis. *Nature Communications* 10, 5029. https://doi.org/10.1038/s41467-019-13036-1

Kapustina, Ž., Medžiūnė, J., Alzbutas, G., Rokaitis, I., Matjošaitis, K., Mackevičius, G., Žeimytė, S., Karpus, L., & Lubys, A. 2021. High-resolution microbiome analysis enabled by linking of 16S rRNA gene sequences with adjacent genomic contexts. *Microbial Genomics* 7, 000624. https://doi.org/10.1099/mgen.0.000624

Khanna, K., Kohli, S., Sharma, N., Kour, J., Devi, K., Bhardwaj, T., Dhiman, S., Singh, A., Sharma, N., Sharma, A., Ohri, P., Bhardwaj, R., Ahmad, P., Alam, P., & Albalawi, T. 2022. Phytomicrobiome communications: novel implications for stress resistance in plants. *Frontiers in Microbiology* 13, 912701. https://doi.org/10.3389/fmicb.2022.912701

Landeta, C., & Mejia-Santana, A. 2021. Union is strength: target-based and whole-cell high-throughput screens in antibacterial discovery. *Journal of Bacteriology* 204, e0047721. https://doi.org/10.1128/jb.00477-21

Lee, A., Chan, C., Wong, L., Yip, C., Lui, W., Cheng, K., Leung, J., Lee, L., Wong, I., Ng, T., Lao, H., & Siu, G. 2023. Identification of microbial community in the urban environment: the concordance between conventional culture and nanopore 16S rRNA sequencing. *Frontiers in Microbiology* 14, 1164632. https://doi.org/10.3389/fmicb.2023.1164632

Lin, T., & Liu, W. 2023. Validation of 16S rRNA gene sequencing and metagenomics for evaluating microbial immigration in a methanogenic bioreactor. *Water Research* 243, 120358. https://doi.org/10.1016/j.watres.2023.120358

Lopes, L., Futrell, S., Bergmeyer, E., Hao, J., & Schachtman, D. (2023). Root exudate concentrations of indole-3-acetic acid (IAA) and abscisic acid (ABA) affect maize rhizobacterial communities at specific developmental stages. *FEMS Microbiology Ecology* 99, fiad019. https://doi.org/10.1093/femsec/fiad019

Manzar, N., Kashyap, A., Goutam, R., Rajawat, M., Sharma, P., Sharma, S., & Singh, H. 2022. *Trichoderma*: advent of versatile biocontrol agent, its secrets and insights into mechanism of biocontrol potential. *Sustainability* 14, 12786. https://doi.org/10.3390/su141912786

Marwa, N., Mishra, N., Singh, N., Mishra, A., Saxena, G., Pandey, V., & Singh, N. (2020). Effect of rhizospheric inoculation of isolated arsenic (As) tolerant strains on growth, As-uptake and bacterial communities in association with *Adiantum capillus-veneris*. *Ecotoxicology and Environmental Safety* 196, 110498. https://doi.org/10.1016/j.ecoenv.2020.110498

Meena, M., Swapnil, P., Divyanshu, K., Kumar, S., Harish., Tripathi, Y., Zehra, A., Marwal, A., & Upadhyay, R. (2020). PGPR-mediated induction of systemic resistance and physiochemical alterations in plants against the pathogens: current perspectives. *Journal of Basic Microbiology* 60, 828–861. https://doi.org/10.1002/jobm.202000370

Minchev, Z., Kostenko, O., Soler, R., & Pozo, M. 2021. Microbial consortia for effective biocontrol of root and foliar diseases in tomato. *Frontiers in Plant Science* 12, 756368. https://doi.org/10.3389/fpls.2021.756368

Mukhopadhyay, M., Mukherjee, A., Ganguli, S., Chakraborti, A., Roy, S., Choudhury, S., Subramaniyan, V., Kumarasamy, V., Sayed, A., El-Demerdash, F., Almutairi, M., Șuțan, A., Dhara, B., & Mitra, A. 2023. Marvels of Bacilli in soil amendment for plant-growth promotion toward sustainable development having futuristic socio-economic implications. *Frontiers in Microbiology* 14, 1293302. https://doi.org/10.3389/fmicb.2023.1293302

Muthappa, D., Lamba, S., Sivasankaran, S., Naithani, A., Rogers, N., Srikumar, S., Macori, G., Scannell, A., & Fanning, S. 2022. 16S rRNA based profiling of bacterial communities colonizing bakery-production environments. *Foodborne Pathogens and Disease* 19, 485–494. https://doi.org/10.1089/fpd.2022.0014

Negi, R., Sharma, B., Parastesh, F., Kaur, S., Khan, S., Kour, D., Singh, S., Rai, A., Rustagi, S., Yadav, N., Shreaz, S., Madouh, T., & Yadav, A. 2024. Microbial consortia mediated regulation of plant defense: A promising tool for sustaining crops protection. *Physiological and Molecular Plant Pathology* 134, 102393. https://doi.org/10.1016/j.pmpp.2024.102393

Nishad, R., Ahmed, T., Rahman, V., & Kareem, A. 2020. Modulation of plant defense system in response to microbial interactions. *Frontiers in Microbiology* 11, 1298. https://doi.org/10.3389/fmicb.2020.01298

Nunes, P., Lacerda, G., Mascarin, G., Guimarães, R., Medeiros, F., Arthurs, S., & Bettiol, W. 2024. Microbial consortium of biological products: do they have a future?. *Biological Control* 188, 105439. https://doi.org/10.1016/j.biocontrol.2024.105439

Omae, N., & Tsuda, K. 2022. Plant-microbiota interactions in abiotic stress environments. *Molecular plant-microbe interactions: MPMI* 35, 511–526. https://doi.org/10.1094/MPMI-11-21-0281-FI

Orozco-Mosqueda, M., Fadiji, A., Babalola, O., & Santoyo, G. 2023. Bacterial elicitors of the plant immune system: an overview and the way forward. *Plant Stress* 7, 100138. https://doi.org/10.1016/j.stress.2023.100138

Pacheco-Moreno, A., Bollmann-Giolai, A., Chandra, G., Brett, P., Davies, J., Thornton, O., Poole, P., Ramachandran, V., Brown, J., Nicholson, P., Ridout, C., DeVos, S., & Malone, J. 2024. The genotype of barley cultivars influences multiple aspects of their associated

microbiota via differential root exudate secretion. *PLOS Biology* 22, e3002232. https://doi.org/10.1371/journal.pbio.3002232
Patil, B., Sridhara, S., Narayanaswamy, H., Hegde, V., & Mishra, A. 2022. Control efficiency and yield response of chemical and biological treatments against fruit rot of arecanut: a network meta-analysis. *Journal of Fungi* 8, 937. https://doi.org/10.3390/jof8090937
Pellan, L., Dieye, C., Durand, N., Fontana, A., Strub, C., & Schorr-Galindo, S. 2021. Biocontrol agents: toolbox for the screening of weapons against *Mycotoxigenic Fusarium*. *Journal of Fungi* 7, 446. https://doi.org/10.3390/jof7060446
Pfeiffer, T., Von Galen, A., Zink, P., Hübner, S., Linkies, A., Felgentreu, D., Drechsel, J., Birr, T., Röder, O., Kotte, M., Dietel, K., Junge, H., Schwarz, E., & Koch, E. 2021. Selection of bacteria and fungi for control of soilborne seedling diseases of maize. *Journal of Plant Diseases and Protection* 128, 1227–1241. https://doi.org/10.1007/s41348-021-00498-z
Poveda, J., & Eugui, D. 2022. Combined use of *Trichoderma* and beneficial bacteria (mainly *Bacillus* and *Pseudomonas*): development of microbial synergistic bio-inoculants in sustainable agriculture. *Biological Control* 176, 105100. https://doi.org/10.1016/j.biocontrol.2022.105100
Prigigallo, M., Staropoli, A., Vinale, F., & Bubici, G. 2023. Interactions between plant-beneficial microorganisms in a consortium: *Streptomyces microflavus* and *Trichoderma harzianum*. *Microbial Biotechnology* 16, 2292–2312. https://doi.org/10.1111/1751-7915.14311
Pršić, J., & Ongena, M. 2020. Elicitors of plant immunity triggered by beneficial Bacteria. *Frontiers in Plant Science* 11, 594530. https://doi.org/10.3389/fpls.2020.594530
Rabari, A., Ruparelia, J., Jha, C., Sayyed, R., Mitra, D., Priyadarshini, A., Senapati, A., Panneerselvam, P., & Mohapatra, P. 2022. Articulating beneficial rhizobacteria mediated plant defenses through induced systemic resistance. *Pedosphere* 33, 556–566. https://doi.org/10.1016/j.pedsph.2022.10.003
Rivera-Vega, L., Grunseich, J., Aguirre, N., Valencia, C., Sword, G., & Helms, A. 2022. A beneficial plant-associated fungus shifts the balance toward plant growth over resistance, increasing cucumber tolerance to root herbivory. *Plants* 11, 282. https://doi.org/10.3390/plants11030282
Romera, F., García, M., Lucena, C., Martínez-Medina, A., Aparicio, M., Ramos, J., Alcántara, E., Angulo, M., & Pérez-Vicente, R. 2019. Induced systemic resistance (ISR) and Fe deficiency responses in dicot plants. *Frontiers in Plant Science* 10, 287. https://doi.org/10.3389/fpls.2019.00287
Rütten, A., Kirchner, T., & Musiol-Kroll, E. 2022. Overview on strategies and assays for antibiotic discovery. *Pharmaceuticals* 15, 1302. https://doi.org/10.3390/ph15101302
Santoyo, G., Gamalero, E., & Glick, B. 2021a. Mycorrhizal-bacterial amelioration of plant abiotic and biotic stress. 5, 672881. https://doi.org/10.3389/fsufs.2021.672881
Santoyo, G., Guzmán-Guzmán, P., Parra-Cota, F., Santos-Villalobos, S., Orozco-Mosqueda, M., & Glick, B. 2021b. Plant growth stimulation by microbial consortia. *Agronomy* 11, 219. https://doi.org/10.3390/AGRONOMY11020219
Sharma, I., Kashyap, S., & Agarwala, N. 2023. Biotic stress-induced changes in root exudation confer plant stress tolerance by altering rhizospheric microbial community. *Frontiers in Plant Science* 14, 1132824. https://doi.org/10.3389/fpls.2023.1132824
Sherpa, M., Sharma, L., Bag, N., & Das, S. 2021. Isolation, characterization, and evaluation of native rhizobacterial consortia developed from the rhizosphere of rice grown in organic

state Sikkim, India, and their effect on plant growth. *Frontiers in Microbiology* 12, 713660. https://doi.org/10.3389/fmicb.2021.713660

Singh, A., Mazahar, S., Chapadgaonkar, S., Giri, P., & Shourie, A. 2023. Phyto-microbiome to mitigate abiotic stress in crop plants. *Frontiers in Microbiology* 14, 1210890. https://doi.org/10.3389/fmicb.2023.1210890

Singh, D., Geat, N., Jadon, K., Verma, A., Sharma, R., Rajput, L., Mahla, H., & Kakani, R. 2024. Isolation and characterization of biocontrol microbes for development of effective microbial consortia for managing *Rhizoctonia bataticola* root rot of cluster bean under hot arid climatic conditions. *Microorganisms* 12, 2331. https://doi.org/10.3390/microorganisms12112331

Sommer, A., Wenig, M., Knappe, C., Kublik, S., Foesel, B., Schloter, M., & Vlot, A. 2024. A salicylic acid-associated plant-microbe interaction attracts beneficial *Flavobacterium* sp. to the *Arabidopsis thaliana* phyllosphere. *Physiologia Plantarum* 176(4), e14483. https://doi.org/10.1111/ppl.14483

Stirk, W., & Van Staden, J. 2022. Bioprospecting for bioactive compounds in microalgae: Antimicrobial compounds. *Biotechnology Advances* 59, 107977. https://doi.org/10.1016/j.biotechadv.2022.107977

Stokes, J., Gutierrez, A., Lopatkin, A., Andrews, I., French, S., Matic, I., Brown, E., & Collins, J. 2019. A multiplexable assay for screening antibiotic lethality against drug-tolerant bacteria. *Nature Methods* 16, 303–306. https://doi.org/10.1038/s41592-019-0333-y

Suwa, S., Ando, M., Nakashima, T., Horii, S., Anai, T., & Takeyama, H. 2024. In Situ Raman hyperspectral analysis of microbial colonies for secondary metabolites screening. *Analytical Chemistry* 96, 14909–14917. https://doi.org/10.1021/acs.analchem.4c02906

Syed, S., Tollamadugu, N., & Lian, B. 2020. *Aspergillus* and *Fusarium* control in the early stages of *Arachis hypogaea* (groundnut crop) by plant growth-promoting rhizobacteria (PGPR) consortium. *Microbiological Research* 240, 126562. https://doi.org/10.1016/j.micres.2020.126562

Tian, L., Lin, X., Tian, J., Ji, L., Chen, Y., Tran, L., & Tian, C. 2020. Research advances of beneficial microbiota associated with crop plants. *International Journal of Molecular Sciences* 21, 1792. https://doi.org/10.3390/ijms21051792

Upamanya, G., Bhattacharyya, A., & Dutta, P. 2020. Consortia of entomo-pathogenic fungi and bio-control agents improve the agro-ecological conditions for brinjal cultivation of Assam. *3 Biotech* 10, 450. https://doi.org/10.1007/s13205-020-02439-3

Vlot, A., Sales, J., Lenk, M., Bauer, K., Brambilla, A., Sommer, A., Chen, Y., Wenig, M., & Nayem, S. 2020. Systemic propagation of immunity in plants. *The New Phytologist* 229, 1234–1250. https://doi.org/10.1111/nph.16953

Wang, Y., Dall'Agnol, R., Bertani, I., Bez, C., & Venturi, V. 2024. Identification of synthetic consortia from a set of plant-beneficial bacteria. *Microbial Biotechnology* 17, e14330. https://doi.org/10.1111/1751-7915.14330

Wu, D., Wang, W., Yao, Y., Li, H., Wang, Q., & Niu, B. 2023. Microbial interactions within beneficial consortia promote soil health. *The Science of the Total Environment* 900, 165801. https://doi.org/10.1016/j.scitotenv.2023.165801

Wu, F., Wang, H., Lin, Y., Feng, S., & Li, X. 2024. Biocontrol mechanisms of antagonistic yeasts on postharvest fruits and vegetables and the approaches to enhance the biocontrol potential of antagonistic yeasts. *International Journal of Food Microbiology* 430, 111038. https://doi.org/10.1016/j.ijfoodmicro.2024.111038

Xiong, C., Singh, B., He, J., Han, Y., Li, P., Wan, L., Meng, G., Liu, S., Wang, J., Wu, C., Ge, A., & Zhang, L. 2021. Plant developmental stage drives the differentiation in ecological role of the maize microbiome. *Microbiome* 9, 171. https://doi.org/10.1186/s40168-021-01118-6

Yang, B., Yang, H., Liang, J., Chen, J., Wang, C., Wang, Y., Wang, J., Luo, W., Deng, T., & Guo, J. 2024. A review on the screening methods for the discovery of natural antimicrobial peptides. *Journal of Pharmaceutical Analysis* 15, 101046. https://doi.org/10.1016/j.jpha.2024.101046

Youseif, S., El-Megeed, F., Humm, E., Maymon, M., Mohamed, A., Saleh, S., & Hirsch, A. 2021. Comparative analysis of the cultured and total bacterial community in the wheat rhizosphere microbiome using culture-dependent and culture-independent approaches. *Microbiology Spectrum* 9, e0067821. https://doi.org/10.1128/Spectrum.00678-21

Yu, Y., Gui, Y., Li, Z., Jiang, C., Guo, J., & Niu, D. 2022. Induced systemic resistance for improving plant immunity by beneficial microbes. *Plants* 11, 386. https://doi.org/10.3390/plants11030386

Zhang, J., Ahmed, W., Dai, Z., Zhou, X., He, Z., Wei, L., & Ji, G. 2022. Microbial consortia: an engineering tool to suppress clubroot of chinese cabbage by hanging the rhizosphere bacterial community composition. *Biology* 11, 918. https://doi.org/10.3390/biology11060918

Zhou, Y., Wei, Y., Zhao, Z., Li, J., Li, H., Yang, P., Tian, S., Ryder, M., Toh, R., Yang, H., & Denton, M. 2022. Microbial communities along the soil-root continuum are determined by root anatomical boundaries, soil properties, and root exudation. *Soil Biology and Biochemistry* 171, 108721. https://doi.org/10.1016/j.soilbio.2022.108721

9

Support of Microbial Formulations to Achieve the Sustainable Development Goal 2 (SDGs)

Monika Singh[1], R.S. Sengar[2], Karuna Singh[1], Rakesh[1], Aastha[1], Shivani Pal[1], and Rocky Kumar[1]*

[1] *Department of Biotechnology, School of Applied and Life Sciences, Uttaranchal University, Dehradun, Uttarakhand, India*
[2] *Department of Plant Biotechnology, Sardar Vallabhbhai Patel University of Agriculture and Technology, Meerut, Uttar Pradesh, India*

* Corresponding author: monikasingh_bhu@yahoo.com

9.1 Introduction

Microbial formulations have emerged as a promising solution to support the United Nations Sustainable Development Goals (SDGs), by having focus on areas such as sustainable agriculture. These formulations utilize beneficial microorganisms, such as bacteria, fungi, and microalgae, to promote sustainability in several fields. These formulations refer to products that contain live microbes or their metabolic byproducts, which are used for specific applications such as improving soil fertility to improve plant growth and development, bioremediation, and disease control. These formulations can be developed as biofertilizers, biopesticides, biochar, and bioremediating agents (Figure 9.1). In this chapter, authors are more focused on SDG2. How microbial formulations support achieving SDG2, which covers sustainable agriculture to achieve food security of the world, which is a basic need of the human population. Some microbial strains and their formulations supported agricultural system in various ways. Some strains such as *Bacillus thuringiensis* and *Trichoderma harzianum* are key strains that act as biological pesticides to protect the crops. These biological pesticides have the potential to contribute significantly to several United Nations Sustainable Development Goals (SDGs), special reference to SDG 2 (Zero Hunger) (Cai & Dimopoulos, 2025).

Recently, crop yield has hampered amid unfavorable climatic scenarios, and decreasing soil health. Synthetic fertilizers harness the capacity of beneficial microbial strains, which ultimately hampers phosphorus (P), zinc (Z), and potassium (K)deterioration, siderophore production, nitrogen fixation, and pathogen protection. Microbial bioformulation is a potential technique for the future of agriculture because of its several benefits, including

Microbial Consortia in Plant Science and Sustainable Agriculture, First Edition. Edited by Ajay Tomar, et al.

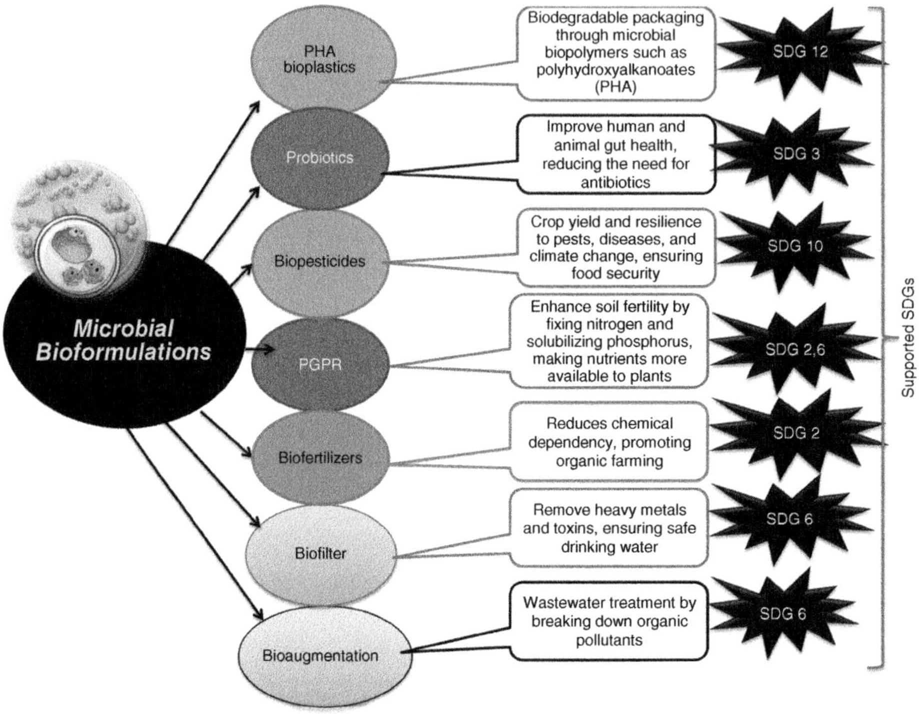

Figure 9.1 Graphical representation of microbial formulations and their supported SDGs other than SDG2.

long-term viability and plant growth-promoting activities. Microbial formulation technology is a cost-effective and eco-friendly approach to sustainable agriculture. Increased food demand is the need of the world; in this reference, microbial formulation applications are essential to boost soil quality and save beneficial soil microbiomes because these microbes ameliorate soil nutritional values, which meets the global food demand. Moreover, the effective application of microbial formulations in the field of crop agriculture is a crucial element to maximize their overall efficiency. This chapter focuses on exploring microbial formulations and their supporting materials, providing insights into their selection. The use of beneficial microorganisms, as plant or inoculants to act as a natural fertilizer and biocontrol agent, or to mitigate abiotic stresses, is one such sustainable approach in agriculture (Singh et al., 2019b; Khan et al., 2023a). In a study, fermented organic fertilizer contained a huge number of cells of *Lactobacillus* spp., actinomycetes, phototrophic bacteria, and yeasts, which are used as microbial inoculants loaded with phosphorus, nitrogen forms, organic acids, and amino acids. Favorable outcomes of the fermented biological fertilizers on soil health and crop productivity will highly depend upon the organic portion, promoting effects of the inoculated microorganisms and supporting effects of microbially synthesized bioactive compounds such as phytohormones and growth regulators (Yamada & Xu, 2001).

Besides fruits and vegetables, legumes are excellent sources of macronutrients and micronutrients, but the crop productivity of leguminous crops is very less because they are available at a high cost in the market. Researchers focused on leguminous crops to enhance crop productivity via microbial formulations, which enrich them with phytochemicals as well as nutritional factors.

Microbial formulation is an eco-friendly approach in agricultural practices to enhance crop productivity. Microbial formulations have several positive effects on various crops. It also enhances biomass content, photosynthetic efficiency, and anti-oxidative response to abiotic stress. In a study, it is reported that microbial formulations augment micronutrients, root and shoot weight, nodulation, and pod yield in kidney beans and soybeans. It also reported that formulated microbes speed up the nutritional contents in sunflower, groundnut, and okra (Naik et al., 2020). In another study, microbial bioformulations and algae-based biostimulants were recognized to support sustainable agriculture. In that study field experiments were conducted in Italy. The results of the experiment concluded that tomato yield and sugar content were enhanced after applying microbial bioformulations (Quintarelli et al., 2025). Several types of microbial formulation and their role in crop productivity are discussed in Table 9.1.

Trypsin inhibitors and phytates are naturally occurring anti-nutritional factors that are primarily present in leguminous and cereal crops. In a study, inhibitors of trypsin and phytates naturally present in edible crops can significantly decrease protein digestion and release of minerals, respectively. The interaction between phenolic and phytase compounds with minerals is essential in cereals and legumes. Germination and fermentation processes are most commonly used to disrupt such interactions and make nutrients and phytochemicals freely available and accessible to digestive enzymes. This study reports on conventional fermentation and germination processes as a source to address diverse interactions via fast activation of endogenous enzymes such as phytase, α-amylase, pullulanase, and other glucosidases. Such enzymes break down complex macronutrients and anti-nutritional factors to their simple and more digestible forms (Nkhata et al., 2018).

9.2 Bioformulation as Plant Growth-promoting Bacteria

Plant growth-promoting rhizobacteria (PGPR) or plant growth-promoting bacteria (PGPB) from the genera *Bacillus, Pseudomonas, Azotobacter, Azospirillum*, etc. are frequently used as a biofertilizer and biocontrol agent in sustainable agriculture. Currently, PGPBs or PGPRs are broadly used as microbial inoculants to enhance crop yield and productivity. These microbial strains are used independently or as consortia for disease management in the plant due to their eco-friendly nature. In this section, authors have explained different aspects of growth promotion such as siderophore production, phosphate solubilization, ammonia production, hydrogen cyanide, and phytohormone (Singh et al., 2019a). da Costa Pereira et al. (2019) studied the *Pseudomonas* Pt13 strain associated with the roots of *Piper tuberculatum*, which shows resistance to infection by *Fusarium solani* f. sp. *piperis*, which causes root rot disease in black pepper (*Piper nigrum* L.). *Pseudomonas* Pt13 was found capable of inhibiting the *in vitro* growth of this root rot pathogen. Thus, results revealed that *pseudomonas* strain Pt13 acts as a biocontrol agent to improve crop production. Results obtained from this study will be useful in future studies of interaction with *F. solani* f. sp. *Piperis*. Another salt-tolerant plant PGPR, *Pseudomonas atacamensis* KSS-6, was selected as a microbial inoculant to enhance the growth

Table 9.1 Microbial formulations for fruit and grain productivity.

Use of Bacterial and viral species as microbial formulations					
Sr. no.	**Microbe**	**Formulation as**	**Uses**	**Crop**	**References**
1	*Bacillus licheniformis* MNNITSR2; *Bacillus velezensis* MNNITSR18	Biofertilizer	Promote plant growth, yield productivity, phosphate solubilization, and salt tolerance.	Rice	Devi et al. (2023)
2	*Azotobacter* sp.	Biofertilizer	Higher production and fruit quality.	Guava	Singh et al. (2020)
3	*Pseudomonas fluorescens*	Biofertilizer	Enhance yield productivity and promote plant growth.	Sweet potato	Fernandez et al. (2021)
4	*Pseudomonas putida*	Biofertilizer	Enhance yield productivity.	Cucumber	Isfahani & Besharati (2012)
5	*Priestia megaterium H1* *Bacillus subtilis I2.*	Biocontrol agent	Antagonistic activity against fire blight	Pear	Din et al. (2024)
6	*Cydia pomonella granulovirus (CpGV)*	Biopesticide	Enhance fruit quality and long-term pest control.	Apple	Cai & Dimopoulos (2025)
7	*Cydia pomonella granulovirus (CpGV)*	Biopesticide	Enhance fruit quality and long-term pest control.	Pear	Cai & Dimopoulos (2025)
8	*Bacillus amyloliquefaciens* Bs006	Biofertilizer	Promote plant growth and yield productivity.	Banana	Gamez et al. (2020)
9	*Pseudomonas palleroniana* Ps006	Biofertilizer	Improve plant growth.	Banana	Gamez et al. (2020)
10	*Peanibacillus polymyxa*	Biofertilizer	Enhance grape productivity and enhance grape quality.	Grapes	Soliman et al. (2024)
11	*Lysinibacillus sphaericus*	Biofertilizer	Improve plant growth and yield productivity.	Grapes	Soliman et al. (2024)
12	*Azospirillium*	Biofertilizer	Improve nutritional uptake, larger food production, and plant height.	Papaya	Dutta et al. (2010)

13	*Pseudomonas fluorescence*	Biofertilizer	Yield productivity, promote plant growth, and improve plant height.	Mango	Shankarappa et al. (2018)
14	*Trichoderma harzianum*	Biofertilizer	Improved nutritional uptake and yield productivity.	Tomato	Coulibaly et al. (2021)
15	*Lasiodiplodia theobromae*	Biopesticide	Reduce disease stress and improve fruit quality.	Mango	Gava et al. (2018)
16	*Bacillus subtilis*	Biofertilizer	Enhance root growth and larger yield productivity	Kiwi	Sharma & Rana (2015)
17	*Bacillus licheniformis*	Biofertilizer	improve root growth and higher yield productivity	Kiwi	Sharma & Rana (2015)
18	*Bacillus velezensis*	Biopesticide	Prevent against plant pathogens and enhance plant growth.	Kiwi	Zhao et al. (2024b)
19	*Bacillus A8a and HA*	Biofertilizer	Promote plant growth and enhance nutrient uptake.	Avocado	Cortazar-Murillo et al. (2023)
20	*Azospirillum brasilemse*	Biofertilizer	Chlorophyll content in leaf and improve plant height.	Plum	Al-Dulaimi & Al-Rawi (2020)
21	*Bacillus amylolequefaciens BChi1*	Biofertilizer	Increase in antioxidant content, higher yield productivity, and enhanced nutrient uptake and fruit quality.	Strawberry	Rahman et al. (2018)
22	*Paraburkholderia fungorum BRRh-4*	Biofertilizer	Higher yield productivity, increase in antioxidant content, and enhanced nutrient uptake and fruit quality.	Strawberry	Rahman et al. (2018)
			Use of fungal Species in microbial formulations		
23	*Trichoderma aureoviride*	Microbial biochar	Promote plant growth, improve soil fertility, and yield productivity.	Watermelon	Medeiros et al. (2020)
24	*Saccharomyces cerevisiae*	Biofertilizer	Improve nutrient content in the roots. Enhance Plant nutrition.	Sugarcane and tomato	Lonhienne et al. (2014)

(Continued)

Table 9.1 (Continued)

Use of Fungal Species in Microbial formulations					
Sr. no.	**Microbe**	**Formulation as**	**Uses**	**Crop**	**References**
25	*Aureobasidium pullulans*	Biocontrol agent	Antagonistic activity against *Aspergillus flavus* pathogen	Tomato	Podgórska-Kryszczuk (2023)
26	*Aspergillus terreus* JF27	Biofertilizer Biocontrol agent	Improve plant growth and yield, Provide Resistance to pathogens	Tomato	Yoo et al. (2018)
27	*Ulocladium chartarum*	Biocontrol agents	Against various pathogens *Botrytis cinerea, Sclerotinia sclerotiorum*, and *Monilinia*	Grapes, berries, citrus fruits, vegetables, legumes, etc.	Thomidis et al. (2015)
28	*Metarhizium anisopliae ICIPE 69*	Biocontrol agents	Against whiteflies, mealy bugs, and thrips	French bean and other legume crops	Nyasani et al. (2015)
29	*Beauveria bassiana*	Biocontrol agents	Against whiteflies, thrips, moths, stinkbugs, red spider mite, and red scale	Peppers, citrus, tomatoes, berries, potatoes, etc.	Erler & Ates (2015)
30	*Verticillium lecanii*	Biocontrol agents	Against thrips, jassids, aphids, whiteflies, and mites	Cereals, vegetables, fruits	Buhroo et al. (2016)
31	*Trichoderma harzianum T-22/ Trichoderma virens G-41*	Biocontrol agents	Against my copathogens	Food and ornamental crops	Javanshir Javid et al. (2016)
32	*Metarhizium anisopliae*	Biocontrol agents	Against termites, beetles, and caterpillars	Cereals, millets, oilseeds, spices, ornamentals, vegetables, fruits, and flowers	Buhroo et al. (2016)
33	*Purpureocillium lilacinum*	Biocontrol agents	Against *Meloidogyne, Radopholussimilis, Heterodera, Globodera, reniformis, Nacobbus, Belonolaimus*, and *Pratylenchus*	Water-dispersible granules	Baidoo et al. (2017), Parajuli et al. (2014)

and yield of rice under salinity conditions. In the whole world, most of the land area is covered by saline soil, which is not suitable for the crops. This *pseudomonas* strain KSS-6 was capable of mitigating salt stress, which is an unfavorable condition for crop growth and development. The talc-based bio-formulation was developed using bacterial strain KSS-6 coupled with organic manure for the growth and yield of rice crops under saline conditions. Co-application of microbial organic manure formulation showed maximum increment in the nutritional content, such as flavonoids, dietary fibers, carbohydrates, proline, antioxidants, and protein, with growth and development. These findings suggest that the application of this microbial formulation could be a promising approach to mitigate the harmful impacts of salt stress, improve the yield and nutritional properties of rice grown in degraded, saline soils, and contribute to achieving zero hunger (Arora et al., 2024). Increasing food demand is becoming a major global concern, which has been raised by India and other developing nations. The demand for food supplies like fruits, grains, and vegetables will surely increase rapidly in the upcoming years. Every living form relies on microorganisms, which are present in every type of climatic and environmental condition. There are several different types of microbes living in earth sphere, including eubacteria, fungi, archaea, actinomycetes, and a few strains of cyanobacteria. Nowadays, these microbes are used as microbial consortiums for sustainable agriculture (Rajput et al., 2024).

Several beneficial plant growth-promoting rhizobacteria (PGPR), such as *Azospirillum, Bacillus, Pseudomonas Burkholderia, Klebsiella, and Rhizobium*, are key strains for sustainable agriculture because they possess plant growth-promoting traits. They have capacity to convert insoluble phosphates to soluble phosphates by the production of organic acids (citric & gluconic acid). Plant growth-promoting microbes are involved in improving nutrient availability, producing phytohormones such as auxins, gibberellins, and cytokinins, providing biocontrol against pathogens, and also increasing abiotic stress tolerance in plants. The necessity of reducing dependency on the use of chemical fertilizers and pesticides, plant growth-promoting microorganisms mitigate environmental impacts and support long-term agricultural productivity via improving soil health. Nitrogen-fixing bacteria possess nitrogenase enzyme complexes, which are capable of reducing atmospheric nitrogen into ammonia. The incorporation of PGPR into biofertilizers and biopesticides offers a sustainable resolution to global agricultural challenges (Yang et al., 2024). In a study, some endophytic bacterial strains, such as *Rhizobium pusense, Bacillus flexus, Bacillus cereus Pseudomonas aeruginosa*, and *Methylophilus flavus*, were applied as inoculants to assess their potential role in the enhancement of tomato crop growth and yield. Results concluded that there was a significant increase observed in the fruit yields in all the treatments; however, *M. flavus* (MS-4) inoculated plants yielded nearly three times higher when compared with untreated plants. Among all, *M. flavus* act as excellent microbial inoculants in tomato, and it might be used in microbial bioformulations as plant growth-promoting bacteria (Singh et al., 2022).

9.2.1 Molecular Approaches in Plant–Microbe Interaction

In the action of microbial inoculums, bacterial strains release exopolysaccharides that form organo-mineral complexes, and fungal strains release mucilaginous exudates composed of extracellular polysaccharides, which improve soil structure and health by promoting the formation of soil aggregates and pores within, making the soil more healthy and fertile. Microbes facilitate increase of organic matter, which boosts the availability of nitrogen, phosphorus, potassium (N, P, K), and iron (Fe) in soil. Microbial inoculation can obtain high crop yield (Rashid et al., 2016). Research is going on Omics technologies in plant–microbe interactions to know how the microbiome interacts with plant aids. Advanced genomic methods such as meta-omics and comparative studies play a crucial role in this aspect. Modern technologies, including metagenomics, microarray, genome sequencing, metatranscriptomics, and comparative genomics strategies can be used to understand the relationships between microbes and plants. Sustainable agricultural techniques, including biological control, bioremediation, and plant growth promotion, can be enhanced by a comprehensive grasp mechanism of host infestation and the role of microbes.

Among all microbial formulations, biofertilizers are basically focused on fulfilling the nitrogen requirement of the plant. The mechanism of nitrogen assimilation is required. Because atmospheric nitrogen is stable and unreactive in nature, it is not absorbable by plants directly. To overcome this concern, microbes such as soil bacteria (rhizobia), which include different genera, such as *Bradyrhizobium, Sinorhizobium, Rhizobium, and Mesorhizobium*, showed a symbiotic relationship with legume roots, which generates nodules. These resulting nodules help fix the atmospheric nitrogen. This action is carried at a molecular level with the help of an enzyme, nitrogenase (Graham & Vance, 2003). Nitrogenase is a complex of protein that makes nitrogen fixation possible in plants. It is composed of enzymes with metal cofactors, namely, dinitrogenase and dinitrogenase reductase (Bulen & LeComte, 1966). Various components of the complex of nitrogenase enzyme are encoded by bacterial *nif* genes also (Kaneko et al., 2000; Galibert et al., 2001; Giraud et al., 2007). Due to the limited bioavailability of nitrogen, a colossal has been created in nitrogen-based fertilizer industries (Dobermann et al., 2007; Westhoff, 2009). Diazotrophic bacteria, by performing biological nitrogen fixation, contribute to nearly 50% of the total nitrogen in crop fields as microbial formulation (Ramírez-Puebla et al., 2019).

Besides nitrogen, phosphorus also plays a vital role in establishing equilibrium between plant and animal. Phosphorus (P) is an essential macro-nutrient for plant growth and development. However, the less availability of phosphorus in soil presents significant challenges for crop yield, especially when crops are facing abiotic stress challenges such as drought and salinity. Phosphorus required by plants helps in photosynthesis, energy transfer, and at the molecular level, it is required for DNA/RNA synthesis (Khan et al., 2023b). Phosphorus contributes approximately 0.2% to the dry weight of plants, making it important for the development and growth of plants (Alori et al., 2017). Phosphorus-solubilizing bacteria are of two different categories, one that mineralizes the organic phosphorus and the other that mineralizes the inorganic phosphorus. The former is known as organic PSB and the latter as inorganic PSB (Cao et al., 2018; Timofeeva et al., 2022). Phosphate solubilization occurs through the process of acidolysis. During this process, pyrroloquinoline quinine (PQQ) and glucose dehydrogenase (GDH) genes facilitate the secretion of organic acids. This process is completed by these two genes. Besides these genes, some chelates are also present in the soil, such as Fe^{3+}, Ca^{2+}, and Al^{3+}, that help to release PO_4^{3+} (Tang & Gao, 2023). The organic phosphorus is not directly used by plants; in this reference, organic phosphorus present in the soil is degraded by enzymatic actions, where *ppxppk, phn*, and some other genes are required for enzymatic actions. These enzymes participate in the hydrolysis of organic phosphorus in the soil, which releases inorganic phosphate. Some of the bacteria play a crucial role in this mechanism (Etesami et al., 2021).

9.3 Biofertilizers and Biopesticides as Microbial Bioformulation

Biofertilizers and biopesticides play a crucial role in sustainable agriculture. These two microbial formulations act as key enhancers in crop yield and productivity. Biofertilizers contain nitrogen-fixing microbes like *Rhizobium, Azotobacter*, and *Azospirillum*, which enhance soil fertility and reduce dependency on chemical fertilizers. In recent years, biological fertilizer has emerged as a key component to fix atmospheric nitrogen, solubilize phosphorus, and finally promote plant growth. These eco-friendly fertilizers have the ability to make the environment clean by the process of bioremediation. These plant growth-promoting microbes, like bacteria, fungi, and algae, promote sustainable development approaches to achieve the zero hunger parameter of the United Nations (Verma & Pandey, 2022). The application of biofertilizers improves soil quality and its nutrients for plant growth. In several research studies, results showed the mechanisms of biofertilizers to improve plant growth, how biofertilizers act on plants, and what their impacts are on the nutritional status of plants. They improves oil environments and the soil microbiome to enhance the plant yield and growth via regulation of phytohormones. The bacterial genera used as biofertilizers included *Pseudomonas, Azospirillum*, and *Bacillus* (Zhao et al., 2024a).

9.4 Bioplastics

Synthetic plastics have emerged as a critical problem in the last few years. They pose a huge pollution problem, impacting soil, water, and plants, and are ultimately harmful to our beautiful planet. Although multifaceted disuses of traditional plastic pollution in agricultural space pose a risk to food security, they ultimately hamper the target to achieve SDG2. Most of the review papers have mentioned different plastic products and their harmful impact on agriculture and discussed different sources of plastic-based pollution, the benefits and disadvantages of utilizing conventional plastic products, and the strategies for removing plastic pollution in agriculture. Moreover, after understanding current plastic applications, their benefits, adverse effects, and risks to soil, plants, and environmental health, researchers have addressed the need for technological regulations, advancements, and social approaches that could contribute to mitigate pollution of our planet, which is caused by conventional plastic (Lakhiar et al., 2024). Widespread use of synthetic plastic has led to accumulation in soil. Synthetic plastic products used in agriculture space pose significant environmental risks, including soil contamination, disruption of ecosystems, and potential harm to human health through food chain contamination. These synthetic plastics negatively impact our earth, in terms of soil and water, which hampers crop growth and development and ultimately threatens the food chain and food security (Hofmann et al., 2023).

Bioplastics are made from renewable resources like cellulose and starch, as well as biodegradable compounds, such as polyhydroxyalkanoates PHA and polylactic PLU. These are eco-friendly acceptable substitutes for conventional, harmful synthetic plastics.

Biodegradation, the natural breakdown of organic materials, is greatly aided by microorganisms such as bacteria, algae, and fungi. The complicated polymer chains of bioplastics can be broken down into simpler, nontoxic compounds by the enzymes that certain microorganisms have evolved. Several research have demonstrated that some bacterial and fungal species, such as *Pseudomonas, Bacillus*, and *Aspergillus* spp., are capable of degrading bioplastics in specific environmental circumstances. These microbial strains have some specific enzymes to degrade bioplastics. The plastic surface is colonized by microorganisms; enzymes are secreted and breakdown products are assimilated as nutrition for the microbes, although the efficiency of this process depends on variables like temperature, moisture content, and the type of microbial community involved in the process. Ongoing research projects endeavor to optimize conditions and identify the most effective microbes for large-scale bioplastic degradation. Microbial degradation may be a significant factor in reducing the environmental impact of bioplastics in near future.

9.4.1 Polyhydroxybutyrates as a Promising Approach to Achieve Zero Hunger

Plastic pollution is a grave environmental threats currently causing several problems for us; it leaves a negative footprint for animal and plant kingdoms. The innovation of bioplastic materials is a boon for mitigating agricultural and environmental pollution and for the reinstatement of the natural things of our incredible planet. Biologically synthesized plastics are emerging as a sustainable alternative of fossil-derived plastics. Among the most promising materials being developed, polyhydroxybutyrate (PHB) is a microbially synthesized polyester of the polyhydroxyalkanoate (PHA) family. This biocompatible and nontoxic polymer is naturally produced and is accumulated by a number of specialized archaeal and eubacterial species. PHB is a prominent substitute for synthetic polymers such as high-density polyethylene (HDPE), which is used globally (McAdam et al., 2020).

In the upcoming years, the whole world might be shifted from HDPE to polyhydroxybutyrates (PHBs) to protect our planet. PHBs are macromolecules synthesized by microbes. Because PHBs rapidly degrade under natural environmental settings, they were considered alternatives for the manufacturing of biodegradable polymers. In a study, more than 15PHB-accumulating strains were isolated. Among all the strains, *Bacillus wiedmannii* AS-02 OK576278 isolated from cattle manure showed excellent activity to degrade pollutants. The isolate *B. wiedmannii* AS-02 OK576278 could possibly be considered an excellent source for industrial PHB production from agricultural wastes (Danialet al., 2021). Synthetic polymers area major threat to the ecosystem, and because of their toxicity, they create a hindrance for agricultural land. Therefore, to assure sustainability, it is important to find an effective substitute for these plastics with bioplastics which is eco-friendly and cost effective as well. In a study, the production of biodegradable polymer was through the bacterial species, i.e. *Bacillus* sp. KE4. The extracted bioplastic was characterized by using various instrumentation techniques, which confirmed the structure of the bioplastic as PHB. The results have shown that the biomaterial is environmentally benign, biocompatible, and biodegradable; therefore, it may have diverse applications involving agriculture, packaging, and allied fields (Acharjee et al., 2024). PHB was also synthesized by microalgae *Chlorella sorokiniana* SVMIICT8 (Kumari et al., 2022). Nowadays plastic is frequently utilized in coverings and as packaging materials. Synthetic plastics' extremely slow degradation and overconsumption create major threats to ecosystems due to their toxic effects, while polyhydroxyalkanoates (PHA) are offering a unique sustainable substitute to synthetic plastics (Ahuja et al., 2024).

Microbial cell-based bioplastics serve as an eco-friendly sustainable approach to overcome the impacts of conventional polymers. Microbial synthesis of biopolymers like polyhydroxyalkanoates, polylactic acid, and polyhydroxybutyrates, by using sustainable feedstocks, as well as industrial wastes, has gained popularity in the recent years. Scientific works on bioplastics have suggested several applications in the field of agriculture and environment. Through these researches, it is concluded that microorganisms, such as *Burkholderia sacchari, Rhizobium* sp., *Alcaligenes latus, Lactobacillus pentosus, Micrococcus* sp., *Klebsiella* sp., *Escherichia* sp., *Bacillus* sp., *Azotobacter* sp., *Enterobacter* sp., *Protomonas* sp., *Cupriavidus* sp., *Halomonas* sp., *Ralstonia* sp., *Pseudomonas* sp., *Kluyveromyces* sp., and *Saccharomyces* sp., are known to synthesize eco-friendly bioplastics. Microorganisms serve as key player in dealing with pollution caused by plastic due to their ability to degrade plastics in natural environments and by forming suitable alternatives as biosynthetic machinery of bioplastics to commonly used petroleum-based plastics. Bioplastics, unlike synthetic plastics, can be produced by using renewable resources and are preferred novel materials of high potential value of the twenty-first century. Bioplastics are also called Green plastics due to their eco-friendly nature; these are either starch or cellulose PLA derivatives. Research on microbially synthesized bioplastics has revealed that several microbes can produce and store bioplastics as a source of ATP and carbon. Depending on microbes, carbon sources, and growth factors, the types and molecular weight of the polymers produced differ significantly. Several bacterial strains, such as *Pseudomonas, Bacillus, Klebsiella, Enterobacter, Escherichia, Citrobacter* sp., and actinomycetes *Rhodococcus* sp., strain BSRT1-1, are some of the well-known PHA/PHB producing bacteria (Trakunjae et al., 2021; Varghese et al., 2022).

The degradation mechanism of bioplastics in soil is closely related to the main components in bioplastics. Synthetic plastics such as polyethylene and polypropylene are used in several fields as a main component for packaging. Plastic wastes are disposed of by incineration and/or landfill treatment. Due to these methods, soil and water pollution has burst out as environmental problems. Several types of bioplastics, such as polylactic acid (PLA), polyhydroxyl alkanoate (PHA), and polybutylene succinate (PBS), which contain biological polymers or natural materials, have been produced. Their use as a replacement for conventional polymers is being considered in the medical fields and agricultural space. Bioplastic degradation through soil microbes has been reported, and several microorganisms, such as *Bacillus* sp. and *Aspergillus* sp., are being isolated and identified as bioplastic degraders from the soil environment. Bioplastics seemed to be efficiently degraded in fertile agricultural land because of rich bacterial biomass. Bioplastic use in agricultural fields is anticipated to increase in the future; thus, the understanding of the influence of bioplastic degradation on materials in the agricultural soil is required. It was concluded that newly synthesized bioplastics having high degradation potential in soil and much safer for the environmental microorganisms should be manufactured in the future (Adhikari et al., 2016). In another study, the isolation of *B. wiedmannii* AS-02 from cattle manure was found capable of efficiently utilizing the fruit peel waste as a nutrient source for PHB production (Danial et al., 2021).

9.5 Microbial Biochar

The growing population of the world poses a global concern about food availability. In this aspect novel eco-friendly fertilizers should be urgently developed to meet the demand of human population. Synthetic fertilizers are proving as a curse for our planet, so the researchers have developed an eco-friendly fertilizer, i.e. biofertilizer or microbial fertilizer. The biochar having a porous nature, has become a promising biological carrier for microbial. However, there is not much discussed or stated about the formulation and mechanisms of the effects of biochar-based microbial fertilizers in the literature. In a study, *Bacillus*, exogenous NPK fertilizers, and biochar were used as raw materials to prepare biochar-based microbial fertilizers by adjusting the preparation steps and the process parameters, and as a result, the releasing patterns of NPK were analyzed. Similarly, a pot experiment was performed to check the effect of biochar-based fertilizer on soil fertility and health, and the results showed that adsorption by biochar helps in maintaining microbial activity, whereas the granulation process decreases the same. The adsorption-granulation process results in increasing the content of total nitrogen and organic matter present in the soil. The use of this bio-based fertilizer has significantly improved the availability of soil nutrients and soil quality, while simultaneously enhancing the presence of bacteria related to carbon and nitrogen metabolism process in the soil (Wang et al., 2023). In another study, the potential of oil palm kernel shell (OPKS) was assessed, which was biochar-based, by using *Bacillus* consortium to form strong biofilm carriers by improving the operating and environmental parameters that influence strong biofilm adhesion. This optimization situation was subjected to storage purposes in different temperatures and *in vitro* soil conditions, which reveals that the *Bacillus* biofilm-adhered OPKS biochar might be stored up to 16 weeks. This work emphasizes the physicochemical and soil stability potential of improved *Bacillus* consortium adhesion on biochar for agricultural purposes (Ajeng et al., 2023).

Biochar is preferred as one of the most promising sustainable solutions for soil quality improvement. In this respect, Vlajkov et al. (2023) developed and characterized a *Bacillus*-loaded biochar for sustainable agriculture. The *Bacillus* sp. BioSol021 was thoroughly analyzed in terms of PGP traits that indicate potential for the production of indole acetic acid (IAA), hydrolytic enzymes, and 1-aminocyclopropane-1-carboxylic acid (ACC) deaminase. This biochar-based Bacillus fertilizer was applied on soybean. The best results were show non maize seed germination and its seedling, root, and shoot growth promotion. Results concluded that it can be used as a beneficial solution for sustainable agricultural practices to mitigate food scarcity.

The European Union, among all the continental unions, is at the top in wheat production worldwide, but wheat productivity depends upon soil fertility. The wheat productivity is directly proportional to soil fertility. Microbial biofertilizers are much preferred over chemical ones, either alone or in combination with biochar. Moreover, with the addition of biofertilizers, specific, plant growth-promoting microbes can also modify and enhance crop productivity. In a study, the two wheat varieties, viz. *Triticum durum* and *Triticum aestivum,* were cultivated under the biochar-based microbial fertilizer field. Results of the field experiment concluded that the effects of microbial biochar were investigated in detail; it enhanced grain yield in both wheat cultivars, but more yield was obtained in *T. durum* when compared to control. The study results indicated that the application of bio-fertilizers and biochar may be beneficial for more wheat grain cultivation and ameliorate the sustainable agriculture (Caldara et al., 2024). Besides the wheat, rice is also a key grain for the human population. Rice (*Oryza sativa* L.) is one of the three major food crops cultivated worldwide. Approximately 50% of the world's population uses rice as a staple food. There are several types of synthetic fertilizers available in the market to enhance crop productivity and yield, but they also reduce the nutritional quality of rice (Zhao & Chen, 2020). The use of biofertilizers among farmers is emerging as an economical and eco-friendly alternative to promote sustainable agricultural practices; it might prove to be a good alternative to chemical fertilizer. Biochar from microalgae/cyanobacteria microbes can be utilized to improve the productivity of rice crops through nitrogen fixation in paddy fields. In a study of rice, it was found that a microbial biochar produced from Spirulina (*Arthrospira platensis*) species enhances the development and seed production of rice plants. These results suggested that biochar derived from *A. platensis* biomass can potentially boost the production of rice seed; it can act as a key player to fulfill the Sustainable Developmental Goal 2 (Minello et al., 2024).

9.6 Bioaugmentation Support to Crop Productivity

Bioaugmentation in agriculture involves introducing specific microorganisms into the soil to enhance the degradation of soil pollutants, synthetic pesticides, and also improve the nutrient availability for plants. Among all microbes, bacteria are excellent bioaugmentation microbes. In a study, a hydroponic experiment was set up for the growth and productivity of two cultivars of *Mentha aquatica* viz. *Mentha aquatica* var. *litoralis* (MAL) and *Mentha aquatica* var. *aquatica* (MAA). The productivity of *mentha* was ameliorated due to the application of microbial consortia of the rhizosphere microbial community, when microbial consortia (Bacteroidetes, Actinobacteria, Planctomycetes, Firmicutes, and Verrucomicrobia) were applied in greenhouse hydroponic setup. They provided significant stimulation for the growth of both cultivars. Results revealed that a significant increase in the dry shoot mass and higher yield were obtained when compared with the uninoculated control (Nanda et al., 2019; Kalniņs et al., 2022). Microbes also have great potential to degrade heavy metals from contaminated sites via the bioaugmentation process. In Zambia, a research study was carried out, in which filamentous fungi are used to bioremediate pollutants such as Cu, Co, Fe, Mn, and Zn from soil. Researchers have isolated indigenous filamentous fungi from heavy metal-contaminated soil to assess their potential for bioremediation purposes through bioaugmentation. The identified filamentous fungal strains are *Aspergillus transmontanensis, Geotrichum candidum,* and *Cladosporium cladosporioides* species, which were helpful in heavy metal remediation. This study has highlighted the highly resistant autochthonous fungal species' potential to remediate soil contaminated from heavy metal pollution, which is crucial for crop productivity and sustainable agriculture (Dusengemungu et al., 2022). Medaura et al. (2021) studied that a microbial consortium of six potentially hydrocarbonoclastic fungi, *Penicillium, Fusarium, Ulocladium,* and *Aspergillus*, is highly capable of degrading PAHs from polluted soil systems. Research reveals that fungi introduction to hydrocarbon-contaminated soil environment stimulates the growth of native microbes that are capable of degrading such contaminants. This fungal bioaugmentation strategy holds a potential, novel remediation approach, specifically in degradation of long-standing, complex, highly toxic and resistant industrial contamination in the soil ecosystem. It supports the SDG2 in an indirect manner. Microorganisms are capable of degrading pollutants, such as oil-degrading *Pseudomonas* species, which helps in environmental cleanup (Diaz-Troya & Huertas, 2024).

9.7 Microbial Formulations as Bio-control Agents

Several pesticides are available in the market to control pests and diseases of agricultural crops, while these are harmful for crops as well as the environment. Chemical pesticides are frequently used by farmers to enhance productivity to meet global food demand. However, farmers are not aware of the bad impacts of these pesticides in the field of agriculture and environment. To overcome the use of these chemical pesticides, researchers have developed microbial-loaded biopesticides, which act as eco-friendly biocontrol agents. Biopesticides are an eco-friendly and cost-effective alternative of synthetic pesticides. Several biopesticides are used in laboratory and field experiments to control insects, pests, and nematodes (Waghunde et al., 2021). Agricultural production is challenged worldwide due to plant pathogens that create deterioration in crop productivity. Innovative approaches are required to mitigate the less crop productivity due to abiotic stress (such as drought, salinity, extreme temperatures, and environmental pollution), which can badly affect plant growth and development via various microbial or nonmicrobial infections and some pests. In this reference, some microbial strains are used as biocontrol agents to overcome the side effects of synthetic insecticides, fungicides, and pesticides.

The Green Revolution has not led to sustainable agriculture, despite improved yields of crops, mostly because of the mismanaged application of chemical-based pesticides. Excess use of synthetic pesticides generates a persistent impact on the ecosystem, especially on agriculture sector. The alternative method for these chemical-based compounds is incorporated as bioprotectants. These bioprotectants might be used to maintain agricultural productivity and plant growth to contribute to decreased use of synthetic pesticides. Bio-pesticides derived from microbes such as bacteria, fungi, and viruses are active against pest species; these biocompounds are nontoxic and nonpathogenic. Now, microbial-based biocontrol agents have been identified, manufactured in large quantities, and successfully prepared for sale in the markets as biopesticides. Some have established themselves as domestic and international market brands (BASF SE, Bayer AG, Syngenta Group, Corteva, and Koppert Biological Systems) (Chaudhary et al., 2024).

9.7.1 Molecular Mechanism of Microbial Cells: As Biocontrol Agents Against Plant Pathogens

Among the microorganisms, bacteria play a crucial role as biocontrol agents for sustainable agriculture. Some bacterial genera such as *Pseudomonas* sp., show antagonist activity against various phytopathogens. In a study, *Pseudomonas* antagonism was revealed against wilt disease-causing plant pathogen, i.e. *Fusarium oxysporium*, via the mechanism of protoplast fusion (a technique where cell walls are removed to fuse cells, which can enhance the antagonistic capabilities). In this study, seven antagonistic potential strains, including four *Enterobacter cloacae*, *Pseudomonas fluorescens P. aeruginosa*, and *Bacillus subtilis*, were examined for the capacity to suppress *Fusarium oxysporum*. The highest antagonism was shown by *P. aeruginosa* followed by *P. fluorescens* and *E. cloacae*. Utilizing *P. aeruginosa and P. fluoresces*, the protoplast fusion process was used to combine several fungal biocontrol mechanisms. Microbial fusants were developed by protoplast fusion. The antagonistic effectiveness of all fusants was higher than that of their parent strains. These fusants are suitable for field use and performed better in suppressing the plant pathogen *F. oxysporum* (Abdel-Salam et al., 2007). Besides this, Microbes are capable of synthesizing various chemicals that can facilitate the process of degradation of the phytopathogen's cell wall. In this reference, plant growth-promoting microbes such as *Trichoderma* sp. synthesize chitinase and β 1,3 glucosidase enzymes; these two enzymes play a crucial role in cell wall degradation of pathogens (Figure 9.2).

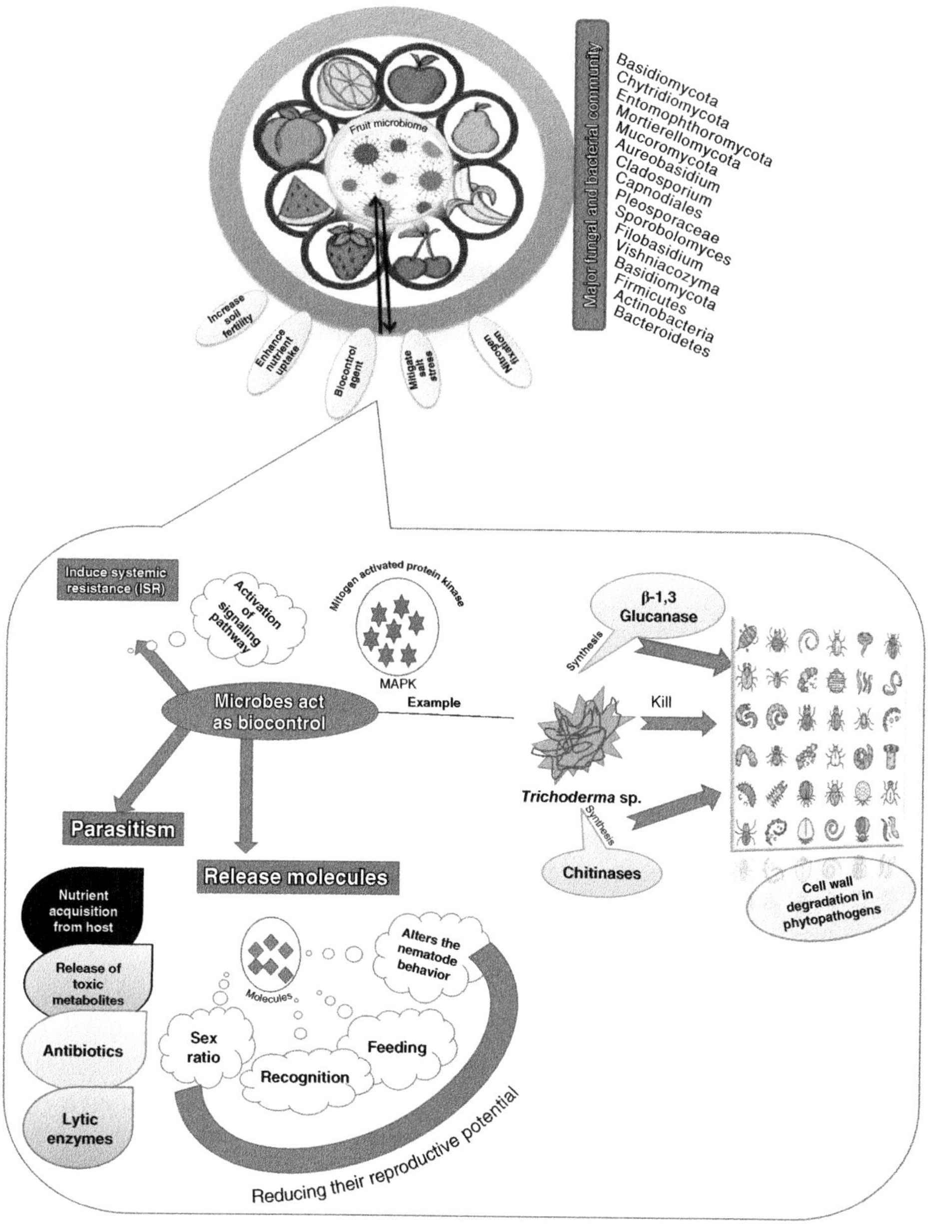

Figure 9.2 Major microbial community of various fruits acting as plant growth-promoting microorganisms, with special reference to biocontrol agent: molecular mechanism.

9.8 Microbial Enzymes Promoting Nutrient Uptake

The increasing population creates the issue of food shortage globally and decreases the availability and accessibility of food grains, eventually demanding food security (Dotaniya et al., 2019; Pandey et al., 2023), along with challenges related to climate change, deforestation, and natural resources exploitation that has made sustainable crop cultivation extremely difficult, making this one of the most focused section of research in agricultural and environmental field across the world (Dotaniya et al., 2019). Conventional farming practices heavily rely on the use of chemical fertilizers and pesticides. Farmers have also been misled this entire time about the overuse of these toxic substances to increase and improve their crop yields, thus further reducing soil biological quality (Dotaniya et al., 2019). Soil is the major component, and its quality is a key factor; these two terms represent the functional state of soil that determines crop growth and plant nutrient availability (Doran & Zeiss, 2000). The use of chemical fertilizers might have enhanced food grain production, but such progress cannot be sustained for long due to overmining of nutrients from the soil apart from those supplied in the form of fertilizers, soil organic matter (SOM) reduction over the time, and climatic factors changes (Meena, 2016; Meena et al., 2016). Plant nutrient availability and transformation rate depend on the SOM; if SOM of any soil composition decreases, the rest will be reduced as well. For sustainable production of crops, modifications in their application rates and climatic conditions area prime requirement for agricultural practices, and also there are different biological and eco-friendly alternatives being introduced as reliable tools for crop productivity enhancement and improvement (Dotaniya et al., 2019).

Plants take up nutrients that are present in available forms, as nutrients present in the soil might be in chemical forms that cannot be directly taken up by the plants. For example, nitrogen is mostly taken in nitrate form, and also sometimes during nitrogen deficiency, ammoniacal nitrogen is preferred by plants. Phosphorous is mostly preferred in the form of orthophosphates or forms that can be converted into insoluble forms (Dotaniya et al., 2019). Microflora play a key role in nutrient transformation as they transform nutrients into mineral forms as a result of their growth and activity, which can be utilized by plants (Dotaniya et al., 2019). The primary nutrients that contribute to the structural and functional molecules of plant growth are C, H, O, N, and K. Secondary nutrients like Ca, Mg, and S and micronutrients such as B, Cl, Cu, Fe, Mn, Mo, and Zn, though comparatively in smaller portions, are also important contributors for plant growth and development (Dotaniya et al., 2019). Microbial enzymes are mainly responsible for the mineralization and transformation of nutrients. These enzymes can be extracellular or endogenous enzymes of microbial cells. As extracellular enzymes are secreted in the soil ecosystem, they degrade or transform the organic or inorganic forms of nutrients into yield-friendly forms (Dotaniya et al., 2017). Collectively, these enzymes are termed soil enzymes and are crucial for soil fertility and health maintenance, also helping in guarding the environment by removing pollutants and remediating the system (Gianfreda & Rao, 2014; Stirling et al., 2017). Enzymes act as biological catalysts by promoting plant growth and crop yield, playing critical role in plant–microbe–soil interactions. Recent studies have shown potential in the involvement of enzyme-based solutions, either in the form of microbial inoculants or direct enzyme formulations, in sustainable agriculture systems for better and improved crop productivity.

This section explains various beneficial enzymes for agriculture, their role in improving and enhancing crop productivity and facilitating resource-friendly farming, which resolves the issue of food security and sustainability.

9.8.1 Soil Enzymes' Role in Agriculture

Soil enzymes influence the maintenance of different nutrient cycles and facilitate soil fertility for healthy plant development (Fang et al., 2010). Soil enzymes accelerate the decomposition rate of organic residues by producing enzymes and influencing the plant nutrient kinetics and interaction between the reactants (microbes, substrate, etc.) (Pankhurst et al., 1995). Rhizosphere, the root-contacted soil areas of a plant, is a unique system where there is a diversified enrichment of microorganisms as compared to other areas of the plant. As a result, enzyme activities in this rhizosphere zone are prominent in mineral transformation and also help to maintain soil–plant interaction in better forms (Gobran et al., 1998). Plant roots are enriched with low molecular weight organic acids, which serve as immediate carbon source for soil microbes, and as a result, microbes showcase priming effect on the production of enzymes, which ultimately speeds up the mineralization of organic substances, thereby releasing simpler forms of inorganic acids into the soil for plant uptake. These inorganic ions further act as chelating agents by forming complexes with different plant nutrients and enzymes; the degradation of these complexes results in the release of enzymes and nutrients in available forms into the soil (Dotaniya et al., 2013). This chelating property helps plants reduce the uptake of toxic, heavy metals from the soil, as it degrades toxic forms into nontoxic forms (Gianfreda & Rao, 2014).

The major enzymes are those involved in the mineralization of C, N, P, and S (Dotaniya et al., 2019). The enzymes involved in C cycling convert complex organic compounds of C into much simpler and utilizable forms, such as organic acids, sugars, etc., as these cannot directly supply nutrients for plant growth. Similarly, enzymes for N, P, and S cycling transform inorganic components into utilizable forms from the soil organic matter for plants and microbes (Dotaniya et al., 2019). There are certain physical factors like pH, EC, and soil composition, which influence enzyme composition, their presence and activities in the soil, indirectly controlling nutrient supply for microbes and plants (Sinsabaugh et al., 2005).

9.8.2 Enzymes Promoting Plant Growth

The major soil enzyme classes are oxidoreductases, hydrolases, and lyases. The oxidoreductases class is mainly involved in the oxidation of various reactants (toxic organic substances) through enzymes such as dehydrogenases, laccases, and other enzymes. Dehydrogenase's critical role involves measurement of microbial metabolic activities in the soil as they are intracellular in nature and are found almost in all living microbes. Also, they are responsible for the H-ion transport and are an important part of respiration and electron transport chains in plants, reflecting their importance in SOM and nutrient uptake (Gianfreda & Rao, 2014). Hydrolases are the dominant class of extracellular enzymes involved in the degradation of macromolecules into forms available for soil microbes and plants (Dotaniya et al., 2019). Enzymes like cellulose and β-glucosidase help in the decomposition of cellulose content in SOM, which is present in the form of crop residues and litter in the soil. These enzymes transform complex, low molecular-weight carbohydrates into simpler sugars like glucose for easy accessibility for the plants, also stimulating microbial activity as the carbon content in soil increases and indirectly enhancing nutrient uptake by the plant (Ajwa & Tabatabai, 1994). β-glucosidase not just serves in nutrient cycling maintenance but also as a natural indicator of soil health and fertility because of its extracellular nature in different soil and climatic conditions (Ajwa & Tabatabai, 1994). In the same way, phosphatases hydrolyze organic phosphorus content into available forms (orthophosphate) that can be easily taken upby plants. Phosphate naturally is present in ore form in the soil, which cannot be directly utilized by the plants; therefore, under acidic environment, microbes and plant roots mineralize phosphates by providing organic acids to their surroundings, which liberates these molecules into easy-to-take forms (Dotaniya et al., 2019). These enzymes are crucial in phosphorus-deficient soils, but excessive acidic conditions can inhibit microbes and plant growth (Dotaniya et al., 2019). Amylase enzyme contributes to nutrient cycling, like C cycling, under situations like residue- or starch-rich soils. They alone do not have that important role as there is not much starch content found in the soil environment that they could be utilized, but they do show their involvement in certain things.

These points indicate that soil enzymes are not just biological catalysts in nutrient availability, but they are equally important tools for learning and understanding about soil health, its fertility rate, and much more, which is gradually affected by physical and chemical factors such as pH, climatic factors, pollution, and crop management strategies. Supporting farming practices by relying on organic farming applications, utilizing various cropping systems, decreasing or preventing chemical use in farming in any form (pesticides or fertilizers), and including enzyme-based activities can potentially enhance and boost soil health. This will automatically improve crop productivity all over and will not only solve quality issues of produce but also create positive impacts toward the Zero Hunger goal.

9.9 Challenges and Future Prospects

Microbial formulations are challenging, but the outcomes supported human welfare in various ways discussed earlier. However, the stability and shelf life of microbial formulations need improvement, and large-scale production faces regulatory and economic hurdles. The most challenging thing is that awareness and adoption by farmers and industries remain limited. In the near future, more advancements in microbial biotechnology, such as genetic engineering, will be needed, which will enhance microbial efficacy. Integration of artificial intelligence and machine learning can optimize microbial formulations for targeted applications. Policies promoting eco-friendly microbial technologies can accelerate global adoption.

9.10 Conclusion

Microbial formulations offer a sustainable, eco-friendly solution to many global challenges. By integrating microbial-based technologies into agriculture management, we can make significant progress toward the SDGs, especially with reference to SDG2 to achieve Zero Hunger. Governments, industries, and researchers of various countries must collaborate to enhance the development, accessibility, and widespread adoption of these technologies for a more sustainable future.

References

Abdel-Salam, M.S., Abd El-Halim, M.M., & El-Hamshary, O.I.M. 2007. Improvement of Pseudomonas antagonism against *Fusarium oxysporum* through protoplast fusion: I-fusants induction. *Res. J. Cell. Mol Biol.* 1(1), 37–41.

Acharjee S.A., Bharali, P., Ramachandran, D., Kanagasabai, V., & Gogoi, M. 2024. Polyhydroxybutyrate (PHB)-Based sustainable bioplastic derived from *Bacillus* sp. KE4 isolated from kitchen waste effluent. *Sustain. Chem. Pharm.* 39, 101507.

Adhikari, D., Mukai, M., Kubota, K., Kai, T., Kaneko, N., Araki, K.S., & Kubo, M. 2016. Degradation of bioplastics in soil and their degradation effects on environmental microorganisms. *J. Agric. Chem. Environ.* 5(1), 23–34. https://doiorg/10.4236/jacen.2016.51003

Ahuja, V., Singh, P.K., Mahata, C., Jeon, J.M., Kumar, G., Yang, Y.H., & Bhatia, S.K. 2024. A review on microbes mediated resource recovery and bioplastic (polyhydroxyalkanoates) production from wastewater. *Microb. Cell. Fact.* 23(1), 187. https://doi.org/10.1186/s12934-024-02430-0

Ajeng, A.A., Abdullah, R., & Ling, T.C. 2023. Biochar-*Bacillus* consortium for a sustainable agriculture: physicochemical and soil stability analyses. *Biochar.* 5(1), 17. https://doi.org/10.1007/s42773-023-00215-z

Ajwa, H.A., & Tabatabai, M.A. 1994. Decomposition of different organic materials in soils. *Biol. Fertil. Soils.* 18, 175–182. https://doi.org/10.1007/BF00647664

Al-Dulaimi, A.S.T., & Al-Rawi, W.A. 2020. Effects of biofertilizers and compost application on vegtative growth of plum transplants. *Plant. Arch.* 20(1), 2215–2220.

Alori, E.T., Glick, B.R., & Babalola, O.O. 2017. Microbial phosphorus solubilization and its potential for use in sustainable agriculture. *Front. Microbiol.* 8, 971. https://doi.org/10.3389/fmicb.2017.00971

Arora, N.K., Mishra, J., Singh, P., & Fatima, T. 2024. Salt-tolerant plant growth-promoting *Pseudomonas atacamensis* KSS-6 in combination with organic manure enhances rice yield, improves nutrient content and soil properties under salinity stress. *J. Basic. Microbiol.* 64(6), 2300767. https://doi.org/10.1002/jobm.202300767

Baidoo, R., Mengistu, T., McSorley, R., Stamps, R.H., Brito, J., & Crow, W.T. 2017. Management of root-knot nematode (*Meloidogyne incognita*) on *Pittosporum tobira* under greenhouse, field, and on-farm conditions in Florida. *J. Nematol.* 49(2), 133.

Buhroo, A.A., Rasheed, F.N., & Khanday, A.L. 2016. An assessment on population density of San Jose scale *Quadraspidiotusperniciosus* (Comstock) and its biological control in Kashmir (Hemiptera: Diaspididae). *Mun. Ent. Zool.* 11, 434–440.

Bulen, W.A., & LeComte, J.R. 1966. The nitrogenase system from Azotobacter: two-enzyme requirement for N2 reduction, ATP-dependent H2 evolution, and ATP hydrolysis. *Proc. Nat. Acad. Sci. USA.* 56(3), 979–986. https://doi.org/10.1073/pnas.56.3.979

Cai, P., & Dimopoulos, G. 2025. Microbial biopesticides: a one health perspective on benefits and risks. *One Health* 20, 100962. https://doi.org/10.1016/j.onehlt.2024.100962

Caldara, M., Gullìa, M., Graziano, S., Riboni, N., Maestri, E., & Matta, M. 2024. Microbial consortia and biochar as sustainable biofertilizers: analysis of their impact on wheat growth and production. *Sci. Total. Environ.* 917, 170168.

Cao, Y., Fu, D., Liu, T., Guo, G., & Hu, Z. 2018. Phosphorus solubilizing and releasing bacteria screening from the rhizosphere in a natural wetland. *Water* 10(2), 195. https://doi.org/10.3390/w10020195

Chaudhary, R., Nawaz, A., Khattak, Z., Butt, M.A., Fouillaud, M., Dufossé, L., & Mukhtar, H. 2024. Microbial bio-control agents: a comprehensive analysis on sustainable pest management in agriculture. *J. Agric. Food. Res.* 18, 101421. https://doi.org/10.1016/j.jafr.2024.101421

Cortazar-Murillo, E.M., Méndez-Bravo, A., Monribot-Villanueva, J.L., Garay-Serrano, E., Kiel-Martínez, A.L., Ramírez-Vázquez, M., & Reverchon, F. 2023. Biocontrol and plant growth promoting traits of two avocado rhizobacteria are orchestrated by the emission of diffusible and volatile compounds. *Front. Microbiol.* 14, 1152597. https://doi.org/10.3389/fmicb.2023.1152597

Coulibaly, P.J., Sawadogo, J., Bambara, Y.A., Ouédraogo, W., Legma, J.B., & Compaoré, E. 2021. Effect of bio-fertilizers on tomato (*Solanum Lycopersicum*) production and on soil physico-chemical properties in Sudan Area of Burkina Faso. *Curr. Agric. Res. J.* 9(1), 43–53. https://doi.org/10.12944/CARJ.9.1.06

da Costa Pereira, A.C., de Castro, G.L.S., Rodrigues, P.C., Silva, G.B., de Oliveira, D.A., & de Souza, C.R.B. 2019. An endophytic *Pseudomonas* sp. of *Piper tuberculatum* promotes growth on *Piper nigrum* through increase of root biomass production. *Physiol. Mol. Plant. Pathol.* 108, 101420. https://doi.org/10.1016/j.pmpp.2019.101420

Danial, A.W., Hamdy, S.M., Alrumman, S.A., Gad El-Rab, S.M., Shoreit, A.A., & Hesham, A.E.L. 2021. Bioplastic production by *Bacillus wiedmannii* AS-02 OK576278 using different agricultural wastes. *Microorganisms* 9(11), 2395. https://doi.org/10.3390/microorganisms9112395

Devi, S., Sharma, S., Tiwari, A., Bhatt, A.K., Singh, N.K., Singh, M., & Kumar, A. 2023. Screening for multifarious plant growth promoting and biocontrol attributes in Bacillus strains isolated from indogangetic soil for enhancing growth of rice crops. *Microorganisms* 11(4), 1085. https://doi.org/10.3390/microorganisms11041085

Diaz-Troya, S., Huertas, M.J. 2024. Green microbes: Potential solutions for key sustainable development goals. *Microb. Biotechnol.* 17(8): e14546. doi: 10.1111/1751-7915.14546.

Din, I.U., Hu, L., Jiang, Y., Wei, J., Afzal, M., & Sun, L. 2024. Bacterial lipopeptides are effective against Pear Fire Blight. *Microorganisms* 12(5), 896. https://doi.org/10.3390/microorganisms12050896

Dobermann, A., 2007. Nutrient use efficiency—measurement and management. In: A. Krauss, K. Isherwood, & P. Heffer (Eds.), *Fertilizer Best Management Practices: General Principles, Strategy for Their Adoption and Voluntary Initiatives Versus Regulations* (pp. 1–28). Paris, France: International Fertilizer Industry Association.

Doran, J.W., & Zeiss, M.R. 2000. Soil health and sustainability: managing the biotic component of soil quality. *Appl. Soil Ecol.* 15(1), 3–11. https://doi.org/10.1016/S0929-1393(00)00067-6

Dotaniya, M.L., Aparna, K., Dotaniya, C.K., Singh, M., & Regar, K.L. 2019. Role of soil enzymes in sustainable crop production. In: M. Kuddus (Ed.), *Enzymes in Food Biotechnology* (pp. 569–589). Academic Press. https://doi.org/10.1016/B978-0-12-813280-7.00033-5

Dotaniya, M.L., Pingoliya, K.K., Meena, H.M., & Prasad, D., 2013. Status and rational use of rock phosphate in agricultural crop production- are view. *Agric. Sustain. Dev.* 1(1), 103–108.

Dotaniya, M.L., Rajendiran, S., Meena, V.D., Saha, J.K., Coumar, M.V., Kundu, S., & Patra, A.K. 2017. Influence of chromium contamination on carbon mineralization and enzymatic activities in Vertisol. *Agric. Res.* 6, 91–96. https://doi.org/10.1007/s40003-016-0242-6

Dusengemungu, L., Gwanama, C., Simuchimba, G., & Mubemba, B. 2022. Potential of bioaugmentation of heavy metal contaminated soils in the Zambian Copper belt using autochthonous filamentous fungi. *Front. Microbiol.* 13, 1045671. https://doi.org/10.3389/fmicb.2022.1045671

Dutta, P., Kundu, S., & Chatterjee, S. 2010. Effect of bio-fertilizers on homestead fruit production of papaya cv. Ranchi. In: *II International Symposium on Papaya* (Vol. 851, pp. 385–388). ISHS. https://doi.org/10.17660/ActaHortic.2010.851.59

Erler, F., & Ates, A.O. 2015. Potential of two entomopathogenic fungi, *Beauveriabassiana* and *Metarhiziumanisopliae* (Coleoptera: Scarabaeidae), as biological control agents against the June beetle. *J. Insec. Sci.* 15(1), 44. https://doi.org/10.1093/jisesa/iev029

Etesami, H., Jeong, B.R., & Glick, B.R. 2021. Contribution of arbuscular mycorrhizal fungi, phosphate–solubilizing bacteria, and silicon to P uptake by plant. *Front. Plant. Sci.* 12, 699618. https://doi.org/10.3389/fpls.2021.699618

Fang, S., Liu, J., Liu, D., & Xie, B. 2010. Enzymatic activity and nutrient availability in the rhizosphere of poplar plantations treated with fresh grass mulch. *Soil Sci. Plant. Nut.* 56(3), 483–491. https://doi.org/10.1111/j.1747-0765.2010.00480.x

Fernández, A.S., Beovides-García, Y., Simó-González, J.E., María C. Penaranda, P., Lopez-Torres., J., Rayas-Cabrera, A., Santos-Pino, A. Basail-Pérez, M. 2021. Effect of a *Pseudomonas fluorescens*-based biofertilizer on sweet potato yield components. *Asian J. Appl. Sci.* 9(2) 105-113. DOI: 10.31031/EAES.2021.08.000692

Galibert, F., Finan, T.M., Long, S.R., Puhler, A., Abola, P., Ampe, F., & Batut, J. 2001. The composite genome of the legume symbiont *Sinorhizobiummeliloti. Science* 293(5530), 668–672. https://www.science.org/doi/10.1126/science.1060966

Gamez, R.M., Ramirez, S., Montes, M., & Cardinale, M. 2020. Complementary dynamics of banana root colonization by the plant growth-promoting rhizobacteria *Bacillus amyloliquefaciens* Bs006 and *Pseudomonas palleroniana* Ps006 at spatial and temporal scales. *Microb. Ecol.* 80(3), 656–668. https://doi.org/10.1007/s00248-020-01571-0

Gava, C.A.T., de Castro, A.P.C., Pereira, C.A., & Fernandes-Júnior, P.I. 2018. Isolation of fruit colonizer yeasts and screening against mango decay caused by multiple pathogens. *Biol. Control.* 17, 137–146. https://doi.org/10.1016/j.biocontrol.2017.11.005

Gianfreda, L., & Rao, M.A. 2014. *Enzymes in Agricultural Sciences.* Foster: OMICS Group International.

Giraud, E., Moulin, L., Vallenet, D., Barbe, V., Cytryn, E., Avarre, J.C., & Sadowsky, M. 2007. Legumes symbioses: absence of Nod genes in photosynthetic bradyrhizobia. *Science* 316(5829), 1307–1312. https://doi.org/10.1126/science.1139548

Gobran, G.R., Clegg, S., & Courchesne, F. 1998. Rhizospheric processes influencing the biogeochemistry of forest ecosystems. In: N. Van Breemen (Ed.), *Plant-induced Soil Changes: Processes and Feedback* (pp. 107–120). Springer. https://doi.org/10.1007/978-94-017-2691-7_6

Graham, P.H., & Vance, C.P. 2003. Legumes: importance and constraints to greater use. *Plant Physiol.* 131(3), 872–877. https://doi.org/10.1104/pp.017004

Hofmann, T., Ghoshal, S., Tufenkji, N., Adamowski, J.F., Bayen, S., Chen, Q., & Wilkinson, K.J. 2023. Plastics can be used more sustainably in agriculture. *Commun. Earth Environ.* 4(1), 332. https://doi.org/10.1038/s43247-023-00982-4

Isfahani, F.M., & Besharati, H. 2012. Effect of biofertilizers on yield and yield components of cucumber. *J. Biol. Earth. Sci.* 2(2), B83–B92.

JavanshirJavid, K., Mahdian, S., Behboudi, K., & Alizadeh, H. 2016. Biological control of *Fusarium oxysporum* f. sp. radicis-cucumerinum by some *Trichoderma harzianum* isolates. *Arch. Phytopath. Plant Protec.* 49(17–18), 471–484. https://doi.org/10.1080/03235408.2016.1242195

Kalniņs, M., Andersone-Ozola, U., Gudra, D., Sieriņa, A., Fridmanis, D., Ievinsh, G., & Muter, O. 2022. Effect of bioaugmentation on the growth and rhizosphere microbiome assembly of hydroponic cultures of *Menthaaquatica. Ecol. Gen. Genomic.* 22, 100107. https://doi.org/10.1016/j.egg.2021.100107

Kaneko, T., Nakamura, Y., Sato, S., Asamizu, E., Kato, T., Sasamoto, S., & Tabata, S. 2000. Complete genome structure of the nitrogen-fixing symbiotic bacterium *Mesorhizobium loti. DNA Res.* 7(6), 331–338. https://doi.org/10.1093/dnares/7.6.331

Khan, A., Singh, A.V., Gautam, S.S., Agarwal, A., Punetha, A., Upadhayay, V.K., & Goel, R. 2023a. Microbial bioformulation: a microbial assisted biostimulating fertilization technique for sustainable agriculture. *Front. Plant Sci.* 14, 1270039. https://doi.org/10.3389/fpls.2023.1270039

Khan, F., Siddique, A.B., Shabala, S., Zhou, M., &Zhao, C. 2023b. Phosphorus plays key roles in regulating plants' physiological responses to abiotic stresses. *Plants* 12(15), 2861. https://doi.org/10.3390/plants12152861

Kumari, P., Kiran, B.R., & Mohan, S.V. 2022. Polyhydroxybutyrate production by *Chlorella sorokiniana* SVMIICT8 under nutrient-deprived mixotrophy. *Bioresour. Technol.* 354, 127135. https://doi.org/10.1016/j.biortech.2022.127135

Lakhiar, I.A., Yan, H., Zhang, J., Wang, G., Deng, S., Bao, R., & Wang, X. 2024. Plastic pollution in agriculture as a threat to food security, the ecosystem, and the environment: an overview. *Agronomy* 14(3), 548. https://doi.org/10.3390/agronomy14030548

Lonhienne, T., Mason, M.G., Ragan, M.A., Hugenholtz, P., Schmidt, S., & Paungfoo-Lonhienne, C. 2014. Yeast as a biofertilizer alters plant growth and morphology. *Crop. Sci.* 54(2), 785–790. https://doi.org/10.2135/cropsci2013.07.0488

McAdam, B., Brennan Fournet, M., McDonald, P., & Mojicevic, M. 2020. Production of polyhydroxybutyrate (PHB) and factors impacting its chemical and mechanical characteristics. *Polymers* 12(12), 2908. https://doi.org/10.3390/polym12122908

Medaura, M.C., Guivernau, M., Moreno-Ventas, X., Prenafeta-Boldu, F.X., &Vinas, M. 2021. Bioaugmentation of native fungi, an efficient strategy for the bioremediation of an aged industrially polluted soil with heavy hydrocarbons. *Front. Microbiol.* 12, 626436. https://doi.org/10.3389/fmicb.2021.626436

Medeiros, E.V., Moraes, M.C., Costa, D.P., Silva, J.S., Oliveira, J.B., Lima J.R., & Hammecker, C. 2020. Biochar and *Trichoderma aureoviride* applied to the sandy soil: effect on soil quality and watermelon growth. *Not. Bot. Horti. Agrobot. Cluj-Napoca.* 48(2), 735–751. https://doi.org/10.15835/nbha48211851

Meena, B.L. 2016. Response of compaction with sulphur fertilization to nutrient content, uptake and economics of barley on highly permeable soil. *Int. J. Agric. Sci.* 8(34), 1719–1722. http://www.bioinfopublication.org/jouarchive.php?opt=&jouid=BPJ0000217

Meena, B.P., Shirale, A.O., Dotaniya, M.L., Jha, P., Meena, A.L., Biswas, A.K., & Patra, A.K. 2016. Conservation agriculture: a new paradigm for improving input use efficiency and crop productivity. In: J. Bisht, V. Meena, P. Mishra, & A. Pattanayak (Eds.), *Conservation Agriculture: An Approach to Combat Climate Change in Indian Himalaya* (pp. 39–69). Springer. https://doi.org/10.1007/978-981-10-2558-7_2

Minello L.V.P., Kuntzler, S.G., Lamb, T.I., Neves, C.O., Berghahn, E., da Paschoa, R.P., Silveira, V., de Lima, J.C., Aguzzoli, C., & Sperotto, R.A. 2024. Rice plants treated with biochar derived from Spirulina (*Arthrospira platensis*) optimize resource allocation towards seed production. *Front. Plant Sci.* 15, 1422935. https://doi.org/10.3389/fpls.2024.1422935

Naik, K., Mishra, S., Srichandan, H., Singh, P.K., & Choudhary, A. 2020. Microbial formulation and growth of cereals, pulses, oilseeds and vegetable crops. *Sustain. Environ. Res.* 30, 1–18. https://doi.org/10.1186/s42834-020-00051-x

Nanda, M., Kumar, V., & Sharma, D.K. 2019. Multimetal tolerance mechanisms in bacteria: The resistance strategies acquired by bacteria that can be exploited to 'clean-up' heavy metal contaminants from water. *Aquatic. Toxic.* 212, 1–10. https://doi.org/10.1016/j.aquatox.2019.04.011

Nkhata, S.G., Ayua, E., Kamau, E.H., & Shingiro, J.B. 2018. Fermentation and germination improve nutritional value of cereals and legumes through activation of endogenous enzymes. *Food. Sci. Nut.* 6(8), 2446–2458. https://doi.org/10.1002/fsn3.846

Nyasani, J.O., Subramanian, S., Poehling, H.M., Maniania, N.K., Ekesi, S., & Meyhöfer, R. 2015. Optimizing western flower thrips management on French beans by combined use of beneficials and imidacloprid. *Insects* 6(1), 279–296. https://doi.org/10.3390/insects6010279

Pandey, A., Kumar, S., Kumar, S., Rajput, M., Singh, M., Pandey, C., & Bhardwaj, N. 2023. Role of microbial enzymes in bioremediation: emerging opportunities and limitations. In: V.K. Sharma, A. Kumar, M.R. Zambrano Passarini, S. Parmar, & V.K. Sing (Eds.), *Microbial Inoculants* (pp. 277–300). Academic Press. https://doi.org/10.1016/B978-0-323-99043-1.00007-4

Pankhurst, C.E., Hawke, B.G., McDonald, H.J., Kirkby, C.A., Buckerfield, J.C., Michelsen, P., & Doube, B.M. 1995. Evaluation of soil biological properties as potential bioindicators of soil health. *Aust. J. Exp. Agric.* 35(7), 1015–1028. https://doi.org/10.1071/EA9951015

Parajuli, G., Kemerait, R., & Timper, P. 2014. Improving suppression of *Meloidogyne* spp. by *Purpureocillium lilacinum* strain 251. *Nematology* 16(6), 711–717.

Podgórska-Kryszczuk, I. 2023. Biological control of *Aspergillus flavus* by the yeast *Aureobasidium pullulans* in vitro and on tomato fruit. *Plants* 12(2), 236. https://doi.org/10.3390/plants12020236

Quintarelli, V., Borgatti, D., Baretta, M., Stazi, S.R., Allevato, E., Pancaldi, S., Baldisserotto, C., Mancinelli, R., Tedeschi, P., Radicetti, E., & Hassine, M.B. 2025. Microbial biofertilizers and algae-based biostimulant affect fruit yield characteristics of organic processing tomato. *J. Sci. Food. Agric.* 105(1), 530–539. https://doi.org/10.1002/jsfa.13851

Rahman, M., Sabir, A.A., Mukta, J.A., Khan, M.M.A., Mohi-Ud-Din, M., Miah, M.G., & Islam, M.T. 2018. Plant probiotic bacteria *Bacillus* and *Paraburkholderia* improve growth, yield and content of antioxidants in strawberry fruit. *Sci. Rep.* 8(1), 2504. https://doi.org/10.1038/s41598-018-20235-1

Rajput, M., Mishra, S., Pandey, A., Basera, N., Rana, V., Singh, M. 2024. Microbial consortia application in the sustainable agricultural practices. In: A. Kumar, J.F. White, & J. Singh (Eds.), *Sustainable Agricultural Practices* (pp. 263–275). Academic Press. https://doi.org/10.1016/B978-0-443-19150-3.00001-1

Ramírez-Puebla, S.T., Ormeño-Orrillo, E., Rogel, M.A., López-Guerrero, M.G., López-López, A., Martínez-Romero, J., & Martínez-Romero, E. 2019. Diversity of native Mexican rhizobia under the light of genomics. *Rev. Mex. Biodivers.* 90, e902681. https://doi.org/10.22201/ib.20078706e.2019.90.2681

Rashid, M.I., Mujawar, L.H., Shahzad, T., Almeelbi, T., Ismail, I.M., & Oves, M. 2016. Bacteria and fungi can contribute to nutrients bioavailability and aggregate formation in degraded soils. *Microbiol. Res.* 183, 26–41. https://doi.org/10.1016/j.micres.2015.11.007

Shankarappa, T.H., Reddy, R.N., Subramanyam, B., Sreenatha, A., & Reddy, N.A. 2018. Biofertilizers for growth and establishment of alphonso mango grafts under nursery condition. *Int. J. Curr. Microbiol. App. Sci.* 7, 5205–5211.

Sharma, N., & Rana, V.S. 2015. Effect of plant growth promoting rhizobacteria and IBA on rooting of cuttings in kiwifruit (*Actinidia deliciosa* Chev.). *J. Horti. Sci.* 10(2), 159–164. https://doi.org/10.24154/jhs.v10i2.123

Singh, A., Kachwaya, D.S., Singh, R. 2020. Effect of Biofertilizers on Growth, Yield and Fruit Quality of Guava (*Psidium guajava* L.) cv. Allahabad Safeda. *Int. J. Curr. Microbiol. App. Sci.* 9(12): 372–378. doi: https://doi.org/10.20546/ijcmas.2020.912.047

Singh, M., Qureshi, K.A., Jaremko, M., Rajput, M., Singh, S.K., & Kaushalendra, K.A. 2022. Bioprospects of endophytic bacteria in plant growth promotion and Ag-nanoparticle biosynthesis. *Plants* 11(14), 1787. https://doi.org/10.3390/plants11141787

Singh, M., Singh, D., Gupta, A., Pandey, K.D., Singh, P.K., &Kumar, A. 2019a. Plant growth promoting rhizobacteria: application in biofertilizers and biocontrol of phytopathogens. In: A.K. Singh, A. Kumar, & P.K. Singh (Eds.), *PGPR Amelioration in Sustainable Agriculture* (pp. 41–66). Woodhead Publishing. https://doi.org/10.1016/B978-0-12-815879-1.00003-3

Singh, V.K., Singh, M., Singh, S.K., Kumar, C., & Kumar, A. 2019b. Sustainable agricultural practices using beneficial fungi under changing climate scenario. In: K.K. Choudhary, A. Kumar, & A.K. Singh (Eds.), *Climate Change and Agricultural Ecosystems* (pp. 25–42). Woodhead Publishing. https://doi.org/10.1016/B978-0-12-816483-9.00002-5

Sinsabaugh, R.L., Gallo, M.E., Lauber, C., Waldrop, M.P., & Zak, D.R. 2005. Extracellular enzyme activities and soil organic matter dynamics for northern hardwood forests receiving simulated nitrogen deposition. *Biogeochemistry* 75, 201–215. https://doi.org/10.1007/s10533-004-7112-1

Soliman, G.M., Mohamed, S.A., Emam, M.T., Hagagg, L.F., Hassan, A.M., & El-Hady, E.S. 2024. Impact of genetically modified bacteria and natural rocks on root-knot nematode biocontrol, improving productivity and chemical properties of grapes. *Egypt. J. Chem.* 67(10), 457–474. https://doi.org/10.21608/ejchem.2024.304638.10023

Stirling, G., Hayden, H., Pattison, T., & Stirling, M. 2017. Soil health, soil biology, soilborne diseases and sustainable agriculture: a guide. *Australasian. Plant Pathol.* 46, 387. https://doi.org/10.1007/s13313-017-0493-0

Tang, D.X., &Gao, Y.Z. 2023. Advances on the strategies of soil phosphate solubilizing microorganisms to promote plan. *Sheng Tai XueBao.* 43, 4390–4399.

Thomidis, T., Pantazis, S., Navrozidis, E., & Karagiannidis, N. 2015. Biological control of fruit rots on strawberry and grape by BOTRY-Zen. New Zealand. *J. Crop. Horticul. Sci.* 43(1), 68–72. https://doi.org/10.1080/01140671.2014.958502

Timofeeva, A., Galyamova, M., & Sedykh, S. 2022. Prospects for using phosphate-solubilizing microorganisms as natural fertilizers in agriculture. *Plants* 11(16), 2119. https://doi.org/10.3390/plants11162119

Trakunjae, C., Boondaeng, A., Apiwatanapiwat, W., Kosugi, A., Arai, T., Sudesh, K., & Vaithanomsat, P. 2021. Enhanced polyhydroxybutyrate (PHB) production by newly isolated rare actinomycetes *Rhodococcus* sp. strain BSRT1-1 using response surface methodology. *Sci. Rep.* 11(1), 1896. https://doi.org/10.1038/s41598-021-81386-2

Varghese, S., Dhanraj, N.D., Rebello, S., Sindhu, R., Binod, P., Pandey, A., & Awasthi, M.K. 2022. Leads and hurdles to sustainable microbial bioplastic production. *Chemosphere* 305, 135390. https://doi.org/10.1016/j.chemosphere.2022.135390

Verma, P., & Pandey, K. 2022. Biofertilizer: an ultimate solution for the sustainable development of agriculture. *Curr. Agric. Res. J.* 10(3), 193–206. http://dx.doi.org/10.12944/CARJ.10.3.04

Vlajkov, V., Pajcin, I., Vucetic, S., Andelic, S., Loc, M., Grahovac, M., & Grahovac, J. 2023. Bacillus-loaded biochar as soil amendment for improved germination of maize seeds. *Plants* 12(5), 1024. https://doi.org/10.3390/plants12051024

Waghunde, R.R., Shinde, C.U., Pandey, P., & Singh, C. 2021. Fungal biopesticides for agro-environmental sustainability. In: A.M. Abdel-Azeem, A.N. Yadav, N. Yadav, & Z. Usmani (Eds.), *Industrially Important Fungi for Sustainable Development:* Volume 1: *Biodiversity and Ecological Perspectives* (pp. 479–508). Springer International Publishing. https://doi.org/10.1007/978-3-030-67561-5_15

Wang, K., Hou, J., Zhang, S., Hu, W., Yi, G., Chen, W., & Zhang, Q. 2023. Preparation of a new biochar-based microbial fertilizer: nutrient release patterns and synergistic mechanisms to improve soil fertility. *Sci. Total. Environ.* 860, 160478. https://doi.org/10.1016/j.scitotenv.2022.160478

Westhoff, P. 2009. The economics of biological nitrogen fixation in the global economy. In: D.W. Emerich & H.B. Krishnan (Eds.), *Nitrogen Fixation in Crop Production* (pp. 309–328). Agronomy Monographs. https://doi.org/10.2134/agronmonogr52.c11

Yamada, K., & Xu, H.L. 2001. Properties and applications of an organic fertilizer inoculated with effective microorganisms. *J. Crop Produc.* 3(1), 255–268. https://doi.org/10.1300/J144v03n01_21

Yang, P., Condrich, A., Scranton, S., Hebner, C., Lu, L., & Ali, M.A. 2024. Utilizing plant growth-promoting rhizobacteria (pgpr) to advance sustainable agriculture. *Bacteria* 3(4), 434–451. https://doi.org/10.3390/bacteria3040030

Yoo, S.J., Shin, D.J., Won, H.Y., Song, J., & Sang, M.K. 2018. *Aspergillus terreus* JF27 promotes the growth of tomato plants and induces resistance against *Pseudomonas syringaepv.* tomato. *Mycobiology* 46(2), 147–153. https://doi.org/10.1080/12298093.2018.1475370

Zhao, G., Zhu, X., Zheng, G., Meng, G., Dong, Z., Baek, J.H., & Jia, B. 2024a. Development of biofertilizers for sustainable agriculture over four decades (1980–2022). *Geogr. Sustain.* 5(1), 19–28. https://doi.org/10.1016/j.geosus.2023.09.006

Zhao, M., Lin, Y. & Chen, H. (2020). Improving nutritional quality of rice for human health. *Theor. Appl. Genet.* 133, 1397–1413. https://doi.org/10.1007/s00122-019-03530-x

Zhao, X., Zhai, Y., Wei, L., Xia, F., Yang, Y., Yi, Y., & Zeng, L. 2024b. Isolation and identification of a novel *Bacillus velezensis* strain JIN4 and its potential for biocontrol of kiwifruit bacterial canker caused by *Pseudomonas syringaepv.* actinidiae. *Front. Plant Sci.* 15, 1513438. https://doi.org/10.3389/fpls.2024.1513438

10

The Synergism of Microbial Interactions Impacting Health and the Environment

Meenakshi Sharma[1], *Raj Kishor Kapardar*[2], *and Rajpal Srivastav*[3*]

[1] *Department of Botany, Daulat Ram College, University of Delhi, Delhi, India*
[2] *The Energy and Resources Institute, New Delhi, India*
[3] *Amity Institute of Biotechnology, Amity University Uttar Pradesh, Noida, Uttar Pradesh, India*

* Corresponding author: rsrivastav2@amity.edu

10.1 Introduction

The microbial activities and their equilibrium determine human health and influence the environment. Microbial activities have direct and indirect associations with factors, including host genetics and environmental factors. The association of the microbial population can exist in various ways, including synergism, commensalism, and parasitism. The synergistic associations are beneficial for the microbial population, and its host gets beneficial effects under certain circumstances, governed by different factors. Health, diet, and living circumstances are some of the factors that affect the synergistic dynamics between the microbes and the host. Understanding the origins and evolution of pathogenic microbes has various implications, such as state policy development related to health, environment, and agriculture. To substantiate this aspect, the genomic information is essential for understanding the pathogenesis and epidemiology of microorganisms.

The pathogenesis of microbes is governed by various factors, including host and environmental factors (Srivastav et al., 2014; Radhakrishnan et al., 2024). The evolution of microbes is being facilitated by gene transfer across populations, and the distribution of antimicrobial resistance genes, for instance, determines the effectiveness of the current drug regimen. Antimicrobial resistance is the prime focus of the research as there is a limited pool of drugs available (Zaura et al., 2015; Madhvi et al., 2017). Antimicrobials are toxic to microbes, and, therefore, microbes modify these chemical entities and expel them from the cell, which subsequently affects the physiological processes of the host (Dethlefsen & Relman, 2011; Srivastav et al., 2013). The microbial dynamics affect the host metabolism associated with the microbiome. It has been found that butyrate production is linked to better outcomes in gut health (Sun et al., 2024). The association of the *Bacteroides fragilis* and *Clostridium difficile* is synergistically unfavoured. The higher number of the first may have protective effects on the gut microbiome, and it avoids colitis (Deng et al., 2018). The extension of

Microbial Consortia in Plant Science and Sustainable Agriculture, First Edition. Edited by Ajay Tomar, et al.

such interactions may have deeper impacts on human health. It is reported that microbial synergistic activities influence the bioremediation of the synthetic polymers (Amelia et al., 2021). Microplastics, a potential hazard to health and the environment, are being generated in various processes, including the degradation of synthetic polymers. They may serve as a long-term reservoir for microbes via adherence and facilitate pathogenesis. Various aspects associated with these are discussed in the following sections.

10.2 Microbial Synergistic Activities and Antimicrobial Resistance

The richness and variety of the microbial communities are affected in the case of dysbiosis. Further, it is a challenging task as probiotics, commensals, and opportunistic interactions can develop resistance through gene transfer, allowing them to survive antimicrobial exposure, hampering biotherapeutics. Although gene therapy and personalized medicine present viable remedies, probiotic-induced resistance mechanisms and their host-specific ramifications require more detailed investigations (Clemente et al., 2012; Grover et al., 2019).

The effects of antimicrobials on gut microbiota have been the subject of numerous studies, which have shown that these molecules drastically modify microbial populations, which affects the physiological processes of the host (Dethlefsen & Relman, 2011; Zaura et al., 2015). The opportunistic bacterium *C. difficile*, which causes antibiotic-associated diarrhea, colonizes more easily when antimicrobials are overused (Looft et al., 2012; Theriot et al., 2014).

The colonization rates, microbial strain diversity, inadequate dose control, and intricate host interactions are measurable factors influencing the efficacy of antimicrobials. These bacteria include *Lactobacillus, Lactococcus, Bifidobacterium*, and *Enterococcus* species (Ljungh & Wadstrom, 2006; Chen et al., 2020). The investigations showed that these microbes may help prevent gut dysbiosis and carcinoma (Legesse et al., 2020) (Figure 10.1).

The role of probiotics cannot be neglected in maintaining gut homeostasis and health. The microbial load and its composition in the gut are one of the factors in microbiome therapeutics. The probiotics can effectively prevent the growth of harmful pathogens, and it is evident that the composition of the gut microbiota is closely linked to the anti-obesity benefits of the probiotics. Probiotic microbes may also be modified to enhance their therapeutic properties. Next-generation probiotics have demonstrated potential in the treatment of metabolic disorders like phenylketonuria, enteric hyperoxaluria, and diabetes. Additionally, they can function as diagnostic tools by producing therapeutic proteins in response to environmental signals. However, problems such as diminished viability in the gut and competition from local gut microorganisms hamper their usefulness (Srivastav et al., 2013; Legesse et al., 2020).

Antibiotic use, cleanliness, disease, age, gender, food, place of origin, and gut probiotics are some of the variables that affect antibiotic resistance genes (ARGs) in the gut. All the ARGs make up the resistome, which reacts differentially to diverse dietary stimuli. According to this study, the vegetarian gut microbiome contains fewer ARGs than that of nonvegetarians. Furthermore, a higher intake of dietary fibers is linked to lower levels of ARG, most likely because of the suppression of facultative anaerobes, which are frequently inflammatory and resistant to antibiotics, and the promotion of anaerobic bacteria. Compared to plant-based diets, animal-based

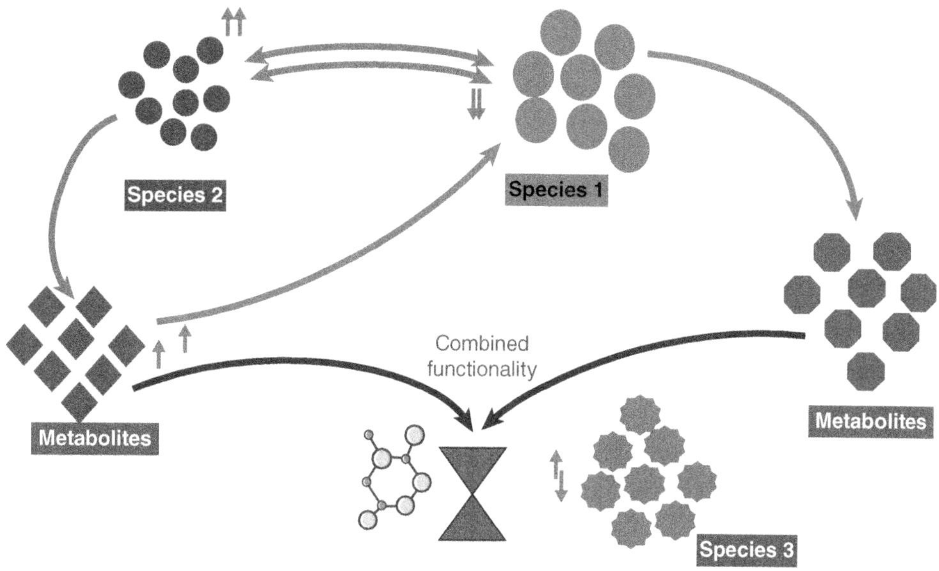

Figure 10.1 The representation of the principle of microbial dynamics and its influences on a specific system.

diets have a higher chance of introducing ARGs into the stomach. From conception to death, several ARGs are essential to the resistome. As antibiotic use has risen over time, these characteristics have changed, resulting in an increase in ARGs and bacteria that are resistant to antibiotics. Health, diet, antibiotic use, and living circumstances are some of the factors that affect the distribution of resistance genes by age. These elements may even lessen age-related resistance development patterns in certain situations (Maestre-Carballa et al., 2022).

10.3 Microbes Alter Metabolism

Bacteroidota contribute to the preservation of gut integrity and the colonization resistance of other species (Shin et al., 2024). Propionate has a role in lipogenesis, while acetate is required for gluconeogenesis. The metabolite butyrate is needed in epithelial cells as an energy source (Turnbaugh et al., 2006; Davis, 2016). It has been observed that Bacteroidota species raise butyrate concentrations (Sofi et al., 2021), suggesting the importance of these species in carbon metabolism. The end products of metabolism are observed to be primarily propionate and acetate (Macfarlane & Macfarlan, 2003; Ríos-Covián et al., 2016). A higher butyrate production is linked to a lower risk of colon cancer development (Scott et al., 2008). Bile salt hydrolases, for instance, are produced by several *Bacteroides* species and deconjugate bile acids that are unable to pass through enterohepatic recirculation (Sorg & Sonenshein, 2008). It is a two-step process. The cholic acid is oxidized to its deoxy form by the enzyme dehydroxylase and finally forms a secondary bile acid. Here, the enzyme dehydroxylase is a key mediator of the bile acid metabolism. It has been demonstrated that deoxycholic acid inhibits the growth of *C. difficile*. In contrast to this, taurocholic acid favors the growth of this microbe (Thanissery et al., 2017).

Table 10.1 Representation of the various microbial populations having importance in health.

S. no.	Microbial strain(s)	Condition(s)	Association(s)	References
1	*Lactobacillus rhamnosus GG*	Atopic dermatitis, gastric	Beneficial	Dempsey and Corr (2022)
2	*Bifidobacterium longum*	Infantitis	Necrotizing enterocolitis	Chichlowski et al. (2020)
3	*Akkermansia muciniphila*	Metabolic disorder	Modulation metabolism	Khalili et al. (2024)
4	*Bacteroides thetaiotaomicron*	Carbohydrate metabolism	Beneficial	Zocco et al. (2007)
5	*Bifidobacterium breve*	Allergies	Beneficial	Gavzy et al. (2023)
6	*Propionibacterium freudenreichii*	Dermatitis	Synergistic	Coronas et al. (2023)
7	*Enterococcus faecium*	Gastric disease	Synergistic	Almeida-Santos et al. (2021)
8	*Lactiplantibacillus plantarum*	Gastric disease	Beneficial	Hao et al. (2024)

Some microbes may have protective effects. For instance, *B. fragilis* has a synergistic effect by regulating gut microbial dynamics. The presence of *B. fragilis* avoids barrier destruction and hence alleviates colitis by pathogenic *Clostridium* species. The use of the infection model has also suggested similar findings. The introduction of *B. fragilis* has not only increased microbial diversity but also enhanced the survival, indicating the beneficial effects of *B. fragilis*. In such analysis, it is also observed that the population of commensalism microbes also increased (Deng et al., 2018). There exists an intricate balance of microbial population, and its dynamics is governed by various host and environmental factors (Srivastav et al., 2019; Gavzy et al., 2023). The relationship of various microbes with the physiological conditions is represented in Table 10.1.

10.4 Microbial Synergism, Biofilms, and Plastisphere

Synthetic polymers like microplastics, with a diameter of less than 5 mm, may serve as a long-term reservoir for antibiotic resistance and colonization during infections (Amaral-Zettler, 2022). Microplastics can be differentiated into two categories such as primary microplastics, which are directly produced and used in toothpaste and cosmetic products (Dhali et al., 2024), while secondary microplastics are a result of degradation. Microbes adhere to microplastics through lipopolysaccharides and pili. The cyanobacteria produce extracellular lipopolysaccharides for protection and aggregation (Laroche, 2022). Specifically, it has been demonstrated that some microplastic surfaces selectively enhance bacterial pathogens. These polymer remains are specifically concentrated in *Vibrio* bacterial species (Zettler et al., 2013).

Aquatic ecosystems and human health are at risk because these molecules act as carriers and colonization sites for microbes (Radhakrishnan et al., 2023). In the early colonization, pioneer microorganisms form biofilms by rapidly covering surfaces and facilitating the attachment of other microbes through extracellular polymeric substances. Over time, diverse communities develop, catalyzing metabolic reactions and transforming the biofilm into a mature plastisphere with stabilized microbial abundance but shifting community structures. Primary producers such as diatoms, cyanobacteria, green algae, and some proteobacteria comprise the early colonizers of aquatic plastispheres (Dhali et al., 2024). Following a day of immersion of microplastics in the marine environment, members of the Oceanospirillaceae, Alteromonadaceae, and Gammaproteobacteria were found. *Alcanivorax, Aestuariicella, Marinobacter*, and members of the Saprospiraceae and Oscillatoriaceae families were among the first organisms to colonize plastic substrates. Later phases of microplastics colonization showed a high abundance of Flavobacteria, Planctomycetaceae, and Phyllobacteriaceae (Amaral-Zettler et al., 2020).

The microbes are attracted to the surfaces through noncovalent interactions like hydrophilic, electrostatic, or Van der Waals interactions. These interactions are mediated by the functional groups present in the polymers. The degradation of these synthetic polymers occurs at a very slow rate in some cases. The enhancement of these reactions acts as a potential area of research in bioremediation. Studies suggest that there are distinct microbial communities on polystyrene and polyvinyl chloride compared to aliphatic plastics. Amaral-Zettler et al. (2020) confirmed microbial diversity differences across various plastics. A meta-analysis also attributed these variations to the chemical composition of plastics, including aromatic functional groups and chemical additives. However, factors like location and salinity seem more influential in shaping microbial communities. Over time, biofilm development equalizes surface chemistry, reducing the impact of plastic type on late-stage colonization. Microorganisms adhere to plastic, using strategies like lipopolysaccharides in gram-negative bacteria and type IV pili in Gammaproteobacteria. Species like cyanobacteria produce extracellular polysaccharides for protection and aggregation. More detailed investigations are required to understand these aspects of microbial activities.

10.5 Conclusive Remarks

The microorganisms form biofilms by rapidly covering surfaces and facilitating the attachment of other microbes through extracellular polymeric substances. These interactions and microbial dynamics are unexplored. Understanding microbial dynamics at the molecular level has various socioeconomic implications. It helps us design biosecurity policies with respect to the epidemiology of the pathogens. Further, these microbial dynamics and synergistic activities have broader applications in healthcare, sustainable environment, and development.

Acknowledgment

RS acknowledges Amity University Uttar Pradesh for infrastructural support during the preparation of this manuscript.

References

Almeida-Santos, A.C., Novais, C., Peixe, L., & Freitas, A.R. 2021. *Enterococcus* spp. as a Producer and Target of Bacteriocins: a double-edged sword in the antimicrobial resistance crisis context. *Antibiotics* 10(10), 1215. https://doi.org/10.3390/antibiotics10101215

Amaral-Zettler, L.A. 2022. Colonization of plastic marine debris: the known, the unknown, and the unknowable. In: A.L. Andrady (Ed.), *Plastics and the ocean: Origin, Characterization, Fate, and Impacts* (pp. 301–316). https://doi.org/10.1002/9781119768432.ch10

Amaral-Zettler, L.A., Zettler, E.R., & Mincer, T.J. 2020. Ecology of the plastisphere. *Nature Reviews Microbiology* 18(3), 139–151.

Amelia, T.S.M., Khalik, W.M.A.W.M., Ong, M.C., Shao, Y.T., Pan, H.-J., & Bhubalan, K. 2021. Marine microplastics as vectors of major ocean pollutants and its hazards to the marine ecosystem and humans. *Progress in Earth and Planetary Science* 8(1), 12. https://doi.org/10.1186/s40645-020-00405-4

Chen, D., Jin, D., Huang, S., Wu, J., Xu, M., Liu, T., Dong, W., Liu, X., Wang, S., Zhong, W., Liu, Y., Jiang, R., Piao, M., Wang, B., & Cao, H. 2020. *Clostridium butyricum*, a butyrate-producing probiotic, inhibits intestinal tumor development through modulating Wnt signaling and gut microbiota. *Cancer Letters* 469, 456–467.

Chichlowski, M., Shah, N., Wampler, J.L., Wu, S.S., & Vanderhoof, J.A. 2020. *Bifidobacterium* longum subspecies *infantis (B. infantis)* in pediatric nutrition: current state of knowledge. *Nutrients* 12(6), 1581. https://doi.org/10.3390/nu12061581

Clemente, J.C., Ursell, L.K., Parfrey, L.W., & Knight, R. 2012. The impact of the gut microbiota on human health: an integrative view. *Cell* 148(6), 1258–1270. https://doi.org/10.1016/j.cell.2012.01.035

Coronas, R., Zara, G., Gallo, A., Rocchetti, G., Lapris, M., Petretto, G.L., Zara, S., Fancello, F., & Mannazzu, I. 2023. Propionibacteria as promising tools for the production of pro-bioactive scotta: a proof-of-concept study. *Frontiers in Microbiology* 14, 1223741. https://doi.org/10.3389/fmicb.2023.1223741

Davis, C.D. 2016. The gut microbiome and its role in obesity. *Nutrition Today* 51(4), 167–174.

Dempsey, E., & Corr, S.C. 2022. *Lactobacillus* spp. for gastrointestinal health: current and future perspectives. *Frontiers in Immunology* 13, 840245. https://doi.org/10.3389/fimmu.2022.840245

Deng, H., Yang, S., Zhang, Y., Qian, K., Zhang, Z., Liu, Y., & Zhi, F. 2018. *Bacteroides fragilis* prevents *Clostridium difficile* infection in a mouse model by restoring gut barrier and microbiome regulation. *Frontiers in Microbiology* 9, 2976.

Dethlefsen, L., & Relman, D.A. 2011. Incomplete recovery and individualized responses of the human distal gut microbiota to repeated antibiotic perturbation. *Proceedings of the National Academy of Sciences of the United States of America* 108(Suppl 1), 4554–4561. https://doi.org/10.1073/pnas.1000087107

Dhali, S.L., Parida, D., Kumar, B., & Bala, K. 2024. Recent trends in microbial and enzymatic plastic degradation: a solution for plastic pollution predicaments. *Biotechnology for Sustainable Materials* 1, 11. https://doi.org/10.1186/s44316-024-00011-0

Gavzy, S.J., Kensiski, A., Lee, Z.L., Mongodin, E.F., Ma, B., & Bromberg, J.S. 2023. *Bifidobacterium* mechanisms of immune modulation and tolerance. *Gut Microbes* 15(2), 2291164. https://doi.org/10.1080/19490976.2023.2291164

Grover, R., Burse, S., Shankrit, S., Aggarwal, A., Kirty, K., Narta, K., Srivastav, R., Ray, A.K., Malik, G., Vats, A., Motiani, R.K., Thukral, L., Roy, S.S., Bhattacharya, S., Sharma, R., Natarajan, K., Mukerji, M., Pandey, R., Gokhale, R.S., & Natarajan, V.T. 2019. Mygl exonuclease couples the nuclear and mitochondrial translational programs through RNA processing. *Nucleic Acids Research* 47, 11, 5852–5866.

Hao, Y., Li, J., Wang, J., & Chen, Y. 2024. Mechanisms of health improvement by *Lactiplantibacillus plantarum* based on animal and human trials: a review. *Fermentation* 10(2), 73. https://doi.org/10.3390/fermentation10020073

Khalili, L., Park, G., Nagpal, R., & Salazar, G. 2024. The role of *Akkermansia muciniphila* on improving gut and metabolic health modulation: a meta-analysis of preclinical mouse model studies. *Microorganisms* 12(8), 1627. https://doi.org/10.3390/microorganisms12081627

Laroche, C. 2022. Exopolysaccharides from Microalgae and Cyanobacteria: diversity of strains, production strategies, and applications. *Marine Drugs* 20(5), 336. https://doi.org/10.3390/md20050336

Legesse, B.T., Feto, T.K., Awoke, K.S., Garedew, A.D., Yifat, F.T., & Birri, D.J. 2020. Probiotics for cancer: alternative prevention and treatment. *Biomedicine & Pharmacotherapy* 129, 110409. https://doi.org/10.1016/j.biopha.2020.110409

Ljungh, A., & Wadström, T. 2006. Lactic acid bacteria as probiotics. *Current Issues in Intestinal Microbiology* 7(2), 73–89.

Looft, T., Johnson, T.A., Allen, H.K., Bayles, D.O., Alt, D.P., Stedtfeld, R.D., Sul, W.J., Stedtfeld, T.M., Chai, B., Cole, J.R., Hashsham, S.A., Tiedje, J.M., & Stanton, T.B. 2012. In-feed antibiotic effects on the swine intestinal microbiome. *Proceedings of the National Academy of Sciences of the United States of America* 109(5), 1691–1696. https://doi.org/10.1073/pnas.1120238109

Macfarlane, S., & Macfarlane, G.T. 2003. Regulation of short-chain fatty acid production. *Proceedings of the Nutrition Society* 62(1), 67–72.

Maestre-Carballa, L., Navarro-López, V., & Martinez-Garcia, M. 2022. A resistome roadmap: from the human body to pristine environments. *Frontiers in Microbiology* 13, 858831. https://doi.org/10.3389/fmicb.2022.858831

Madhvi, A., Hingane, S., Srivastav, R., Joshi, N., Subramani, C., Muthumohan, R., Khasa, R., Varshney, S., Kalia, M., Vrati, S., Surjit, M., & Kumar, C.T.R. 2017. A screen for novel hepatitis C virus RdRp inhibitor identifies a broad-spectrum antiviral compound. *Scientific Reports* 7, 5816. https://doi.org/10.1038/s41598-017-04449-3

Radhakrishnan, A., Sharma, C., Malviya, V.N., & Srivastav, R. 2024. Deciphering the molecular functionality of Cdc45 in replisomal complex. *Biochemistry and Biophysics Reports* 37, 101643.

Radhakrishnan, A., Balaganesh, P., Vasudevan, M., Natarajan, N., Chauhan, A., Arora, J., Ranjan, A., Rajput, V.D., Sushkova, S., Minkina, T., Basniwal, R.K., Kapardar, R. & Srivastav, R. 2023. Bioremediation of hydrocarbon pollutants: recent promising sustainable approaches, scope, and challenges. *Sustainability* 15(7), 5847.

Scott, K.P., Duncan, S.H., & Flint, H.J. 2008. Dietary fibre and the gut microbiota. *Nutrition Bulletin* 33(3), 201–211.

Shin, J.H., Tillotson, G., MacKenzie, T.N., Warren, C.A., Wexler, H.M., & Goldstein, E.J. 2024. *Bacteroides* and related species: the keystone taxa of the human gut microbiota. *Anaerobe* 85, 102819. https://doi.org/10.1016/j.anaerobe.2024.102819

Sofi, M.H., Wu, Y., Ticer, T., Schutt, S., Bastian, D., Choi, H.J., & Yu, X.Z. 2021. A single strain of *Bacteroides fragilis* protects gut integrity and reduces GVHD. *JCI Insight* 6(3), e136841.

Sorg, J.A., & Sonenshein, A.L. 2008. Bile salts and glycine as cogerminants for *Clostridium difficile* spores. *Journal of Bacteriology* 190(7), 2505–2512.

Srivastav, R., Kumar, D., Grover, A., Singh, A., Manjasetty, B., Sharma, R., & Taneja, B. 2014. Unique subunit packing in mycobacterial nanoRNase leads to alternate substrate recognitions in DHH phosphodiesterases. *Nucleic Acids Research* 42(12), 7894–7910.

Srivastav, R., Sharma, R., Tandon, S., & Tandon, C. 2019. Role of DHH superfamily proteins in nucleic acids metabolism and stress tolerance in prokaryotes and eukaryotes. *International Journal of Biological Macromolecules* 127, 66–75.

Srivastav, R., Singh, A., Jangir, P.K., Kumari, C., Muduli, S., & Sharma, R. 2013. Genome sequence of *Staphylococcus massiliensis* strain S46 isolated from surface of healthy human skin. *Genome Announcements* 1(4), e00553-13.

Sun, J., Chen, S., Zang, D., Sun, H., Sun, Y., & Chen, J. 2024. Butyrate as a promising therapeutic target in cancer: from pathogenesis to clinic (Review). *International Journal of Oncology* 64(4), 44. https://doi.org/10.3892/ijo.2024.5632

Thanissery, R., Winston, J.A., & Theriot, C.M. 2017. Inhibition of spore germination, growth, and toxin activity of clinically relevant *C. difficile* strains by gut microbiota derived secondary bile acids. *Anaerobe* 45, 86–100.

Theriot, C.M., Koenigsknecht, M.J., Carlson, P.E. Jr., Hatton, G.E., Nelson, A.M., Li, B., Huffnagle, G.B., Li, J.Z., & Young, V.B. 2014. Antibiotic-induced shifts in the mouse gut microbiome and metabolome increase susceptibility to *Clostridium difficile* infection. *Nature Communications* 5, 3114. https://doi.org/10.1038/ncomms4114

Turnbaugh, P.J., Ley, R.E., Mahowald, M.A., Magrini, V., Mardis, E.R., & Gordon, J.I. 2006. An obesity-associated gut microbiome with increased capacity for energy harvest. *Nature* 444(7122), 1027–1031. https://doi.org/10.1038/nature05414

Zaura, E., Brandt, B.W., Teixeira de Mattos, M.J., Buijs, M.J., Caspers, M.P., Rashid, M.U., Weintraub, A., Nord, C.E., Savell, A., Hu, Y., Coates, A.R., Hubank, M., Spratt, D.A., Wilson, M., Keijser, B.J., & Crielaard, W. 2015. Same exposure but two radically different responses to antibiotics: resilience of the salivary microbiome versus long-term microbial shifts in feces. *mBio* 6(6), e01693-15. https://doi.org/10.1128/mBio.01693-15

Zettler, E.R., Mincer, T.J., & Amaral-Zettler, L.A. 2013. Life in the "plastisphere": microbial communities on plastic marine debris. *Environmental Science & Technology* 47(13), 7137–7146.

Zocco, M.A., Ainora, M.E., Gasbarrini, G., & Gasbarrini, A. 2007. *Bacteroides thetaiotaomicron* in the gut: molecular aspects of their interaction. *Digestive and Liver Disease* 39, 707–712. https://doi.org/10.1016/j.dld.2007.04.003

11

Encouraging Probiotics as Plant Stimulants for Sustainable Agriculture Through Microbial Consortium

Deepika Sharma[1] and Sakshi Singh[2]*

[1] *Amity Institute of Organic Agriculture, Amity University, Noida, Uttar Pradesh, India*
[2] *Young Professional-II, ICAR-Central Potato Research Institute, Shimla, Himachal Pradesh, India*

* Corresponding author: dpkasharma44@gmail.com

11.1 Introduction

A different strategy for using synthetic chemicals is needed to comply with international standards as modern agriculture moves toward sustainability, which is accomplished with the use of green and eco-friendly resources. Plant biostimulants are microorganisms or compounds that, when introduced inside the soil's rhizosphere, encourage the symbiotic processes that enhance the quality of the crop by boosting the uptake of nutrients as well as efficiency, consequently enabling the plants to withstand biotic and abiotic challenges.

The innovative technologies need to be explored to allow an increase in agricultural productivity under uniform and convenient circumstances to attain sustainability in agriculture and the environment. Biostimulants and other bio-based products are the most effective and sustainable technologies to ensure the achievement of agricultural sustainability. Biostimulants are generally of microbial origin, which can be used significantly for open-field and greenhouse conditions, thereby involving fruits, vegetables, and ornamentals (Basile et al., 2020).

Plant pathogens are typically managed using traditional management techniques that include physical, chemical, and cultural approaches. However, new greener technologies for disease control are required because of the limitations of current approaches, which include the development of resistance, low effectiveness, high initial cost, and adverse environmental effects. The use of living organisms, which are ingested by plants in appropriate amounts, promotes substantial growth regulation for the plants along with the host's capacity to uphold biotic and abiotic stresses. Thus, to maintain the balance between pathogenic and beneficial bacteria, probiotics use organic extracts, biostimulants, plant growth-promoting rhizobacteria (PGPRs), mycorrhizal fungi, endophytes, and their consortia (Kanmani et al., 2013; Servin, 2004).

Microbial Consortia in Plant Science and Sustainable Agriculture, First Edition. Edited by Ajay Tomar, et al.

The application of beneficial microorganisms to reduce dependence and agriculture's reliance on pesticides and, consequently, the adverse effects of those chemicals on the environment has shown increasing tendencies. It is thought that plants naturally associated with advantageous microbes play a significant role in improved crop health and stress tolerance. Plant stimulants are the rhizospheric microbes that affect plant growth and development via inducing resistance against stress in plants (Abhilash et al., 2016). Organic plant extracts, bioformulations, and complex preparations that incorporate consortia that stimulate the metabolic and physiological processes in plants are examples of biostimulants, which are bioactive substances that promote plant development (Yakhin et al., 2017). The scientific community frequently uses the term "biostimulant" to describe a variety of compounds and how they work (Atero-Calvo et al., 2024). Biostimulants are known as multipurpose compounds that help plants by improving the nutrient intake from the soil. Indirect activity against various biotic stressors involving harmful microorganisms is another aspect of biostimulants. The condition of the target host plant is improved by bioformulations that contain combinations of fulvic acid, amino acids, and plant extracts, in combination with certain inorganic components and beneficial fungi or bacteria (Du Jardin, 2015).

Plant growth promotion and nutrient uptake, intake, and cycling of nutrients are also associated with beneficial and well-established interactions between microorganisms and plants. Utilizing the nonpathogenic strains of *Pseudomonas, Bacillus, Azotobacter, Trichoderma, Serratia*, and *Azospirillum*, known as PGPRs, is another way that allows improvement in soil health and productivity. Defense responses involving the activation of induced systemic resistance and systemic acquired resistance are triggered by interactions between plants and biostimulants. Because of their interaction, the pretreated plants have demonstrated improved seed germination with higher yield potential.

Plant defense mechanisms trigger the use of many enzymatic compounds like proline, superoxide dismutase, and aminocyclopropane-1-carboxylate activity (Yakhin et al., 2017). Utilizing microbial consortia in conjunction with seaweed extracts, polymers, botanicals, and inorganic chemicals with animal products increased the reliability of using beneficial plant products (Beckett & Van Staden 1989). Different studies and researchers have shown that rhizobacteria, rhizofungi, and microbial consortia have potential as agricultural probiotics (Kong et al., 2018).

Studies of plant–microbes and their interactions are important for the formation of microbial consortia that have been found through field research and *in vitro* screening. Understanding the benefits and drawbacks of microbial inoculants and the plants is thereby crucial for the formation of successful consortia (De Souza et al., 2020). Cooperation, commensalism, mutualism, rivalry, and parasitism are among the symbiotic relationships between organisms, which are mostly described by the various interactions among microbes. The foundation of cooperative relationships between these microorganisms and plants is often the result of interactions that increase the effectiveness and efficiency of bacteria that are connected to one another. The interactions help the microorganisms produce organic molecules, which in turn activate their metabolic processes. Another kind of microbial interaction is commensalism, which refers to community-based interactive actions that benefit other community members. Another phenomenon where species benefit from one another is mutualism.

It is necessary to investigate these many kinds of interactions since they are advantageous for the development of microbial consortia as well as various antagonistic interactions between them, such as competition and parasitism (Ghoul & Mitri, 2016). Since they are crucial in identifying the traits of microorganisms to be employed in consortium composition, the term "soil microbiome" refers to the population of microbes in a complex community with all the potential interactions among them. Utilizing multiple microbial species to create bioformulations is a more promising strategy in agriculture than utilizing only one. As their capacity to colonize various ecological niches of the plant rhizosphere likewise improves, the microbial consortium exhibits no discernible hostile activity toward each other (Wei et al., 2015).

Soil microbiome, which is known as the population of microbes within a complex community with potential interactions between them that are crucial for identifying the traits of microbes used for consortium composition, and the interactions between microorganisms themselves must both be investigated. Several microbial consortia species that have antagonistic activity against one another are used in agriculture to colonize various ecological niches (Wei et al., 2015).

11.2 Plant Stimulants as Probiotics

Soil–plant interaction involving the use of beneficial microorganisms has varying capacities for promoting plant growth, thereby mobilizing nutrients. These probiotics are bacteria that are found on the surface of plants and have proactive biological actions because of their biological efficacy. When added to soil, these probiotic microbes initiate the production of phytohormones, such as auxin, gibberellins, and osmolyte metabolites, that regulate the ACC deaminase activity, thereby benefiting plants. The probiotic-based bioformulations with high colonization qualities enhance plant nutrition and health. The harmful effects of synthetic chemical formulations comprising sulfur, copper, and zinc cause environmental deterioration, which also affects beneficial creatures that are not the intended target. To mitigate the negative impacts on the ecosystem, eco-friendly methods use beneficial microorganisms more widely. Several findings claim that using natural substances as biostimulants boosts agricultural productivity and lessens the severity of different types of stress (De Pascale et al., 2017).

To control plant diseases, microbes associated with the phyllosphere and rhizosphere of the plants work in coordination to encourage growth. When present in sufficient quantities, plant biostimulants offer the host a number of health benefits and advantages through the use of PGPRs, which have the ability to solubilize phosphate and have an antagonistic effect on pathogens (Figure 11.1).

Beneficial microbes have both direct and indirect impacts, such as the generation of indole-3-acetic acid (IAA), gibberellins, ethylene, and siderophores; antibiotic synthesis; solubilization of vital nutrients; and induced systemic resistance (Tripathi et al., 2024). Plants with higher vegetative growth and quality have exhibited natural defense responses when exposed to probiotic chitosan (Salachna & Zawadzinska, 2014).

The idea of a plant biostimulant emerged in the 1930s as a significant concoction to lessen the negative effects of ecological imbalances (Sharma et al., 2014). These naturally

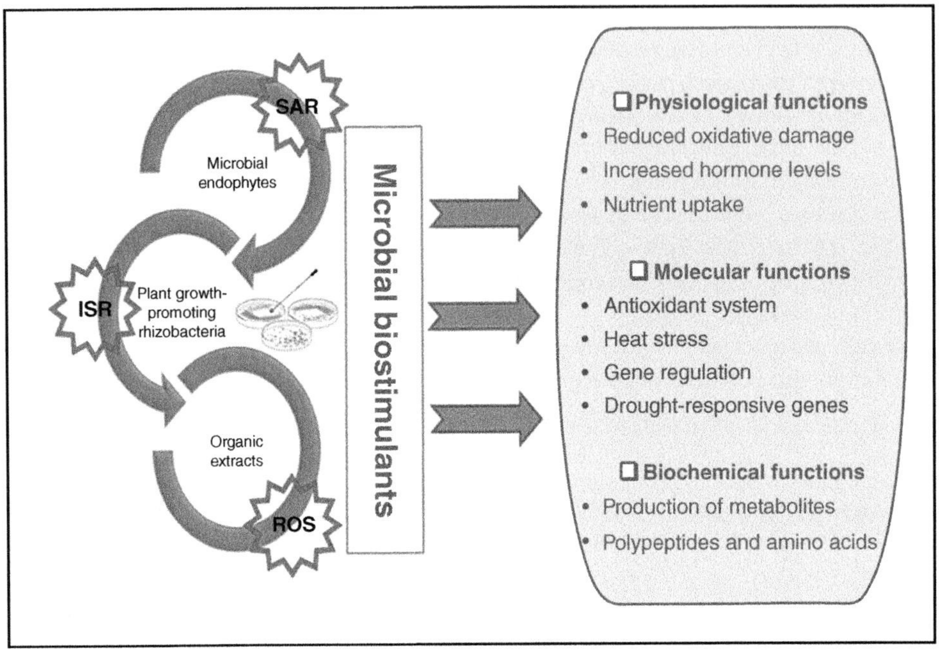

Figure 11.1 The image illustrates the role of microbial consortia involving biostimulants like microbial endophytes, PGPRs, and organic extracts in triggering defense mechanisms in plants. As depicted, the microbial endophytes are involved in activating the systemic acquired resistance (SAR) by signal transduction pathway because of salicylic acid accumulation and activity of pathogenesis-related proteins. PGPRs are linked with the response of induced systemic resistance (ISR) because of which the induction of reactive oxygen species (ROS) occurs, which is being regulated by the usage of botanical extracts. Different functions of microbial biostimulants are illustrated in the figure that includes physiological, molecular, and biochemical aspects involving reduced oxidative damage, enhancement of antioxidant systems, and activation of drought-responsive genes. Production of metabolites is also regulated with the synthesis of polypeptides and amino acids, which are being regulated as biochemical functions of plant biostimulants. Created with BioRender.com.

occurring compounds or bacteria are known as biostimulants, and they are used to improve the different characteristics of the growing medium. As the biostimulants synthesized from plants aid in plant development, they have a special way of working at the cellular and molecular levels (Lie et al., 2021).

With careful consortia selection, microbial biostimulants provide a high-nutrient agricultural output, which aids in the better utilization of these products. Root surface area typically improves and increases by the symbiotic mycorrhizal connections, which makes the absorption of other minerals and nutrients easier. Endophytic rhizobacteria are important because they lower biotic stress, which in turn encourages plant growth. With a decrease in abscisic acid (ABA) production and a rise in hormone levels, a relationship between hormone accumulation and promoting plant development was seen. Thus, at concentrations of 0.1 and 0.5%, plant extracts such as guava, aloe vera, garlic, and alfalfa have successfully helped to enhance the growth of rice (Ei et al., 2024). There have also been reports of using *Trichoderma* strains to activate tomatoes' *in vitro* defensive and metabolic interactions (De Palma et al., 2016).

11.2.1 Different Biostimulants with Different Modes of Action

Different crops are grown in unfavorable environmental conditions, necessitating the use of sustainable organic resources to preserve biodiversity and guarantee food security and the use of biostimulants that are beneficial for improving crop nutrition and, as a result, using organic methods to address environmental concerns (Bulgari et al., 2015). Probiotics have increased the growth and development of plants while also improving quality measures. The application of *Ascophyllum nodosum*, a plant probiotic, together with around 50% of the recommended fertilizer dosage, has been shown to improve the quality attributes and microbial population as well.

Applying biostimulants as a spray treatment to chickpeas was reported to increase the accumulation of dry matter content (Sharanya et al., 2022). Tejada et al. (2011) reported that soil modified with biostimulants, which improve the soil's texture and arability, also exhibited soil enzymatic activity. According to reports, applying *Trichoderma* as a biostimulant can increase soil fertility and nutrient availability (Sani et al., 2020).

11.2.2 Organic Extracts

Plant protection plants are made using botanicals, which are compounds that are taken from plants. Much research has been done on their potential as biostimulants (Kirn et al., 2010). Humic chemicals, which are found in organic materials, help bacteria break down plant leftovers and improve their metabolic activity. These humic compounds are categorized as heterogeneous molecules, including humins and humic acid, according to their molecular weight and solubility. Thus, these compounds function by interacting with other natural molecules through their combination (Table 11.1).

11.2.3 Microbial Endophytes

Seaweed extracts are utilized as biostimulants and biofertilizers, both of which have economic applications. Plant growth promotion is positively impacted by seaweed extracts and their breakdown products, including polysaccharides, laminarin, and alginates (Khan et al., 2009). Other species of algae, including *Ascophyllum, Fucus*, and *Laminaria*, can also be effectively used as biostimulants, in addition to botanicals, which are plant-based products with therapeutic value (Seiber et al., 2014).

A free-living species of PGPRs like *Azotobacter* influences the nitrogen levels in soil, which are crucial for plant growth and can also be used as a biostimulant. These bacteria's capacity to withstand heat allows them to fight off many abiotic stressors, such as drought and salinity, while also fostering advantageous relationships with other microbes (Beckett & Van Staden 1989; Kalimuthu et al., 2024). because they can be effectively utilized with suitable consortia involving actinomycetes, symbiotic fungi, and other bacteria that release phytohormones, vitamins, and antifungal compounds under stress conditions (Viscardi et al., 2016).

Plants also harbor endophyte populations, which are microbial communities that live inside plants and contribute to plant growth, in addition to displaying the visible symptoms of plant infection. Many endophytic microorganisms are commonly reported by

Table 11.1 Biostimulants and their utilization in plant growth and protection.

Biostimulant		Stimulative effect on crops	References
Plant growth-promoting Rhizobacteria	*Bacillus* sp.	Lettuce	Vetrano et al. (2020)
	Burkholderia ambifaria	Onion	Pellegrini et al. (2020)
	Pseudomonas putida	Tomato	He et al. (2019)
	Bacillus subtilis	Tomato	Chandrasekaran et al. (2019)
	Azospirillum brasilense	Cucumber	Islam et al. (2016)
	Pseudomonas sp.	Apple	Aslanta et al. (2007)
	Pseudomonas putida	Grape	Sabir (2013)
Seaweed	*Laminaria digitata*	Tomato	Esserti et al. (2017)
	Ascophyllum nodosum	Capsicum	Khan et al. (2009)
	Kappaphycus alvarezii	Soybean	Shukla et al. (2018)
	Kappaphycus alvarezii	Potato	Prajapati et al. (2016)
	Gracilaria sp.	Rice	Layek et al. (2018)
	Ulva lactuca	Wheat	Shah et al. (2013)
Endophytes	*Bacillus subtilis*	Chickpea	Abd-Allah et al. (2018)
	Piriformospora indica	Barley	Baltruschat et al. (2008)
	Burkholderia phytofirmans	Maize	Naveed et al. (2014)
Organic extracts	Fulvic acid	Capsicum	Aminifard et al. (2012)
	Fulvic acid	Broad bean	Shahid et al. (2012)
	Humic acid	Rice	Garcia et al. (2012)
	Ginger extract	Tomato	Hyder et al. (2024)

various workers, including bacteria such as *Pseudomonas, Pantoea, Stenotrophomonas,* and *Actinobacter*. Glomeromycota, Ascomycota, and Basidiomycota are examples of eukaryotic endophytes that are categorized into distinct phyla (Hardoim et al., 2015; Zhang et al., 2019).

11.2.4 Role of PGPRs

The plant rhizobacteria, commonly referred to as PGPRs, are linked to rhizosphere colonization because they adhere closely to the soil surface. The nitrogen-fixing bacteria like *Rhizobium, Azotobacter* spp., and *Frankia* are among the different categories of PGPRs that can be used as biostimulants (Kalimuthu et al., 2024). By operating in different ways and producing phytohormones, HCN, siderophores, minerals, and other compounds, the utilization of various soil-resident bacteria plays a role in regulating and promoting the plant growth that exhibits a variety of beneficial activities. According to Niu et al. (2018), several species of bacteria, including *Bacillus, Pseudomonas, Azospirillum*, and *Serratia*, as PGPR consortia of *Bacillus subtilis, B. thurigiensis*, and *B. megaterium* also have a role to play in enhancing the metabolic activities in plants like chlorophyll content, riboflavin, and asparagine that even under stress conditions in crops like wheat and chickpea, along with plant growth modulation (Khan et al., 2019; Meenakshi et al., 2019).

Stimulatory effects of the soil microorganisms as biostimulants with plant growth promotion properties are subjected to extensive research. Development regulators such as cytokinins, gibberellic acid (GA), and IAA contribute to the germination of plants with increased vegetative development, adventitious roots, and increased metabolite production (Jha & Saraf, 2015). Because probiotics like *Burkholderia* sp., *Gluconacetobacter* sp., and *Azospirillum* sp. are the most effective and environmentally viable, using them to carry out the nitrogen needs of plants without using commercial fertilizers is both viable and eco-friendly (Ramakrishna et al., 2019).

11.3 Role of Biostimulants

By improving abiotic stress tolerance and raising quality and yield indices, natural and synthetic compounds referred to as biostimulants are essential in influencing plant growth. With an increment in the growth of plants and a reduction in abiotic stress, these biostimulants may provide a novel method for stimulating several processes in plants. Plant metabolism and photosynthetic processes are altered by biostimulants, directly and indirectly (Carillo et al., 2019). Minerals, including phosphorus, sulfur, calcium, magnesium, potassium, and iron, are absorbed as a result of the action of biostimulants in various crops, including maize, tomato, and other cultivated crops (Rady & Rehman, 2016; Ertani et al., 2011).

According to reports, biostimulants can also change the number of sugars in plants by altering enzymes, including polygalacturonase, starch synthase, and dehydrogenase (Rouphael et al., 2017). To adapt to different stress conditions, oxidative mechanisms, along with the signaling system, stimulate plants. According to reports, beans and sweet peppers benefit from the use of biostimulants during drought stress. Under stress, enzymatic defense activities involving catalase, superoxide dismutase, and peroxidase activity also increase plants' antioxidant systems (Zhou et al., 2021).

Applying seaweed extracts that are biostimulants has also been shown to increase blooming in tomato plants (Dookie et al., 2021) (Figure 11.2). Likewise, the application of microalgae *Ecklonia maxima* has been shown to have a biostimulant effect on the yield of the lettuce crop (Di Mola et al., 2020). Fruit and vegetable crops that get biostimulants as foliar treatments or root dips have their protein hydrolysate activated, which triggers the defensive mechanism and causes stimulated absorption with multilayer signaling (Baglieri et al., 2014; Ronga et al., 2019).

Onion, along with garlic plants, thereby benefits from humic compounds by absorbing nitrogen and sulfur, which results in a notable increase in output (Gemin et al., 2019).

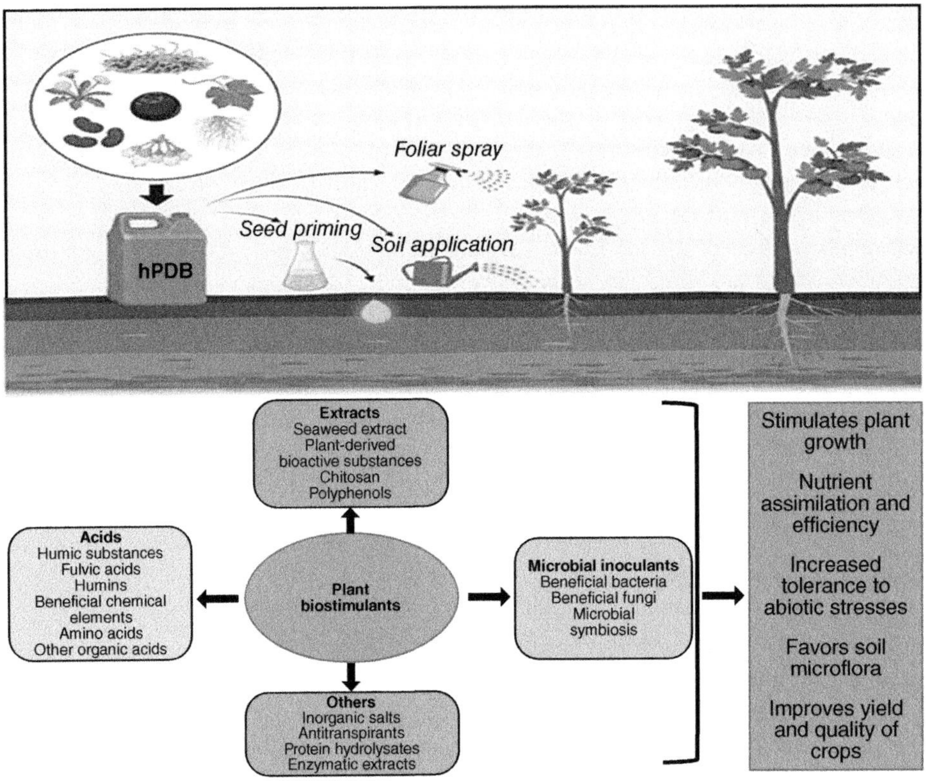

Figure 11.2 Application methods of higher plant-derived biostimulants, which include soil application, seed priming application method, and foliar application that improve the capability of plants to fight against stress. The different biostimulants having stimulative action in plants include organic extracts like seaweed, chitosan extracts, and humic and fulvic acids, along with microbial inoculants including fungi and bacteria. These microbial inoculants have a stimulative effect on plants and play an important role in stimulating plant growth, higher assimilation of nutrients, increased tolerance to abiotic stress, improving soil microflora, and also increasing the yield and quality of the crops (Kumari et al., 2022; Martínez-Lorente et al., 2024).

Chitosan and related biopolymers were widely used in many horticulture crops to help plants resist biotic stress. Biostimulants are thus administered in a number of ways, involving foliar spraying, soil application, as well as direct coating of different vegetable products (Pichyangkura & Chadchawan, 2015). Furthermore, a new tactic called "rhizosphere engineering" surfaced, which involves integrating microbial inoculants with other biological networks to encourage the reemergence of beneficial microbial communities (Kirn et al., 2010). According to Gopalkrishnan et al. (2015), they include the synthesis of exopolysaccharides, nitrogen fixation, phosphate solubilization, siderophores, and phytohormones. Utilizing plant microbiome engineering entails identifying possible bacteria that promote plant growth by thoroughly analyzing their constituent parts and assessing the microorganisms' compatibility (Berg et al., 2014; Kong et al., 2018).

11.4 Microbial Consortium-mediated Protection

The presence of different synthetic compounds of phenols and nitrophenolates is a characteristic of biostimulants. As per the compatibility of different microorganisms, integration of the plant microbiome entails cultivating putative PGPRs (Berg et al., 2014; Kong et al., 2018). Numerous investigations have been conducted about the use of *Azotobacter* and *Trichoderma* as a microbial consortium in conjunction with another rhizosphere microbiota (Ventorino et al., 2018).

Beneficial strains of fungi such as *Trichoderma* sp. prime or activate plants before pathogens attack them, which increase the abundance of microbial consortia by inducing systemic acquired resistance (SAR) and ISR (Manganiello et al., 2018; Song et al., 2024). The use of microbial consortia stimulates processes that are known to promote the growth of plants and thereby offer defense against stress conditions, along with functions such as nitrogen fixation, siderophore production of phytohormones, and exopolysaccharide production (Gopalkrishnan et al., 2015; Ventorino et al., 2014; El Hamss et al., 2023). Fore genuinely useful and effective plant products, the microbial consortium consists of *Trichoderma, Azotobacter*, seaweed, botanical and inorganic substances, polymers, and animal-derived products (Woo & Pepe, 2018; Kredics et al., 2024).

Using microbial consortia to fight biotic stress has been very beneficial for various crops (Wen et al., 2023; Singh et al., 2024). Thus, new possibilities for plant biostimulants can be used to create a sustainable agroecosystem, improving crop quality and fueling the expansion of the biofertilizer business.

Researchers have demonstrated that the drought tolerance of a consortium of four bacterial species – *Stenotrophomonas rhizophila, Xanthomonas retroflexes, Microbacterium oxydans*, and *Paenibacillus amylolyticus* – in *Arabidopsis* promotes plant growth and disease resistance (Yang et al., 2021). The application of a consortium of *Ochrobactrum* sp., *Pantoea* sp., and *Enterobacter* sp., along with *Bacillus* sp., has been reported to have significantly improved the growth of plants (Bairey et al., 2016). For microorganisms to be effectively used as probiotics, they must be compatible with organic extracts and have the best viable microbial units available, with the crucial quantity of bioinoculant active units required to achieve the intended outcomes and efficacy (Sanjuan et al., 2023).

11.5 Conclusion

Because of its many benefits, the use of probiotic bacteria has been connected to enhancements in quality and development traits. As biopesticides and biofertilizers that aid in plant growth and development, these microbial inoculants were used. The use of these probiotics guarantees the sustainability and productivity needed for the future, and these classes of plant stimulants function as growth promoters for plants. Therefore, these plant probiotics enable the other bacteria to trigger their defense and help plants overcome a range of stressors, enhancing their output and general well-being.

Strains of microorganisms have been researched for their potential application and potential utility as a useful tool in biological management. However, because they favor chemical methods for managing diseases, farmers are less likely to be employed in microbial-based goods. By combining different microorganisms with distinct modes of action, microbial consortia can decrease the need for chemicals and improve the longevity of plant protectants.

References

Abd-Allah, E.F., Alqarawi, A.A., Hashem, A., Radhakrishnan, R., Al-Huqail, A.A., Al-Otibi, F.O.N., Malik, J.A., Alharbi, R.I., & Egamberdieva, D. 2018. Endophytic bacterium *Bacillus subtilis* (BERA 71) improves salt tolerance in chickpea plants by regulating the plant defense mechanisms. *Journal of Plant Interactions* 13(1), 37–44.

Abhilash, P.C., Dubey, R.K., Tripathi, V., Gupta, V.K., & Singh, H.B. 2016. Plant growth-promoting microorganisms for environmental sustainability. *Trends in Biotechnology* 34(11), 847–850.

Aminifard, M.H., Aroiee, H., Nemati, H., Azizi, M., & Jaafar, H.Z.E. 2012. Fulvic acid affects pepper antioxidant activity and fruit quality. *African Journal of Biotechnology* 11(68), 13179–13185.

Aslantas, R., Ramazan, C., & Sahin, F. 2007. Effect of plant growth promoting rhizobacteria on young apple tree growth and fruit yield under orchard conditions. *Scientia Horticulturae* 111(4), 371–377.

Atero-Calvo, S., Izquierdo-Ramos, M.J., García-Huertas, C., Rodríguez-Alcántara, M., Navarro-Morillo, I., & Navarro-León, E. 2024. An evaluation of the effectivity of the green leaves biostimulant on lettuce growth, nutritional quality, and mineral element efficiencies under optimal growth conditions. *Plants* 13(7), 917. https://doi.org/10.3390/plants13070917

Baglieri, A., Cadili, V., Monterumici, C.M., Gennari, M., Tabasso, S., Montoneri, E., & Negre, M. 2014. Fertilization of bean plants with tomato plants hydrolysates. Effect on biomass production, chlorophyll content and N assimilation. *Scientia Horticulturae* 176, 194–199.

Bairey, E., Kelsic, E.D., & Kishony, R. 2016. High-order species interactions shape ecosystem diversity. *Nature Communications* 7(1), 12285.

Baltruschat, H., Fodor, J., Harrach, B.D., Niemczyk, E., Barn, B., Gullner, G., Janeczko, A., Kogel, K.H., Schäfer, P., Schwarczinger, I., & Zuccaro, A. 2008. Salt tolerance of barley induced by the root endophyte *Piriformospora indica* is associated with a strong increase in antioxidants. *New Phytologist* 180(2), 501–510.

Basile, B., Rouphael, Y., Colla, G., Soppelsa, S., & Andreotti, C. 2020. Appraisal of emerging crop management opportunities in fruit trees, grapevines and berry crops facilitated by the application of bio-stimulants. *Scientia Horticulturae* 267, 109330.

Beckett, R.P., & Van Staden, J. 1989. The effect of seaweed concentrates on the growth and yield of potassium stressed wheat. *Plant and Soil* 116, 29–36.

Bulgari, R., Cocetta, G., Trivellini, A., Vernieri, P., & Ferrante, A. 2015. Bio stimulants and crop responses: A review. *Biological Agriculture and Horticulture* 31(1), 1–17.

Carillo, P., Colla, G., Fusco, G.M., Dell'Aversana, E., El-Nakhel, C., Giordano, M., & Rouphael, Y. 2019. Morphological and physiological responses induced by protein hydrolysate-based bio stimulant and nitrogen rates in greenhouse spinach. *Agronomy* 9(8), 450.

Chandrasekaran, M., Chun, S.C., Oh, J.W., Paramasivan, M., Saini, R.K., & Sahayarayan, J.J. 2019. *Bacillus subtilis* CBR05 for tomato (*Solanum lycopersicum*) fruits in South Korea as a Novel Plant Probiotic Bacterium (PPB): implications from total phenolics, flavonoids, and carotenoids content for fruit quality. *Agronomy* 9(12), 838.

De Palma, M., D'Agostino, N., Proietti, S., Bertini, L., Lorito, M., Ruocco, M., Caruso, C., Chiusano, M.L., & Tucci, M. 2016. Suppression subtractive hybridization analysis provides new insights into the tomato (*Solanum lycopersicum* L.) response to the plant probiotic microorganism *Trichoderma longibrachiatum* MK1. *Journal of Plant Physiology* 190, 79–94.

De Pascale, S., Rouphael, Y., & Colla, G. 2017. Plant biostimulants: innovative tool for enhancing plant nutrition in organic farming. *European Journal of Horticultural Science* 82(6), 277–285.

De Souza, R.S.C., Armanhi, J.S.L., & Arruda, P. 2020. From microbiome to traits: designing synthetic microbial communities for improved crop resiliency. *Frontiers in Plant Science* 11, 1179.

Di Mola, I., Cozzolino, E., Ottaiano, L., Giordano, M., Rouphael, Y., El-Nakhel, C., & Mori, M. 2020. Effect of seaweed (*Ecklonia maxima*) extract and legume-derived protein hydrolysate bio stimulants on baby leaf lettuce grown on optimal doses of nitrogen under greenhouse conditions. *Australian Journal of Crop Science* 14(9), 1456–1464.

Dookie, M., Ali, O., Ramsubhag, A., & Jayaraman, J. 2021. Flowering gene regulation in tomato plants treated with brown seaweed extracts. *Scientia Horticulturae* 276, 109715.

Du Jardin, P. 2015. Plant biostimulants: definition, concept, main categories and regulation. *Scientia Horticulturae* 196, 3–14.

Ei, E., Park, H.H., & Kuk, Y.I. 2024. Effects of plant extracts on growth promotion, antioxidant enzymes, and secondary metabolites in rice (*Oryza sativa*) plants. *Plants* 13(19), 2727. https://doi.org/10.3390/plants13192727

El Hamss, H., Radouane, N., Belabess, Z., & Lahlali, R. 2023. Microbial consortia: an approach to enhance the effectiveness of beneficial soil microbes. In: *Novel Biological and Biotechnological Applications in Plant Nematode Management* (pp. 133–166). Singapore: Springer Nature Singapore.

Ertani, A., Schiavon, M., Altissimo, A., Franceschi, C., & Nardi, S. 2011. Phenol-containing organic substances stimulate phenylpropanoid metabolism in *Zea mays*. *Journal of Plant Nutrition and Soil Science* 174(3), 496–503.

Esserti, S., Smaili, A., Rifai, L.A., Koussa, T., Makroum, K., Belfaiza, M., Kabil, E.M., Faize, L., Burgos, L., Alburquerque, N., & Faize, M. 2017. Protective effect of three brown seaweed extracts against fungal and bacterial diseases of tomato. *Journal of Applied Phycology* 29, 1081–1093.

García, A.C., Berbara, R.L.L., Farías, L.P., Izquierdo, F.G., Hernández, O.L., Campos, R.H., & Castro, R.N. 2012. Humic acids of vermicompost as an ecological pathway to increase resistance of rice seedlings to water stress. *African Journal of Biotechnology* 11(13), 3125–3134.

Gemin, L.G., Mógor, Á.F., Amatussi, J.D.O., & Mógor, G. 2019. Microalgae associated to humic acid as a novel biostimulant improving onion growth and yield. *Scientia Horticulturae* 256, 108560.

Ghoul, M., & Mitri, S. 2016. The ecology and evolution of microbial competition. *Trends in Microbiology* 24(10), 833–845.

Gopalakrishnan, S., Sathya, A., Vijayabharathi, R., Varshney, R. K., Gowda, C.L., & Krishnamurthy, L. 2015. Plant growth promoting rhizobia: challenges and opportunities. *3 Biotech* 5, 355–377.

Hardoim, P.R., Van Overbeek, L.S., Berg, G., Pirttilä, A.M., Compant, S., Campisano, A., Döring, M., & Sessitsch, A. 2015. The hidden world within plants: ecological and evolutionary considerations for defining functioning of microbial endophytes. *Microbiology and Molecular Biology Reviews* 79(3), 293–320.

He, Y., Pantigoso, H.A., Wu, Z., & Vivanco, J.M. 2019. Co-inoculation of *Bacillus* sp. and *Pseudomonas putida* at different development stages acts as a biostimulant to promote growth, yield, and nutrient uptake of tomato. *Journal of Applied Microbiology* 127(1), 196–207.

Hyder, S., Gondal, A.S., Sehar, A., Khan, A.R., Riaz, N., Rizvi, Z.F., & Rizwan, M. 2024. Use of ginger extract and bacterial inoculants for the suppression of *Alternaria solani* causing early blight disease in Tomato. *BMC Plant Biology* 24(1), 131.

Islam, S., Akanda, A.M., Prova, A., Islam, M.T., & Hossain, M.M. 2016. Isolation and identification of plant growth-promoting rhizobacteria from cucumber rhizosphere and their effect on plant growth promotion and disease suppression. *Frontiers in Microbiology* 6, 1360.

Jha, C.K., & Saraf, M. 2015. Plant growth-promoting rhizobacteria (PGPR): a review. *Journal of Agricultural Research and Development* 5(2), 108–119.

Kalimuthu, K., Kumutha, K., Sabarinathan, K.G., Vijaya, P.P., Mini, M.L., & Amutha, R. 2024. Unraveling the intriguing potential of protein-rich microbial biostimulants for horticultural crops. *Plant Science Today* 11(4), 864–879

Kanmani, P., Satish Kumar, R., Yuvaraj, N., Paari, K.A., Pattukumar, V., & Arul, V. 2013. Probiotics and its functionally valuable products—a review. *Critical Reviews in Food Science and Nutrition* 53(6), 641–658.

Khan, N., Bano, A., Rahman, M.A., Guo, J., Kang, Z., & Babar, M.A. 2019. Comparative physiological and metabolic analysis reveals a complex mechanism involved in drought tolerance in chickpea (*Cicer arietinum* L.) induced by PGPR and PGRs. *Scientific Reports* 9(1), 2097.

Khan, W., Rayirath, U.P., Subramanian, S., Jithesh, M.N., Rayorath, P., Hodges, D.M., & Prithiviraj, B. 2009. Seaweed extracts as biostimulants of plant growth and development. *Journal of Plant Growth Regulation* 28, 386–399.

Kirn, A., Kashif, S.R., & Yaseen, M. 2010. Using indigenous humic acid from lignite to increase growth and yield of okra (*Abelmoschus esculentus* L.). *Soil Environment* 29(2), 187–191.

Kong, W., Meldgin, D.R., Collins, J.J., & Lu, T. 2018. Designing microbial consortia with defined social interactions. *Nature Chemical Biology* 14(8), 821–829.

Kredics, L., Büchner, R., Balázs, D., Allaga, H., Kedves, O., Racić, G., & Sipos, G. 2024. Recent advances in the use of *Trichoderma*-containing multicomponent microbial inoculants for pathogen control and plant growth promotion. *World Journal of Microbiology and Biotechnology* 40(5), 162.

Kumari, M., Swarupa, P., Kesari, K.K., & Kumar, A. 2022. Microbial inoculants as plant bio stimulants: a review on risk status. *Life* 13(1), 12.

Layek, J., Das, A., Idapuganti, R.G., Sarkar, D., Ghosh, A., Zodape, S.T., Lal, R., Yadav, G.S., Panwar, A.S., Ngachan, S., & Meena, R.S. 2018. Seaweed extract as organic bio-stimulant improves productivity and quality of rice in eastern Himalayas. *Journal of Applied Phycology* 30, 547–558.

Li, H., Yue, H., Li, L., Liu, Y., Zhang, H., Wang, J., & Jiang, X. 2021. Seed biostimulant *Bacillus* sp. MGW9 improves the salt tolerance of maize during seed germination. *AMB Express* 11, 1–15. https://doi.org/10.1186/s13568-021-01237-1

Manganiello, G., Sacco, A., Ercolano, M.R., Vinale, F., Lanzuise, S., Pascale, A., Napolitano, M., Lombardi, N., Lorito, M., & Woo, S.L. 2018. Modulation of tomato response to *Rhizoctonia solani* by *Trichoderma harzianum* and its secondary metabolite harzianic acid. *Frontiers in Microbiology* 9, 1966

Martínez-Lorente, S.E., Martí-Guillén, J.M., Pedreño, M.Á., Almagro, L., & Sabater-Jara, A.B. 2024. Higher plant-derived biostimulants: mechanisms of action and their role in mitigating plant abiotic stress. *Antioxidants* 13(3), 318.

Meenakshi, Annapurna, K., Govindasamy, V., Ajit, V., & Choudhary, D.K. 2019. Mitigation of drought stress in wheat crop by drought tolerant endophytic bacterial isolates. *Vegetos* 32(4), 486–493.

Naveed, M., Mitter, B., Reichenauer, T.G., Wieczorek, K., & Sessitsch, A. 2014. Increased drought stress resilience of maize through endophytic colonization by *Burkholderia phytofirmans* PsJN and *Enterobacter* sp. FD17. *Environmental and Experimental Botany* 97, 30–39.

Niu, X., Song, L., Xiao, Y., & Ge, W. 2018. Drought-tolerant plant growth-promoting rhizobacteria associated with foxtail millet in a semi-arid agroecosystem and their potential in alleviating drought stress. *Frontiers in Microbiology* 8, 2580.

Pellegrini, M., Spera, D., Ercole, C., & del Gallo, M. 2020. *Allium cepa* L. seed inoculation with a consortium of plant growth-promoting bacteria: effects on plant growth and development and soil fertility status and microbial community. *In Proceedings* 66(1), 20

Pichyangkura, R., & Chadchawan, S. 2015. Biostimulant activity of chitosan in horticulture. *Scientia Horticulturae* 196, 49–65.

Prajapati, A., Patel, C.K., Singh, N., Jain, S.K., Chongtham, S.K., Maheshwari, M.N., Patel, C.R., & Pate, R.N. 2016. Evaluation of seaweed extract on growth and yield of potato. *Environment & Ecology* 34(2), 605–608.

Rady, M.M., & Rehman, H. 2016. Supplementing organic biostimulants into growing media enhances growth and nutrient uptake of tomato transplants. *Scientia Horticulturae* 203, 192–198.

Ramakrishna, W., Yadav, R., & Li, K. 2019. Plant growth-promoting bacteria in agriculture: two sides of a coin. *Applied Soil Ecology* 138, 10–18.

Ronga, D., Biazzi, E., Parati, K., Carminati, D., Carminati, E., & Tava, A. 2019. Microalgal biostimulants and biofertilisers in crop productions. *Agronomy* 9(4), 192.

Rouphael, Y., Colla, G., Giordano, M., El-Nakhel, C., Kyriacou, M.C., & De Pascale, S. 2017. Foliar applications of a legume-derived protein hydrolysate elicit dose-dependent increases of growth, leaf mineral composition, yield and fruit quality in two greenhouse tomato cultivars. *Scientia Horticulturae* 226, 353–360.

Sabir, A. 2013. Improvement of grafting efficiency in hard grafting grape Berlandieri hybrid rootstocks by plant growth-promoting rhizobacteria (PGPR). *Scientia Horticulturae*, 164, 24–29.

Salachna, P., & Zawadzińska, A. 2014. Effect of chitosan on plant growth, flowering and corms yield of potted freesia. *Journal of Ecological Engineering* 15(3), 97–102.

Sani, M.N.H., Islam, M.N., Uddain, J., Chowdhury, M.S.N., & Subramaniam, S. 2020. Synergistic effect of microbial and nonmicrobial bio stimulants on growth, yield, and nutritional quality of organic tomato. *Crop Science* 60(4), 2102–2114.

Sanjuán, J., Nápoles, M.C., Pérez-Mendoza, D., Lorite, M.J., & Rodríguez-Navarro, D.N. 2023. Microbials for agriculture: why do they call them bio-stimulants when they mean probiotics? *Microorganisms* 11, 153.

Seiber, J.N., Coats, J., Duke, S.O. 2014. Biopesticides: state of the art and future opportunities. *Journal of Agricultural and Food Chemistry* 62, 11613–11619.

Servin, A.L. 2004. Antagonistic activities of lactobacilli and bifidobacteria against microbial pathogens. *FEMS Microbiology Reviews* 28(4), 405–440.

Shah, M.T., Zodape, S.T., Chaudhary, D.R., Eswaran, K., & Chikara, J. 2013. Seaweed sap as an alternative liquid fertilizer for yield and quality improvement of wheat. *Journal of Plant Nutrition* 36(2), 192–200.

Shahid, M., Dumat, C., Silvestre, J., & Pinelli, E. 2012. Effect of fulvic acids on lead-induced oxidative stress to metal sensitive *Vicia faba* L. plant. *Biology and Fertility Soils* 48, 689–697.

Sharanya, B.R., Gowda, A.M., & Srinivasappa, K.N. 2022. Impact of bio-stimulants on growth and yield of cowhage (*Mucuna pruriens* L.). *The Pharma Innovation Journal* 11(11), 399–404.

Sharma, H.S., Fleming, C., Selby, C., Rao, J.R., & Martin, T. 2014. Plant biostimulants: a review on the processing of macroalgae and use of extracts for crop management to reduce abiotic and biotic stresses. *Journal of Applied Phycology* 26, 465–490.

Shukla, P.S., Shotton, K., Norman, E., Neily, W., Critchley, A.T., & Prithiviraj, B. 2018. Seaweed extract improve drought tolerance of soybean by regulating stress-response genes. *AoB Plants* 10(1), plx051.

Singh, D., Jadon, K.S., Verma, A., & Kakani, R.K. 2024. Harnessing nature's defenders: unveiling the potential of microbial consortia for plant defense induction against *Alternaria* blight in cumin. *Folia Microbiologica* 3: 1–24.

Song, J., Xi, X., Qian, X., Li, J., Tao, Y., Li, L., & Zhou, G. 2024. Revealing the microbial composition changes and relationship with Fusarium caused by rot disease in the *Crocus sativus* L. *Annals of Microbiology* 74(1), 24.

Tejada, M., Benítez, C., Gómez, I., & Parrado, J. 2011. Use of bio stimulants on soil restoration: effects on soil biochemical properties and microbial community. *Applied Soil Ecology* 49, 11–17.

Tripathi, A., Pandey, V.K., Jain, D., Singh, G., Brar, N.S., Taufeeq, A., Pandey, I., Dash, K.K., Samrot, A.V., & Rustagi, S. 2024. An updated review on significance of PGPR-induced plant signalling and stress management in advancing sustainable agriculture. *Journal of Agriculture and Food Research* 16, 101169.

Ventorino, V., Sannino, S., Piccolo, A., Cafaro, V., Carotenuto, R., & Pepe, O. 2014. *Methylobacterium populi* VP2: plant growth-promoting bacterium isolated from a highly polluted environment for polycyclic aromatic hydrocarbon (PAH) biodegradation. *The Scientific World Journal* 2014(1), 931793.

Vetrano, F., Miceli, C., Angileri, V., Frangipane, B., Moncada, A., & Miceli, A. 2020. Effect of bacterial inoculum and fertigation management on nursery and field production of lettuce plants. *Agronomy* 10(10), 1477.

Wei, Z., Yang, T., Friman, V.P., Xu, Y., Shen, Q., & Jousset, A. 2015. Trophic network architecture of root-associated bacterial communities determines pathogen invasion and plant health. *Nature Communications* 6(1), 8413.

Wen, T., Xie, P., Liu, H., Liu, T., Zhao, M., Yang, S., & Yuan, J. 2023. Tapping the rhizosphere metabolites for the prebiotic control of soil-borne bacterial wilt disease. *Nature Communications* 14(1), 4497.

Woo, S.L., & Pepe, O. 2018. Microbial consortia: promising probiotics as plant bio-stimulants for sustainable agriculture. *Frontiers in Plant Science* 9, 1801.

Yakhin, O.I., Lubyanov, A.A., Yakhin, I.A., & Brown, P.H. 2017. Bio-stimulants in plant science: a global perspective. *Frontiers in Plant Science* 7, 2049.

Yang, N., Nesme, J., Røder, H.L., Li, X., Zuo, Z., Petersen, M., Burmølle, M., & Sørensen, S.J. 2021. Emergent bacterial community properties induce enhanced drought tolerance in *Arabidopsis*. *NPJ Biofilms Microbiomes* 7(1), 82.

Zhang, Z., Zhang, J., Xu, G., Zhou, L., & Li, Y. 2019. Arbuscular mycorrhizal fungi improve the growth and drought tolerance of *Zenia insignis* seedlings under drought stress. *New Forests* 50, 593–604.

Zhou, X., Joshi, S., Patil, S., Khare, T., & Kumar, V. 2021. Reactive oxygen, nitrogen, carbonyl and sulfur species and their roles in plant abiotic stress responses and tolerance. *Journal of Plant Growth Regulation* 2: 1–24.

12

Commercialization Aspects of Microbial Biocontrol Agents

Saurabh Sharma[1], *Justine Sanasam*[2], *Umalaxmi Thingujam*[3], *Krishnakumari Yendrembam*[4], *and Guneshori Maisnam*[5*]

[1] *Department of Agricultural Extension Education, Nagaland University, Dimapur, Nagaland, India*
[2] *(Agriculture) Agronomy, Amity Institute of Organic Agriculture, Noida, Uttar Pradesh, India*
[3] *Department of Soil Science and Agricultural Chemistry, PSB, Visva Bharati, West Bengal, India*
[4] *Lovely Professional University, Jalandhar, Punjab, India*
[5] *Department of Rural Development and Planning, Nagaland University, Medziphema Campus, Nagaland, India*

[*] Corresponding author: gunemaisnam@gmail.com

12.1 Introduction

The increasing demand for sustainable and eco-friendly agricultural practices has led to the growing interest in microbial biocontrol agents (MBAs) as alternatives to synthetic pesticides. When it comes to managing plant diseases, biocontrol agents (BCAs) are a more environmentally friendly option than chemical pesticides. An outline of the several strategies used by BCAs to inhibit plant diseases and pests is provided in this chapter. Direct hostility, competition, secondary metabolite synthesis, induced systemic resistance (ISR), and microbial interactions are some of these mechanisms. A better comprehension of these processes provides information about how to use them in sustainable agriculture and integrated pest management (IPM). MBAs, including beneficial bacteria, fungi, viruses, and protozoa, play a crucial role in suppressing plant pathogens and pests while minimizing environmental and human health risks. Unlike chemical pesticides, these biological agents offer high specificity, reduced toxicity, and lower chances of pest resistance, making them an integral part of IPM strategies. Despite their advantages, the commercialization of MBAs faces several challenges, such as mass production difficulties, formulation stability, regulatory approvals, and market acceptance (Glare & O'Callaghan, 2000). Ensuring microbial viability during storage and application is a key concern, requiring advancements in formulation technologies such as encapsulation, freeze-drying, and liquid fermentation. Additionally, regulatory frameworks for biopesticides vary across different regions, often requiring extensive safety testing and registration procedures that can slow down product development and market entry (Lugtenberg & Kamilova, 2009). The success of MBAs in the commercial market depends on cost-effective production, efficient distribution

Microbial Consortia in Plant Science and Sustainable Agriculture, First Edition. Edited by Ajay Tomar, et al.

networks, and farmer awareness programs. Adoption rates among farmers remain a challenge due to factors like limited knowledge, inconsistent field performance, and competition with well-established chemical pesticides. However, advancements in biotechnology, government incentives, and increasing consumer demand for organic and residue-free food are driving the expansion of the biocontrol industry. This chapter explores the key commercialization aspects of MBAs, highlighting technological innovations, regulatory considerations, market dynamics, and strategies to enhance adoption. By addressing these challenges, MBAs have the potential to revolutionize pest management and contribute to a more sustainable global agricultural system.

12.1.1 History of BCAs

The use of natural creatures, such as viruses, parasites, and predators, to manage disease and pest populations is known as biological control or biocontrol. Biocontrol has become a viable substitute for chemical pest control over time, providing a sustainable, eco-friendly, and efficient way to manage agricultural diseases and pests (Kairo, 2000). Although there is evidence that biocontrol was used in early civilizations thousands of years ago, scientific study did not provide the groundwork for contemporary biocontrol methods until the late nineteenth and early twentieth centuries. The history of BCAs is covered in this chapter, from their first uses to the current usage of natural enemies in IPM systems. The long-standing human desire to manage illnesses and pests in a sustainable and ecologically friendly way is demonstrated by the history of BCAs. From traditional farming methods to contemporary IPM techniques, biocontrol has developed into an essential weapon in the battle against pests (Kaiser, 2007). Biocontrol is a viable route toward more environmentally friendly farming methods as the world's environmental problems worsen, guaranteeing food security while reducing ecological effects.

12.1.2 Mechanisms of Action

Antibiotics, enzymes, and other toxic chemicals that directly prevent pathogen development or virulence are examples of antagonism. For instance, chitinases and glucanases produced by *Trichoderma* species break down plant pathogen cell walls (Madhavan et al., 2015).

Competition for Space and Nutrients: BCAs prevent pathogens from establishing themselves by competing with them for scarce nutrients in the rhizosphere or phyllosphere. For instance, *Pseudomonas fluorescens* produces siderophores to compete for iron, limiting the amount of this vital nutrient that pathogens can access (Glick et al., 2007).

12.1.3 Indirect Mechanisms

1. **ISR:** By causing ISR, a persistent, all-around resistance to infections, BCAs can promote plant defense mechanisms. For instance, *Pseudomonas* species and *Bacillus subtilis* cause ISR in plants by activating signaling pathways and plant immune receptors (Van Loon et al., 2014).
2. **Production of Secondary Metabolites:** BCAs have the ability to produce secondary metabolites that inhibit the growth of pathogens, such as antifungal peptides and

volatile organic compounds. For instance, *Streptomyces* species generate antibiotics that inhibit a variety of plant diseases, such as streptomycin and actinomycin (Sundar, 2015). To control pests and diseases, BCAs employ a variety of strategies, ranging from simple predation to more intricate relationships like creating toxic compounds or causing plant resistance. The particular pest or disease, the surrounding environment, and the intended crop management result all influence the choice of biocontrol technique. Particularly in horticulture and agriculture, BCAs are organisms or their by-products that are employed to manage infections, pest populations, or unwanted species. The effects of BCAs can be produced through a variety of different processes and methods. Here are several important methods of action along with illustrations:

a. **Competition for Resources**: In competition for vital nutrients and space, BCAs may be able to outcompete infections or pests. They lessen the resources available to the dangerous species by colonizing the habitat more quickly than the target organisms. For instance, fungi called *Trichoderma* spp. outcompete plant diseases like *Fusarium* and *Rhizoctonia* for nutrients and space on plant roots, hence reducing the establishment and disease-causing activity of these pathogens.
b. **Antibiosis (Production of Antimicrobial Compounds):** Mechanisms include toxic compounds, like antibiotics, that limit the growth or kill pests and diseases can be produced by BCAs. For instance, the antimicrobial substances subtilin and bacilysin, which are produced by *B. subtilis*, prevent the growth of some plant diseases, such as *Pythium* and *Fusarium* species.
c. **The Mechanism of the Parasite:** Pathogens or pests can be parasitized by BCAs. In this instance, the target organism is harmed or killed by the BCA that resides on it. As an illustration, hymenopteran parasitoids, such as *Trichogramma* species, deposit their eggs inside those of *Lepidoptera* species and other insect pests. The insect eggs are killed by the developing larvae, stopping additional infection.
d. **Predation Mechanism:** Certain BCAs naturally prey on pests, lowering their numbers by feeding on them directly. Aphids are naturally preyed upon by ladybird beetles, such as *Coccinella septempunctata*. They drastically lower aphid numbers by feasting on them, which prevents aphids from harming plants by sucking sap.
e. **Systemic Acquired Resistance and ISR Mechanism**: By inducing the plant's immunological response, biocontrol chemicals can increase the plant's resistance to future pathogen infections (Sivapalan & Singh, 2020). The entire plant may be impacted by this resistance, which may be systemic or local. For instance, the rhizobacterium *P. fluorescens* causes ISR in plants, which strengthens their defenses against diseases like *Rhizoctonia* and *Pythium*.
f. **Toxin Biodegradation:** By breaking down or neutralizing toxic substances that infections or pests produce, certain BCAs might lessen the negative effects they have on plants. Toxins produced by *Bacillus thuringiensis* are poisonous to insect larvae but not to humans or other animals. It is frequently used to manage insect pests such as flies, mosquitoes, and caterpillars.
g. **Excessive Parasites:** To improve its own capacity to manage pest populations, a BCA may parasitize or target other BCAs or pests' natural enemies. For instance, several species of ladybird beetles in the genus *Hyperaspis* function as hyperparasites, targeting aphid parasitoids such as *Aphidius ervi* and interfering with the aphids' normal biocontrol mechanisms.

h. **Modification of Habitat:** Certain biocontrol chemicals aid in altering the habitat to make it less conducive to infections or pests. They may have an impact on the microenvironment that reduces the ability of dangerous organisms to survive or reproduce. For instance, soil-dwelling insect larvae are managed by nematodes like *Steinernema* spp. By boosting soil moisture and aeration, these nematodes change the environment, which has a detrimental effect on pest survival.
i. **Symbiosis-Based Biological Control:** To shield plants from diseases and pests, certain BCAs develop symbiotic partnerships with plants or other species. The BCA may benefit from these interactions as well. As an illustration, mycorrhizal fungi, such as *Glomus* species, establish symbiotic connections with plants. By enhancing plant health and nutrient intake, these fungi can increase a plant's resilience to illnesses like *Fusarium* wilt.
j. **Genetic Engineering:** BCAs can be improved to become more effective against particular pests or diseases by selective breeding or genetic engineering. For instance, *B. thuringiensis* (Bt) has been genetically manipulated to create poisons that are extremely efficient against particular pest species, such as the European maize borer. To control pests and diseases, BCAs employ a variety of strategies, ranging from direct predation to more intricate relationships, such as creating plant resistance or producing dangerous materials. The particular pest or disease, the surrounding environment, and the intended crop management result all influence the choice of biocontrol technique.

12.1.4 Biocontrol Microorganisms

The detrimental effects of chemical pesticides on the environment and human health have led to heightened scrutiny of agricultural practices in recent decades. These substances endanger human health, damage beneficial organisms, and cause pesticide resistance. Consequently, the use of microorganisms for biological control has drawn interest as a viable and environmentally beneficial substitute. To manage pests, diseases, and other unwanted species while maintaining the integrity of ecosystems, biocontrol microorganisms are essential.

There are many different types of biocontrol microorganisms, such as bacteria, fungi, yeasts, and viruses, and each has a special way of fighting off illnesses and pests. These microbes have a number of tactics at their disposal, such as direct hostility, resource competition, and toxin manufacturing. The kinds of microorganisms utilized in biocontrol, their mechanisms of action, and noteworthy instances of biocontrol success are reviewed in this manuscript.

12.1.5 Types of Biocontrol Microorganisms

1. **Bacteria:** One of the most often employed microorganisms for managing pathogens and pests is bacterial BCAs. Through a variety of strategies, including direct predation, nutritional competition, and the synthesis of antimicrobial chemicals, these bacteria can reduce pest populations.

Examples:

***B. thuringiensis*:** One of the most well-known biocontrol bacteria is *B. thuringiensis* (Bt), which makes proteins that are poisonous to some insect larvae, including the European corn borer. Products made from Bt have been used extensively in agriculture to manage pests without endangering beneficial insects.

***P. fluorescens*:** This bacterium outcompetes harmful microorganisms, creates antibiotic chemicals, and causes plant resistance. It is employed to manage root diseases like as *Fusarium* and *Pythium*.

2. **Fungi:** Because they may target a wide variety of plant diseases and pests, fungus BCAs are used extensively. Fungi can work by parasitizing pest organisms or by generating mycotoxins that stop infections from growing.

   Examples:

   ***Trichoderma* spp.:** *Trichoderma* species are well-known for their capacity to shield plants against soilborne diseases such as *Fusarium* and *Rhizoctonia*. They can create enzymes that break down the infections' cell walls or outcompete them for nutrients and space.

   ***Beauveria bassiana*:** This entomopathogenic fungus infects termites, aphids, and whiteflies, among other insect pests. Spores of *B. bassiana* adhere to the insect's cuticle and enter its body, killing it.

3. **Viruses:** Insect pests can be effectively managed with viral BCAs. Particularly, baculoviruses have been thoroughly researched for their potential to manage pests. Due to their extreme specificity – they frequently exclusively infect one type of insect – these viruses have a minimal effect on creatures that are not their intended targets.

   Example: Baker's yeast or *Saccharomyces cerevisiae*: Although best known for its use in baking, *S. cerevisiae* has also been studied for its ability to inhibit plant pathogens such as *Botrytis cinerea* (grey mold) by creating a protective biofilm on plant surfaces that stops the pathogen from adhering.

12.1.6 Applications in Agriculture

Numerous agricultural contexts employ biocontrol microbes, such as the following:

1. **Field Crops:** Diseases like blights and molds, as well as pests like aphids, beetles, and caterpillars, are controlled with BCAs. For instance, Bt-based products are used to keep insect pests away from crops like cotton and maize.
2. **Greenhouse Crops:** Applying BCAs more precisely is made possible by the controlled atmosphere of greenhouses. Microorganisms such as *Trichoderma* and *Pseudomonas* species aid in the control of root rot and soilborne illnesses in greenhouse plants.
3. **Fruit and Vegetable Crops:** Aphids and root diseases are managed in fruit orchards and vegetable farms by using *Trichoderma* and *B. bassiana*.
4. **Weed Management:** By infecting and destroying undesirable plants, fungus pathogens like *Colletotrichum* species have demonstrated the ability to control weeds.

12.1.7 Success Stories in Biocontrol

A number of success stories demonstrate how beneficial biocontrol bacteria are the following:

1. **Bt. in Corn and Cotton:** The application of Bt, namely *B. thuringiensis* products, has transformed corn and cotton pest management. By lowering the demand for chemical pesticides, Bt crops have reduced environmental contamination and encouraged sustainable farming methods.
2. ***Trichoderma* spp. in Tomato and Cucumber Production:** The application of *Trichoderma* species has greatly decreased soilborne illnesses such as *Fusarium* wilt in tomato and cucumber cultivation, improving crop yields and lowering the need for pesticides.
3. ***B. bassiana* in Insect Pest Control:** In several areas, *B. bassiana* has been successfully used to control pests like the African cotton bollworm, which lessens the need for chemical insecticides.

12.1.8 Challenges and Future Directions

Although biocontrol microbes have a lot of promise, there are still a number of obstacles to overcome:

1. **Regulation and Commercialization**: Strict regulatory frameworks and a dearth of standardized procedures for assessing efficacy frequently impede the commercialization of BCAs.
2. **Environmental Variability**: Temperature, humidity, and soil composition are some examples of environmental factors that might impact the performance of BCAs and limit their efficacy in particular areas.
3. **Cost and Scalability**: Large-scale production of BCAs can be costly, which may prevent their widespread use, especially in environments with limited resources.

Integrating biocontrol microorganisms into IPM programs, where they can be employed in conjunction with other control strategies, including chemical and cultural approaches, is where their future rests. Biotechnology developments, such as genetically modifying biocontrol chemicals to increase their effectiveness, are probably going to play a vital part in getting past present obstacles. In agriculture, biocontrol microorganisms are a viable and sustainable substitute for chemical pesticides, offering efficient ways to control diseases and pests. These microbes have the potential to be crucial to the development of sustainable agriculture in the future if further study, research, and commercialization are conducted.

12.1.9 Advantages and Disadvantages of BCAs

The technique known as biological control, or biocontrol, uses diseases, parasites, predators, and competitors as natural enemies to manage pest species. Because of its possible economic and environmental advantages, it is being explored more as a substitute for chemical control techniques. The usefulness and limitations of BCAs in pest management are reviewed in this publication, along with their benefits and drawbacks. Biological control has many benefits, including long-term pest management, environmental sustainability, and wildlife and human safety. However, before putting biocontrol ideas into practice, issues including sluggish action, nontarget consequences, and environmental dependency must be carefully taken into account. A thorough understanding of the pest, the BCAs, and the application environment is necessary for successful biocontrol. Biocontrol is anticipated to grow in importance as a technique for IPM as research progresses.

12.1.9.1 Advantages of BCAs

Eco-friendly: In contrast to chemical pesticides, BCAs are frequently regarded as environmentally sustainable. In general, they have negligible detrimental effects on ecosystems, soil health, and nontarget species. Because BCAs usually only target particular pests, they don't harm the environment broadly or contaminate the air, soil, or water – two main issues with chemical control methods. For instance, less broad-spectrum pesticides are required when *Trichogramma* species (parasitoid wasps) are used to manage lepidopteran pests like moths (Bigler et al., 2006).

Long-Term Pest Control: By establishing populations in the environment, BCAs can offer long-term pest control, potentially reducing insect populations for longer periods of time. These agents can naturally proliferate and spread after being injected, negating the need for further treatments. For instance, *Cactophagus* weevil was effectively introduced to Australia to manage invasive prickly pear cacti, resulting in a long-term decrease of the cactus population (Julien & Griffiths, 1998).

Decreased Development of Resistance: Biocontrol treatments are less likely to cause resistance in target pest populations than chemical pesticides, which can cause resistance over time. This is due to the fact that certain pests are targeted by natural enemies in a manner that makes it challenging for the pest to adapt. For instance, pests are less likely to develop resistance when *Aphidius* parasitoids are used to manage aphids because the natural predators maintain a balanced control strategy (Heimpel & Mills, 2017).

Safety for People and Animals: In general, BCAs are less harmful to people, animals, and wildlife than chemical pesticides. Because of this, they are a safer choice for controlling pests, particularly in agricultural systems where food safety is an issue. For instance, unlike many synthetic pesticides, the use of *B. thuringiensis* (Bt) as a biopesticide is thought to be harmless for both beneficial creatures and people (Glare et al., 2012).

Cost-Effective: Since biocontrol eliminates the need for frequent pesticide applications, it may end up being less expensive than chemical control over time. Many BCAs require little extra funding once they are established. For instance, it has been demonstrated that using *Neoseiulus* mites to manage spider mites in greenhouses lowers the expenses related to the usage of pesticides (Van Houten et al., 2020).

12.1.9.2 Disadvantages of BCAs

Slow Action: Compared to chemical pesticides, BCAs may take longer to lower insect populations. When quick pest management is required, like during an invasive species outbreak or in highly competitive agricultural environments, this delayed action may prove troublesome. For instance, it may take weeks or even months for *P. fluorescens* to start producing discernible effects when used to suppress soilborne diseases (Kremer et al., 2012).

Nontarget Effects: Despite the fact that BCAs are typically more selective than chemical pesticides, nontarget effects are always possible. A BCA's introduction could inadvertently impact non-pest species through competition or predation, which may upset regional ecosystems. For instance, native plant species were inadvertently harmed when the *Cactophagus* weevil was introduced to control prickly pears (Hoffmann, 2000).

Dependency on Environmental Conditions: Environmental elements like temperature, humidity, and habitat quality frequently affect how effective BCAs are. The effectiveness or survivability of BCAs can be diminished by extreme weather, which reduces their dependability in a variety of environmental conditions. For instance, *Aphidius* parasitoids may find it difficult to develop in areas with exceptionally hot or dry temperatures, which would restrict their application in some agricultural areas (Van Lenteren, 2000).

Lack of Agent Control: Because BCAs are living things, their actions are not always precisely regulated like chemical pesticides can. They might not always target the desired pest species, or they might lose some of their effectiveness as a result of shifting population dynamics or pest behavior. For instance, *B. thuringiensis* distribution in some agricultural systems has had inconsistent outcomes because some pests either become immune to the virus or steer clear of infected areas (Hajek et al., 2008).

High Initial Cost and Complexity: Biocontrol program development and implementation can be costly and intricate. Some small-scale or resource-poor farmers may find it difficult to invest in and possess the technical know-how necessary for the research, development, and distribution of BCAs. For instance, the initial high expense of introducing *Trichogramma* in large-scale agricultural systems is one example of how meticulous planning is necessary to build successful biocontrol programs (Gurr et al., 2016).

12.1.10 Challenges in Adopting BCAs

1. **Regulatory Barriers:**

   Absence of Standardized Rules: The discharge of BCAs is subject to a lack of standardized rules in many nations. Guidelines differ between nations, which may cause delays in approval and execution.

   Approval Process Complexity: The regulatory procedures for BCA testing and approval are drawn out and intricate. Extensive testing standards for non-native species might cause major delays in some areas.

   Cost of Registration: Because the expenses associated with developing, testing, and registering BCAs are frequently substantial, the financial burden of doing so may deter farmers or small businesses from utilizing them.

2. **Effectiveness and Reliability:**

   Variable Efficacy: Particularly in outdoor settings, BCAs may not always offer the appropriate degree of control. Environmental factors, host resistance, and other competing species can all contribute to this unpredictability.

   Species Specificity: A large number of BCAs are extremely species-specific, which means that they might only be effective against a limited number of pests. This may restrict their use in situations where pest populations are heterogeneous or fluctuating.

Insufficient Knowledge of Ecological Interactions: Unexpected ecological outcomes may result from a lack of understanding of how BCAs interact with other organisms in the environment, including nontarget species.

3. **Economic and Market Barriers:**

   Cost of Implementation: Although BCAs are frequently thought of as being more environmentally friendly, their cost can be higher than that of chemical alternatives, especially when taking into account the agents' manufacture, use, and upkeep.

   Farmer Reluctance: A lot of farmers are reluctant to use BCAs because they don't trust them, are unsure about them, or have had bad experiences with improperly administered or ineffective products in the past.

   Infrastructure and Support: In many areas, the infrastructure needed for the manufacture, delivery, and use of BCAs is currently insufficient, which restricts their uptake. Furthermore, farmers who are not familiar with these techniques frequently lack technical assistance and training.

4. **Biological and Environmental Limitations:**

   Climate Sensitivity: It might be challenging to deploy BCAs in specific climates or during specific seasons since they are frequently sensitive to environmental factors, including temperature, humidity, and soil quality.

   BCA Persistence: BCAs frequently have a short field life, meaning they might not be effective for long enough to effectively control pests. Certain agents might not reproduce well in the intended environment or may disappear too soon.

   Competition with Native Species: Non-native BCAs have the potential to disrupt local ecosystems or outcompete native species when they are introduced.

5. **Knowledge and Training Gaps:**

   Lack of Expertise and Knowledge: Farmers and pest control specialists frequently lack the know-how to successfully incorporate biocontrol chemicals into their pest management plans. BCAs may be applied incorrectly without training, which would decrease their efficacy.

6. **Integration with Other Control Methods:** BCAs frequently need to be employed in conjunction with other pest control techniques (such as chemical treatments or cultural activities), necessitating an understanding of and proficiency with IPM. Farmers who are not accustomed to this method may find it challenging.

7. **Problems with the Supply Chain and Distribution:**

   Limited Access to BCAs: In certain regions, it may be difficult to get trustworthy and efficient BCAs, or the items might not be available when pests are most active.

   Storage and Shelf Life: BCAs may need particular storage conditions and have a short shelf life, making them challenging to transport and store efficiently.

8. **Perception and Trust by the Public:**

   Public Safety Concerns: Some farmers and the general public have reservations about the safety of releasing BCAs into the environment, especially when it comes to non-native species. Adoption may be impeded by misconceptions or anxieties over possible adverse ecological effects.

9. **Lack of Education and Awareness:** Even if BCAs are shown to be successful by science, the general population may not completely comprehend the advantages of biocontrol techniques, which could lead to opposition to adoption.

12.1.11 Government Regulations on BCAs

Utilizing natural creatures like parasitoids, viruses, and predators to control pest populations is known as biological control or biocontrol. Biocontrol has drawn more and more attention globally as an environmentally responsible and sustainable substitute for chemical pesticides. To guarantee their efficacy and safety, the government has strict laws governing the introduction and use of BCAs. These rules are essential for reducing any ecological hazards and guaranteeing that BCAs don't damage nontarget species or disturb ecosystems.

Structure of Regulation for BCAs

1. **Global Standards:** International organizations such as the World Trade Organization and the Food and Agriculture Organization (FAO, 2019) have an impact on the introduction of BCAs.

   Guidelines for the safe use of BCAs are provided by the FAO Code of Conduct on the Use of BCAs to reduce hazards to human health and biodiversity. Furthermore, a framework for the prudent use of BCAs is established by the Convention on Biological Diversity, especially when taking into account the threats to native species and ecosystems.
2. **Rules at the National Level:** The regulatory bodies in charge of regulating the implementation and use of BCAs vary by nation. For example, in the United States, the importation of BCAs is governed by the Animal and Plant Health Inspection Service, which is a division of the Department of Agriculture. They guarantee that these substances are not invasive and do not endanger the environment or agriculture. The safety of BCAs used in pest management for people, animals, and the environment is another responsibility of the Environmental Protection Agency (EPA).

   European Union: Regulation (EC) No. 1107/2009, which regulates the approval of plant protection products, including BCAs, established a comprehensive regulatory framework. Before being put on the market, the rule makes sure that BCAs undergo extensive testing for safety, effectiveness, and environmental impact.

   Australia: BCA registration and regulation are supervised by the Australian Pesticides and Veterinary Medicines Authority in Australia. A thorough risk assessment procedure that incorporates assessments of ecological, human, and environmental safety is part of their framework.
3. **Regulatory Procedures:** There are multiple steps in the BCA approval process. The foundation of the regulatory process is risk assessment. The possible harm that a BCA could do to nontarget species, the ecology, and human health is evaluated by regulatory bodies.

   Efficacy Testing: To make sure the BCA can effectively manage the pest without causing unforeseen repercussions, its efficacy is also assessed.

   Post-Release Monitoring: To identify any unanticipated ecological effects, regulators may demand ongoing monitoring of the BCAs after approval.

12.1.12 Case Studies of Biocontrol Regulation

1. **The Australian Cane Toad's Introduction:** The 1935 introduction of the Cane Toad (*Rhinella marina*) to Australia to control the sugarcane beetle is among the most well-known (and contentious) examples of biocontrol. But the toad turned into an

invasive species that spread quickly and seriously harmed native species' ecological well-being. This case emphasizes the significance of thorough risk evaluations as well as the possible long-term effects of biocontrol measures.
2. **The *Tamarixia triozae* Wasp's Effective Use:** On the other hand, the Asian citrus psyllid has been successfully managed in the United States through the employment of the *T. triozae* wasp. Following extensive testing, the wasp was brought in and has successfully decreased psyllid populations without endangering other species. This instance demonstrates the beneficial effects of properly regulated BCAs.

Government rules are essential for guaranteeing the safe and efficient use of biocontrol chemicals. The legal frameworks governing the use of biocontrol must change to accommodate new issues and advancements in the field, as it continues to gain traction as a practical substitute for chemical pest control. Protecting ecosystems, agriculture, and human health all depend on BCAs being well tested for safety and effectiveness.

12.1.13 Biocontrol Industries

Growing interest in biocontrol methods is a result of worries about the dangers that chemical pesticides pose to human health and the environment. The biocontrol businesses, which use living creatures like diseases, parasitoids, predators, and biochemicals to give natural pest management solutions, are essential in tackling the world's agricultural problems. The development, innovations, and effects of the biocontrol industries are examined in this book, with a focus on the technologies employed, market dynamics, and potential future developments. Biocontrol has the potential to transform pest management systems globally when combined with other sustainable techniques.

Natural enemies are used in biological control, or biocontrol, an environmentally friendly method of managing pests. As a substitute for the widespread use of synthetic insecticides, it has been a crucial component of integrated pest control (IPM) techniques. Biocontrol businesses are becoming more popular as a safer, more sustainable alternative to conventional pesticides due to growing worries about their detrimental impacts on ecosystems, biodiversity, and human health. The market for biocontrol has grown significantly in recent years due to consumer demand for organic products, regulatory restrictions, and biotechnology advancements. Utilizing helpful species like parasitoids, predators, and microbiological agents like bacteria, viruses, and fungi to manage detrimental pests in horticultural, forestry, and agricultural contexts is known as biocontrol.

12.1.14 Types of BCAs

There are three primary categories of BCAs:

1. **Predators:** These organisms consume pest species directly. Predators frequently target a variety of pests because they are generalists. Aphids, for instance, are commonly managed with ladybirds (Coccinellidae).
2. **Parasitoids:** These organisms lay their eggs inside or on top of a pest, finally killing the host as the larvae grow. *Trichogramma* species, which are employed to combat lepidopteran pests, are among the best-known examples.

3. Microbial agents are organisms that infect and kill pest species, such as bacteria, fungi, viruses, and nematodes. The bacterium *B. thuringiensis* (Bt) is commonly employed to combat a variety of insect larvae, whereas the fungal infection *B. bassiana* targets a broad spectrum of insects.
4. **Biochemicals:** These are naturally occurring substances that interfere with the biological functions of the pest, such as mating disruption and pheromones for pest capture.

12.1.15 Evolution and Development of Biocontrol Industries

Although the industry started to grow more methodically in the twentieth century, the first documented use of biological control dates to the late nineteenth century. As industrialized agriculture, which frequently depended heavily on synthetic pesticides, grew, so did the biocontrol sector.

The following are important factors propelling the biocontrol businesses' expansion:

- **Pesticide Regulations:** Stricter rules and prohibitions on dangerous compounds have resulted from growing worries about the effects chemical pesticides have on the environment. For instance, the widespread prohibition of DDT in the 1970s increased interest in biocontrol techniques as a substitute.
- **Developments in Biotechnology:** Molecular biology and genetic engineering have improved the specificity and effectiveness of BCAs. For example, Bt has been incorporated into crops through genetic engineering, which lessens the requirement for outside applications.
- **Demand for Organic Products:** The biocontrol sector has developed products that are appropriate for organic farming in response to the growing customer demand for food that is cultivated sustainably and organically.
- **Sustainability in Agriculture:** IPM techniques that use BCAs have become more and more popular as a result of the drive for more environmentally friendly farming methods.

12.1.16 Key Players and Market Dynamics

In recent years, the worldwide biocontrol market has grown significantly. According to a Grand View Research analysis, the worldwide biocontrol market was estimated to be worth USD 5.6 billion in 2022 and is expected to increase at a 14.2% compound annual growth rate between 2023 and 2030. Product development advancements, government assistance, and the growing use of sustainable farming methods are the main factors propelling the market.

Among the major companies in the biocontrol sector are the following:

- **Syngenta:** It is a well-known international agribusiness that has made significant investments in biocontrol goods, especially microbial solutions.
- **Badische Anilin- und Soda-Fabrik (BASF):** BCAs have been incorporated into BASF's wider pest management techniques. BASF is well known for its biological crop protection solutions.

- **Valent BioSciences:** It focuses on creating and promoting BCAs, especially biological insecticides.
- **Certis USA:** Certis provides a variety of biocontrol solutions for different crops and is a significant player in the biological pesticides market.

12.1.17 Production and Formulation Challenges in the Commercialization of MBAs

MBAs have emerged as effective alternatives to synthetic pesticides, offering environmentally friendly solutions for pest and disease management in agriculture. However, their large-scale commercialization faces significant obstacles, particularly in the areas of production and formulation (Zhou & Li, 2019). These challenges impact the efficacy, stability, and overall market success of MBAs, making it essential to address them through technological advancements and strategic approaches. One of the primary challenges in the commercialization of MBAs is the selection and optimization of microbial strains. The chosen strains must exhibit strong biocontrol activity, high adaptability to different environmental conditions, and genetic stability over successive generations. Many microbial strains that perform well in controlled laboratory conditions often fail to deliver consistent results in field applications due to variations in temperature, soil composition, and microbial competition (Malar & Yuvakkumar, 2021). Additionally, ensuring that MBAs do not negatively impact beneficial soil microbiota or crop growth is crucial. Research efforts focus on enhancing strain efficacy through methods such as mutation breeding, metabolic engineering, and adaptive evolution (Eilenberg & Hajek, 2001). However, these techniques require significant investment in biotechnology and regulatory approval, which can delay commercialization. The next major challenge lies in the mass production of MBAs (Tabashnik et al., 2013). Large-scale production is typically achieved through solid-state fermentation (SSF) or liquid fermentation, each with distinct advantages and limitations. SSF is commonly used for fungal BCAs like *Trichoderma* and *B. bassiana*, utilizing solid substrates such as rice husk or wheat bran. While cost-effective, SSF presents challenges related to aeration control, contamination risks, and scalability. On the other hand, liquid fermentation, used for bacteria and some fungi (*B. subtilis, P. fluorescens*), allows better control over microbial growth parameters but requires expensive bioreactors and nutrient media (Khan & Mohammad, 2017). The transition from laboratory-scale to industrial-scale production often leads to reduced microbial efficacy due to changes in growth conditions, making process optimization a critical aspect of commercialization (AI-Dosary and Khan (2021).

Formulation is another crucial factor influencing the success of MBAs in the market. Unlike chemical pesticides, MBAs are living organisms that require stable formulations to maintain viability during storage and transportation. One of the biggest challenges is ensuring a long shelf life while preserving microbial efficacy (Baird, 2020; Cloyd, 2019). Factors such as temperature, humidity, and exposure to ultraviolet (UV) light can drastically reduce microbial viability, leading to inconsistent field performance. To address this, advanced formulation techniques such as freeze-drying, spray-drying, and encapsulation are employed to enhance stability. However, these methods can be costly and may sometimes alter the biological activity of the microorganisms (Vidal & Soberón, 2020). The selection of an appropriate carrier material is also essential, as it must protect the microbes while allowing their effective release upon application (Prat & Diatloff, 2015). Apart from stability, the efficiency of delivery systems and application methods remains a concern. MBAs can be applied through seed coatings, foliar sprays, or soil amendments, but ensuring even distribution and adherence to plant surfaces poses challenges (Sharma & Meena, 2021). The survival of MBAs after application is influenced by environmental conditions such as soil pH, moisture levels, and microbial competition (Van Lenteren, 2018). Incompatibility with other agricultural inputs, such as chemical fertilizers and pesticides, further complicates their adoption by farmers, who often use IPM approaches (Ghazavi & Faramarzi, 2020). Developing formulations that allow MBAs to function effectively in diverse farming systems is a key area of ongoing research. Beyond technical challenges, the high cost of production and regulatory barriers hinder the large-scale commercialization of MBAs. The production of microbial biopesticides involves complex processes that increase manufacturing costs, making them less competitive compared to chemical pesticides (Cottrell & Wraight, 2011). Additionally, stringent regulatory requirements demand extensive safety evaluations, environmental impact assessments, and field trials before MBAs can be approved for commercial use (Chandler & Baize, 2016). These regulatory processes vary across countries and can take years, delaying market entry and increasing financial risks for biocontrol companies. Overall, the successful commercialization of MBAs depends on overcoming these production and formulation challenges through continuous research, technological advancements, and policy support (Sharma & Kaur, 2020). Innovations in fermentation techniques, formulation technologies, and regulatory frameworks will play a crucial role in making MBAs more accessible and effective for large-scale agricultural use (Baker, 1990). Addressing these challenges will not only enhance the adoption of microbial biocontrol solutions but also contribute to the global shift toward sustainable and environmentally responsible farming practices (Vannini & De, 2013).

12.1.18 Regulatory and Market Barriers in the Commercialization of MBAs

The commercialization of MBAs faces significant regulatory and market barriers that hinder their widespread adoption and integration into mainstream agricultural practices. These barriers are primarily related to the complexities of regulatory approval processes, competition from chemical pesticides, and challenges in farmer acceptance. Overcoming these obstacles is essential to unlock the full potential of MBAs in sustainable pest and disease management (Feng & Gao, 2017).

12.1.19 Regulatory Barriers

The regulatory landscape for MBAs is one of the most significant challenges in their commercialization. Unlike chemical pesticides, which are regulated as chemicals, MBAs are treated as living organisms, which complicates their regulatory approval. Each country or region has its own regulatory framework for the approval of biopesticides, often governed by environmental and public health concerns. For instance, in the United States, the EPA oversees the approval of microbial biopesticides, while in Europe, similar regulations are handled by the European Food Safety Authority. The approval process for MBAs typically requires extensive data on their toxicity, environmental impact, and nontarget effects. Companies must demonstrate that the microbial agent is safe for humans, animals, beneficial organisms, and the environment. This can involve field trials, laboratory experiments, and comprehensive safety evaluations. These procedures are time-consuming and costly, with the potential for delays in market entry. For smaller companies or startups, the financial burden of meeting regulatory requirements can be overwhelming, creating a barrier to entry. In addition to safety and environmental concerns, some countries may lack clear or standardized regulations for MBAs, leading to uncertainties in the approval process. The absence of harmonized international regulations also complicates the export and trade of MBAs. Companies often face challenges when attempting to register their products in multiple regions, requiring separate approvals, which further extend the commercialization timeline.

12.1.20 Market Barriers

In addition to regulatory challenges, market barriers hinder the adoption of MBAs in agriculture. The primary market challenge is the strong competition from conventional chemical pesticides, which have a well-established presence in the agricultural market. Chemical pesticides are often preferred due to their proven efficacy, ease of application, and longer shelf life. They also tend to be cheaper to produce, making them more attractive to farmers in terms of cost-effectiveness. MBAs, on the other hand, face higher production costs due to the complexities involved in their mass production, formulation, and storage. These factors make MBAs more expensive than synthetic pesticides, which may deter farmers, especially those with limited financial resources. The higher cost is often attributed to the need for specialized production facilities, longer development timelines, and more intricate formulation processes to ensure microbial viability. As a result, the price point of MBAs can be a significant barrier to their adoption, particularly in regions where cost is a primary consideration for farmers. Another significant market barrier is farmer skepticism and lack of awareness. Many farmers are unfamiliar with the benefits and applications of MBAs, and there is often a reluctance to adopt new pest management practices, particularly when chemical pesticides are deeply ingrained in farming systems. MBAs may also be perceived as less effective or more labor-intensive, especially in cases where their field performance is inconsistent or influenced by environmental conditions. This skepticism is compounded by limited access to education and training on the use of MBAs. In many cases, farmers may not see the value in switching from traditional chemical methods to biocontrol solutions, especially if they are unsure of the results.

Additionally, infrastructure and distribution networks for MBAs are often underdeveloped, particularly in rural or developing regions. Unlike chemical pesticides, which are widely available through established distribution channels, MBAs may not have the same level of market penetration or logistical support. Without efficient distribution systems, the availability and accessibility of MBAs are limited, further hindering their widespread use.

12.1.21 Addressing Regulatory and Market Barriers

Overcoming these barriers requires coordinated efforts from various stakeholders, including governments, regulatory bodies, research institutions, and the private sector. Regulatory bodies could work toward harmonizing standards for MBAs across regions, which would simplify the approval process and reduce the time and costs associated with registration. Additionally, government incentives such as subsidies or tax breaks for companies developing MBAs could help offset the high production costs and make biocontrol solutions more affordable for farmers. On the market side, increasing farmer education and outreach programs can help build trust and demonstrate the efficacy and long-term benefits of MBAs (Van Lenteren, 2003). Demonstration trials, farmer field schools, and extension services can play a crucial role in showing farmers how MBAs can be integrated into their existing pest management strategies. Furthermore, partnerships with agricultural cooperatives and large farming organizations can increase the visibility and reach of MBAs. Finally, investment in supply chain infrastructure is essential to improve the accessibility of MBAs. Developing robust distribution networks and improving storage and handling technologies can ensure that MBAs are available in a timely manner, even in remote areas. In conclusion, while the commercialization of MBAs is hindered by significant regulatory and market barriers, these challenges can be overcome with strategic collaboration, research advancements, and targeted policy interventions (Liu & Chen 2019). By addressing these barriers, MBAs can become a mainstream solution in global efforts toward sustainable agriculture (Herrero et al., 2019).

12.1.22 Strategies for Successful Commercialization of MBAs

The commercialization of MBAs offers promising opportunities for sustainable pest management, but it requires careful planning and strategic approaches to overcome production, regulatory, and market barriers. To ensure the successful market adoption of MBAs, companies must focus on a variety of strategies that address the challenges of product development, market acceptance, and scalability. The following are some key strategies for the successful commercialization of MBAs.

1. **Collaboration and Public–Private Partnerships:**
   One of the most effective strategies for successful MBA commercialization is fostering public–private partnerships (PPPs). Collaboration between private companies, research institutions, and government agencies can provide the necessary resources for both research and market entry (He & Li, 2021). Public sector involvement can help subsidize research costs, improve access to funding, and streamline regulatory approval processes. Governments can also create policies that encourage sustainable agriculture and incentivize the use of BCAs over chemical pesticides (Simberloff & Stiling, 2015). Partnerships with universities and research organizations can contribute to strain optimization, product formulation, and testing, while private companies can focus on production, marketing, and distribution. By combining expertise from both sectors, PPPs enable a more efficient path to commercialization and help mitigate the financial risks associated with developing new biopesticide products.
2. **Product Differentiation and Innovation:**
   For MBAs to stand out in a market that is still largely dominated by synthetic chemical pesticides, product differentiation and continuous innovation are critical (Van Oers, 2016). Companies must focus on developing unique and effective microbial formulations that can compete with conventional products in terms of performance, cost, and ease of use. Investing in biotechnology advancements – such as genetic engineering,

strain improvement, and nanoformulation technologies – can increase the efficacy, stability, and shelf life of MBAs. For instance, bioengineered strains of bacteria or fungi that are more resistant to environmental stressors or are optimized for better pest control could offer a competitive edge (Lacey & Shapiro-Ilan, 2008). Additionally, innovative delivery systems (e.g. controlled-release formulations or soil-adhesion technologies) can improve the ease of application and ensure greater effectiveness in the field.

3. **Education and Farmer Outreach:**
   Farmer education and awareness are vital to the adoption of MBAs (Hajek & Leger, 2016). Many farmers are not familiar with the benefits of using MBAs, and there is often skepticism about their efficacy compared to traditional chemical pesticides. Demonstration trials, farmer field schools, and extension services can help educate farmers about the practical use of MBAs and demonstrate their success in real-world conditions. Field trials and case studies that highlight successful applications of MBAs can build confidence in their performance. Additionally, providing farmers with comprehensive training programs on the integration of MBAs into IPM systems can facilitate smoother adoption. Farmers should be shown not just the effectiveness of MBAs but also their economic benefits, such as reduced pesticide resistance, lower environmental impact, and long-term sustainability (Barbosa, 1998). Educational campaigns must also address concerns regarding the cost-effectiveness of MBAs, emphasizing how they can be integrated into existing farming practices to improve overall farm productivity.
4. **Streamlining Regulatory Processes:**
   Regulatory hurdles often delay the commercialization of MBAs, making it crucial to work on streamlining regulatory processes for microbial biopesticides (European Commission, 2020; Dey and Pradhan, 2019). Companies should engage with regulatory bodies early in the product development process to ensure that they meet safety and efficacy standards while minimizing time and cost spent on regulatory approvals. Efforts to harmonize regulatory standards across different regions (such as the EU, United States, and Asia) can reduce the complexity of the approval process for multinational companies and allow quicker market access. Advocating for fast-track approval systems for microbial products, particularly those that have been shown to be safe and effective, could also help speed up their commercialization (Lacey et al., 2015). Moreover, collaboration with industry associations and biopesticide industry groups can help lobby for favorable regulatory frameworks, creating a more predictable and efficient path to market for MBAs.
5. **Building Robust Distribution Networks:**
   Developing a robust and efficient distribution system is essential for ensuring the widespread availability of MBAs (Mendelsohn et al., 2003; Zhang et al., 2019). Unlike chemical pesticides, microbial agents require specific storage and transportation conditions to maintain their viability and effectiveness. Establishing an efficient logistics infrastructure that supports the safe and timely distribution of MBAs is key to preventing product degradation during transit. Working with distributors, agricultural cooperatives, and large-scale farming organizations can help ensure that MBAs are accessible to a broad range of farmers (Holt et al., 2004). Companies must also consider partnering with established agrochemical companies that have existing distribution networks

and customer bases. This could help overcome the infrastructure gaps that might otherwise hinder market penetration.

Additionally, e-commerce platforms for the sale of biocontrol products can offer more flexible, direct-to-farmer solutions, particularly for small-scale farmers in remote areas. Online sales can reduce costs related to distribution and make MBAs more accessible to a larger customer base.

6. **Offering Competitive Pricing and Value-Added Services:**
   To compete with established chemical pesticides, MBAs must be priced competitively. Although MBAs often involve higher production costs, companies can reduce prices by improving production efficiency and increasing scale. Additionally, subsidies, incentives, and discounts for early adoption or bulk purchases can help make MBAs more affordable for farmers. In addition to competitive pricing, offering value-added services – such as technical support, pest management consultancy, and customized solutions – can differentiate a company's MBA product in the market. Providing a package of services that help farmers integrate MBAs into their existing practices will improve customer loyalty and promote long-term relationships.

12.1.23 Commercialization of BCAs

As an alternative to chemical pesticides, BCAs are gaining a lot of attention due to the increased need for sustainable agricultural methods. Eco-friendly pest management options are offered by BCAs, which include microbial agents, natural enemies, and other biological products. Notwithstanding its potential, a number of obstacles stand in the way of BCA commercialization, including manufacturing, legal restrictions, and customer acceptance. The commercialization of BCA, its main factors, obstacles, and potential growth paths in agriculture and pest control are all examined in this book.

The hunt for sustainable substitutes for conventional pest management techniques has accelerated due to growing worries about pesticide resistance, environmental damage, and food safety. long shelf life while preserving for IPM systems, BCAs have drawn a lot of interest among these alternatives. Predators, parasitoids, and diseases are examples of BCAs -organisms that naturally control pest populations (Aeschlimann, 1990). They also comprise agents that are based on microorganisms, such as viruses, bacteria, and fungi. However, overcoming a number of technological, financial, and legal obstacles is necessary to make the shift from lab research to extensive commercial usage.

12.1.24 Challenges in Commercialization of BCAs

1. **Production and Formulation:** The mass production and formulation of BCAs are two of the main obstacles to their commercialization. Effective and economical techniques that preserve the BCAs' viability and efficacy in a range of environmental circumstances are necessary for large-scale production. Certain microbial agents, for instance, need growing conditions, and storage or transit may jeopardize their stability (Vargas et al., 2013).
2. **Efficiency and Consistency:** BCAs must function consistently in a variety of settings and under a range of pest stressors. Farmers may be discouraged by variations in BCA efficacy brought on by ecological and climatic conditions.

3. **Regulatory Approval:** Before approving BCAs for usage, regulatory bodies need comprehensive information on their efficacy and safety. These procedures can be expensive and time-consuming, which slows down the introduction of new products to the market.
4. **Distribution and the Supply Chain:** Another challenge is creating trustworthy distribution channels for BCAs. BCAs need specific handling, storage, and transportation conditions to guarantee their effectiveness when delivered to end users, in contrast to synthetic chemicals that have established distribution networks.
5. **Market Acceptance:** Due to a lack of awareness, a perception that using BCAs is difficult, or a lack of knowledge about the advantages over traditional pest control techniques, farmers and consumers may be reluctant to use them.

12.2 Future Prospects

With recent developments in microbial engineering, genetics, and biotechnology opening up new avenues, the biocontrol sector appears to have a bright future. The following are important trends influencing biocontrol going forward:

- **Gene Editing:** By making BCAs more durable and targeted, technologies like clustered regularly interspaced short palindromic repeats are being investigated to increase their effectiveness.
- **Biological Pesticide Development:** More choices for IPM are being made available by the increased interest in creating microbial-based solutions that target a wider variety of pests.
- **Automation and AI:** It is anticipated that the automated application of BCAs and the use of artificial intelligence (AI) in pest monitoring would save labor costs and boost productivity.
- **Global Expansion:** Biocontrol companies are anticipated to go beyond their conventional markets in Europe and North America into emerging regions as the need for sustainable agriculture solutions increases globally. Because they provide a sustainable and eco-friendly substitute for chemical pesticides, biocontrol industries are vital to the direction of pest management in the future (Gallardo & DeBach, 1990). The biocontrol sector is well-positioned for future success due to continuous developments in biotechnology and the rising desire for organic products from consumers, despite obstacles like production scalability and regulatory approval. To create resilient agricultural systems that can satisfy the demands of an expanding global population, biocontrol technologies must be integrated with other sustainable practices.

12.3 Conclusion

The growing demand for ecologically friendly farming methods and the reduction in the usage of conventional pesticides bode well for the commercialization of BCA (Gurr & Wratten, 2000). However, advancements in formulation, production, and regulatory procedures are essential if biocontrol drugs are to become a widely used solution. Overcoming current obstacles will require strong collaborations between government agencies, business, and research organizations. Furthermore, BCAs might emerge as a key component of integrated and sustainable pest management systems globally as consumer preferences and technological advancements change.

12.3.1 The Future of the Commercialization of BCAs: Opportunities, Challenges, and Pathways Forward

Over the past few decades, the commercialization of BCAs has changed dramatically due to the increased need for environmentally friendly farming methods, food security, and sustainable agriculture. A viable substitute for chemical pesticides is BCAs, which include diseases, parasites, natural predators, and other advantageous species. The present state, new developments, and prospects for the commercialization of BCAs are examined in this study. Technological advancements, market dynamics, regulatory frameworks, and the incorporation of BCAs into IPM systems are important subjects (Horsfall & Cowling 2001). Along with highlighting industry obstacles like cost, scalability, and efficacy issues, the report makes suggestions on how to get past these obstacles and achieve broad use of biocontrol technologies. As worries about the effects of chemical pesticides on the environment grow, BCAs have become a practical and environmentally responsible substitute. BCAs manage pest populations in agricultural systems by using natural enemies, including bacteria, viruses, fungi, and insects. The demand for BCAs has increased significantly as the global agriculture market moves more toward ecologically friendly and sustainable practices.

12.4 Conclusion

The commercialization of MBAs represents a significant opportunity for transforming pest and disease management in agriculture. As demonstrated by successful case studies like *B. thuringiensis* (Bt) and *Trichoderma* spp., microbial BCAs offer sustainable, environmentally friendly alternatives to chemical pesticides, responding to the increasing demand for eco-conscious farming solutions. These agents have proven effective in controlling a range of agricultural pests and diseases, gaining acceptance particularly within organic farming and IPM systems. However, challenges remain in scaling production, ensuring product stability, and navigating regulatory hurdles. The complexities of microbial formulation and the lengthy approval processes continue to hinder broader market adoption. Despite these obstacles, ongoing technological advancements, such as improved formulation methods and genetic engineering, hold promise for overcoming these barriers and enhancing the efficacy and stability of BCAs. Looking forward, the prospects for microbial BCAs are optimistic, driven by global trends toward sustainability and the reduction of chemical pesticide use. As innovations in biotechnology continue to evolve and regulatory frameworks become more conducive to biological product approval, microbial BCAs are likely to play an increasingly prominent role in modern agriculture. The integration of microbial agents with other pest management practices, coupled with growing consumer demand for environmentally safe solutions, signals a bright future for the commercialization of MBAs, offering immense potential for the agricultural industry worldwide.

In conclusion, the continued investment in research, product development, and regulatory support will be crucial in unlocking the full potential of microbial BCAs. As the industry matures, MBAs have the potential to revolutionize pest management, making agriculture more sustainable, cost-effective, and resilient to the challenges of the future.

References

Aeschlimann, J.P. 1990. *Biological Control in Agriculture and Forestry*. Springer-Verlag.

Al-Dosary, M.S., & Khan, M.A. 2021. Biocontrol agents and their commercial applications. *Agricultural Systems*.

Baird, R. 2020. The commercial potential of biocontrol in integrated pest management. *Pest Management Science*.

Baker, A.J. 1900. The Vedalia Beetle and its Work in California. California Agricultural *Experiment Station Circular* 94, 1–32.

Baker, R., & Sinha, R. 2020. Challenges in scaling microbial biocontrol agents for commercial use. *Microbial Biotechnology* 13(5), 1121–1136.

Barbercheck, M.E., & Dickinson, K. 2010. Commercializing microbial biocontrol agents for insect management. *Annual Review of Entomology* 55, 165–188.

Barbosa, P. 1998. *Conservation Biological Control*. Academic Press.

Bigler, F., Babendreier, D., & Kuhlmann, U. 2006. Environmental impact of commercial biocontrol agents. In: *Biological control in IPM systems in Africa* (pp. 45–66). Springer.

Chandler, D., & Baize, M. 2016. Regulatory approaches for microbial biocontrol agents. *Crop Protection* 76, 57–65.

Cloyd, R.A. 2019. The role of biocontrol agents in integrated pest management. *Journal of Pest Science* 92(4), 1151–1160.

Cottrell, T., & Wraight, S.P. 2011. Regulatory considerations for commercializing microbial biocontrol agents. *Biocontrol Science and Technology* 21(5), 595–613.

Davidson, R., & Walters, M. 2018. Market trends for biocontrol and biopesticides. *International Journal of Pest Management* 64(5), 411–422.

Dey, P., & Pradhan, S. 2019. Challenges in the commercialization of microbial biocontrol agents in developing nations. *Journal of Biological Control* 32(4), 1–12.

Eilenberg, J., & Hajek, A.E. 2001. Does biological control work in the real world? *Trends in Ecology & Evolution* 16(8), 409–410.

European Commission 2020. The European Green deal retrieved from Harman, G.E., 2004. "*Trichoderma* spp. as biocontrol agents". *Bio Control* 49(4), 47–68.

FAO 2019. The role of biological control in integrated pest management. *Food and Agriculture Organization of the United Nations*.

FAO Code of Conduct on the Use of Biological Control Agents. Food and agriculture Organization of the United Nations.

Feng, M., & Gao, L. 2017. Microbial biocontrol: from laboratory to commercialization. *Biocontrol Science and Technology* 27(8), 1064–1075.

Gallardo, R.K., & DeBach, P. 1990. Biological control in pest management systems of plants. Academic Press.

Ghazavi, M., & Faramarzi, M.A. 2020. Commercialization of biocontrol agents: challenges and opportunities. *Agricultural Biology Research* 2(2), 75–88.

Glare, T.R., & O'Callaghan, M. 2000. Microbial biopesticides. In: *Biocontrol: A Practice Guide for Farmers and Growers* (pp. 161–180). Springer.

Glare, T.R., O'Callaghan, M., & Baird, R. 2012. *Biological Control: Measures of Success.* Springer.

Glick, B. R., Todorovic, B., Czarny, J., Cheng, Z., Duan, J., & McConkey, B. 2007. The role of *Pseudomonas fluorescens* in plant growth promotion and biological control. *Journal of Applied Microbiology*, 113(3), 407–419.

Grand View Research 2023. Biocontrol Market Size, Share & Trends Analysis Report by Product (Microbials).

Gurr, G. M., & Wratten, S.D. 2000. *Biological Control: Measures of Success.* Kluwer Academic Publishers.

Gurr, G.M., Wratten, S.D., & Luna, J.M. 2016. *Biocontrol: A Global Perspective.* Springer.

Gurr, G.M., Mazzi, D., Calanca, P., Nikolaus, K.D., & Pellissier, L. 2016. *Biological Control and Pest Management: Opportunities and Challenges.* Springer.

Hajek, A.E., & St. Leger, R.J. 2016. Microbial biocontrol agents for agriculture. *Annual Review of Entomology* 61, 347–367.

Hajek, A.E., & Leger, R.J. (2008). *Biological Control of Pests: Ecology and Evolution.* Cambridge University Press.

He, L., & Li, W. 2021. The role of microbial biocontrol agents in integrated pest management. *Biocontrol Science and Technology* 31(10), 1383–1395.

Heimpel, G.E., & Mills, N.J. 2017. *Biological Control of Insect Pests in the 21st Century: Risks and Rewards.* Springer.

Herrero, S., Borja, M., & Ferre, J. 2019. Advances in *Bacillus thuringiensis* for pest control: development and optimization of genetically engineered strains. *Microbial Biotechnology* 12(1), 25–38.

Hoffmann, J.H. (2000). The biological control of prickly pear in Australia: progress and challenges. *Biological Control* 17(2), 168–175.

Holt, J., Brown, M., & Warner, P. 2004. *Biological Control: A Guide to Natural Enemies in North America.* Oxford University Press.

Horsfall, J.G., & Cowling, E.B. (2001). *Biological Control of Plant Diseases.* Academic Press.

Julien, M.H., & Griffiths, M. (1998). *Biological Control of Weeds: A World Catalogue of Agents and Their Target Weeds* (4th ed.). CABI Publishing.

Kairo, M.T.K. 2000. The potential of biological control of invasive pests in tropical regions. *Biocontrol Science and Technology* 10(3), 387–401.

Kaiser, W. 2007. The role of natural predators in ancient agriculture. *Agricultural History* 81(2), 249–267.

Khan, F.A., & Mohammad, F. (2017). *Microbial Biocontrol of Plant Diseases.* Springer.

Kremer, R.J., Neuenschwander, P., Borgemeister, C., & Langewald. J. 2012. Biocontrol of soilborne diseases. In: *Biological Control in IPM Systems in Africa* (pp. 23–43). Springer.

Kumar, S., & Chandra, A. 2021. Exploring market opportunities for microbial biocontrol in global agriculture. *World Journal of Microbiology & Biotechnology* 37(3), 68.

Lacey, L.A., & Shapiro-Ilan, D.I. (2008). Commercialization of microbial biocontrol agents in the United States. *Biocontrol Science and Technology* 18(8), 879–898.

Lacey, L.A., Frutos, R., Kaya, H.K., & Vail, P. 2015. *Insect Pathogens: Functional Biology of Insect-Pathogenic Fungi, Nematodes, and Microorganisms*. Springer.

Liu, Z., & Chen, L. 2019. Microbial biocontrol in agriculture: market challenges and opportunities. *Agriculture* 9(7), 154.

Lugtenberg, B., & Kamilova, F. 2009. Plant growth-promoting rhizobacteria. In: *Microbial Biocontrol: A Sustainable Alternative to Chemical Pesticides* (pp. 137–168). Springer.

Microbial, Biochemicals, by Application (Crop Protection, Turf & Ornamental), by Region, and Segment Forecasts, 2023–2030. *Grand View Research*.

Madhavan, M., Friedman, P.A., & Lennon, R.J. 2015. Mechanisms of biological control in *Trichoderma*. *World Journal of Microbiology and Biotechnology*.

Malar, A., & Yuvakkumar, R. 2021. Commercialization of microbial biocontrol agents in crop protection. *Biocontrol Science and Technology* 31(4), 417–434.

Mendelsohn, M., Kough, J., Vaituzis, Z., & Matthews, K. 2003. The effects of Bt crops on non-target organisms. *Nature Biotechnology* 21(11), 1474–1478.

Mishra, A., & Rajeswari, S. 2019. *Microbial Biocontrol Agents for Sustainable Agriculture and Commercialization Strategies*. Springer.

Prat, S., & Diatloff, E. 2015. Strategic and economic aspects of microbial biocontrol in agriculture. *Agricultural Economics Review* 46(3), 65–77.

Regulation (EC) No. 1107/2009 of the European Parliament and of the Council of 21 October 2009. European Union.

Roberts, D.W., & Campbell, R. 2004. *Biological Control and Microbial Biopesticides: Applications in Agriculture*. Elsevier.

Sharma, P., & Kaur, G. 2020. Biocontrol and sustainable agriculture: market opportunities for microbial agents. *Journal of Biopesticides* 13(1), 55–70.

Sharma, P.K., & Meena, V.S. (2021). Market dynamics and challenges in microbial biocontrol. *Environmental Science and Pollution Research* 28(32), 43500–43512.

Simberloff, D., & Stiling, P. 2015. How well do biological control agents perform? A review of the evidence. *BioScience* 65(5), 465–474.

Sivapalan, S., & Singh, R. 2020. Development of biocontrol products for agricultural applications. *Critical Reviews in Microbiology* 46(5), 629–649.

Sundar, S. 2015. Secondary metabolites from Streptomyces as biocontrol agents. *Journal of Microbial and Biochemical Technology* 7(6), 331–335.

Sundararajan, K., & Parvez, M. 2020. Regulation and commercialization of biocontrol agents. *Microbial Biotechnology for Sustainable Agriculture* 14, 82–101.

Tabashnik, B.E., Carrière, Y., Crowder, M., & Dennehy, S. 2013. Resistance to Bt crops and the future of transgenic pest control. *Nature Biotechnology* 31(2), 197–199.

van den Bosch, R., & Mumford, J.D. 2021. Economic impacts of biocontrol strategies in crop protection. *Crop Protection Journal* 28(4), 512–552.

Van Houten, Y.M., Bigler, F., Barratt, B.I.P., Moran, V.C., & van der Geest, L.C. 2020. Biocontrol of greenhouse pests with natural enemies: costs, benefits, and research needs. *Biological Control* 148, 104356.

Van Lenteren, J.C. 2000. Integrated pest management and biological control. *Crop Protection* 19(8), 641–646.

Van Lenteren, J.C. 2003. Commercial availability of biological control agents in the European Union. *Biological Control* 28(1), 5–11.

van Lenteren, J.C. 2018. Commercialization of biological control agents: current status and future prospects. *Biological Control* 121, 23–33.

Van Lenteren, J.C. 2018. Commercializing biological control in agriculture. *Biological Control* 129, 80P–90P.

Van Lenteren, J.C., Bigler, F., Barratt, B.I.P., & Moran, V.C. 2018. *Biological Control: Measures of Success*. Springer.

Van Lenteren, J.C., Bigler, F., Barratt, B.I.P., & Moran, V.C. 2018. The role of biocontrol in the sustainability of Agriculture. *Agricultural Systems* 165, 1–5.

Van Loon, L.C., Bakker P.A.H.M., & Pieterse. C.M.J. 2014. Induced systemic resistance by rhizobacteria. *Annual Review of Phytopathology*, 52, 347–375.

van Oers, M.M. (2016). Genetic engineering of baculoviruses for insect pest control. *Viruses* 8(10), 282.

Vannini, A., & De, F.A. 2013. *Microbial Biocontrol in Crop Protection and Commercialization Perspectives*. Springer Series in Sustainable Development.

Vargas, R., de Souza, J.A.C.C., Rivera-Pedroza, L.F., Jahnke, M.F., García, L.F., & Jahnke S.M. 2013. Production and application of biocontrol agents for pest management. *Biological Control* 66, 45–59.

Vidal, S., & Soberón, M. (2020). Microbial biocontrol agents and their commercialization in developing countries. *Microorganisms* 8(8), 1173.

Zhang, H., Duan, X., & Liu, C. 2019. *Biotechnological Advances in Biocontrol of Agricultural Pests*. Springer.

Zhang, H., Duan, X., & Liu, C. 2021. Genetic enhancement of virulence and adaptability in entomopathogenic fungi. *Journal of Applied Microbiology* 130(5), 1234–1247.

Zhou, Z., & Li, Y. 2019. *Microbial Biocontrol Agents for Plant Disease Control*. Springer International Publishing.

13

Management of Soil-borne Diseases with *Trichoderma* spp.

Manoj Kumar Maurya[1], Vivek Singh[2], and Ajay Tomar[3]*

[1] *Department of Vegetable Science, Acharya Narendra Dev University of Agriculture and Technology, Kumarganj, Ayodhya, India*
[2] *Department of Plant Pathology, Acharya Narendra Dev University of Agriculture and Technology, Kumarganj, Ayodhya, India*
[3] *Amity Institute of Organic Agriculture, Amity University, Noida, Uttar Pradesh, India*

* Corresponding author: drmanoj.pathovs@nduat.org

13.1 Introduction

Soilborne diseases remain a major challenge in agriculture because many pathogens can persist in the soil for extended periods. Conventional farming practices, which often involve the excessive application of chemical fertilizers and pesticides, along with the continuous monoculture of crops, have significantly disrupted the ecological balance of farmland. This has resulted in increased plant diseases, pest infestations, pesticide residues in crops, and environmental pollution of soil and water (Bardin et al., 2015). *Trichoderma* is a beneficial fungus that mostly occurs in soils, in the air, and on plant surfaces and is widely recognized for its ability to manage a broad range of phytopathogens that incite diseases in crops (Wang et al., 2022). It is particularly effective in combating soilborne pathogens and can also help control certain foliar and spike diseases (Abbas et al., 2022). Besides disease suppression, *Trichoderma* contributes to plant health by enhancing growth, improving nutrient uptake, boosting plant defenses, and mitigating agrochemical pollution (Tilocca et al., 2020; Fontana et al., 2021). Various species, such as *Trichoderma harzianum, Trichoderma hamatum, Trichoderma longibrachiatum, Trichoderma koningii, Trichoderma viride, Trichoderma polysporum*, and *Trichoderma asperellum*, have been extensively researched for their biocontrol potential (Di Marco et al., 2022). Several studies indicate that key species of *Trichoderma* produce secondary metabolites and enzymes that degrade cell walls, which help in suppressing plant pathogens and promoting overall plant vigor (Kubicek et al., 2019). *Trichoderma* spp. have strong potential as BCAs due to their multifaceted roles in disease control and the development of plant growth (Harman et al., 2004).

Microbial Consortia in Plant Science and Sustainable Agriculture, First Edition. Edited by Ajay Tomar, et al.

13.2 Soilborne Pathogens and Their Impact

Soilborne phytopathogens pose a serious challenge to agriculture worldwide, leading to substantial crop losses (Janvier et al., 2007; Ampt et al., 2019). They cause several diseases such as wilting, damping-off, and root rot, which adversely affect plant health by retarded growth, decreasing survival, and interfering with water and nutrient absorption (Baligar et al., 2001). Major diseases caused by fungal pathogens are given in Table 13.1.

13.2.1 Economic Consequences

The economic toll of soilborne diseases is considerable worldwide. These pathogens can cause severe yield losses, sometimes reducing crop production by 50–75%, depending on the crop. For instance, *Fusarium* wilt leads to significant crop damage, while *Pythium* particularly affects plants at the seedling or nursery stage. Infected plants often exhibit reduced root development and stunted growth, and, in severe cases, die prematurely, all of which contribute to lower yields and productivity. To manage these diseases, farmers often rely on costly control methods such as fungicides, soil fumigation, or biopesticides, further increasing the cost of crop cultivation.

Table 13.1 Key soilborne phytopathogens cause diseases and affect crops.

S. no.	Fungal pathogen	Disease caused	Commonly affected crops
1.	*Fusarium oxysporum*	*Fusarium* wilt	Tomato, banana, chickpea, pea, lentil, cotton, chilli
2.	*Rhizoctonia solani*	Root rot, damping-off, sheath blight, black scurf	Potato, beans, lettuce, rice
3.	*Pythium* spp.	Damping-off, root rot	Cucumber, pepper, tobacco, chilli, tomato, brinjal, cabbage, cauliflower, papaya
4.	*Phytophthora* spp.	Root rot, crown rot, blight	Tomato, potato, cocoa, citrus
5.	*Sclerotium rolfsii*	Southern blight, collar rot	Tomato, carrot, beans, groundnut, chickpea
6.	*Verticillium dahliae*	*Verticillium* wilt	Cotton, eggplant, sunflower
7.	*Macrophomina phaseolina*	Charcoal rot	Soybean, sorghum, maize, sesame
8.	*Gaeumannomyces graminis*	Take-all	Wheat, barley
9.	*Sclerotinia sclerotiorum*	Watery soft rot, *Sclerotinia* stem rot	Brinjal, peas

13.3 *Trichoderma* spp.

Trichoderma spp. are filamentous, free-living fungi mostly found in soil and around plant roots. They interact closely with plant roots, soil environments, and occasionally leaf surfaces. These fungi produce secondary metabolites that trigger both local and systemic defense mechanisms in plants. Once they colonize the roots, *Trichoderma* can promote root development, improve nutrient uptake, increase crop yields, and enhance the plant's resistance to biotic and abiotic stresses. They primarily reproduce asexually through conidia, while sexual reproduction (rarely) through ascospores is associated with the genus *Hypocrea*.

13.3.1 Taxonomy of *Trichoderma*

Trichoderma, as a genus of asexual filamentous fungi, is closely related to *Hypocrea*, which represents its sexual form. The taxonomic classification is as follows:

- **Asexual Stage (Conidial)**
  - Subdivision: Deuteromycotina
  - Class: Hyphomycetes
  - Order: Moniliales
  - Family: Moniliaceae
  - Genus: *Trichoderma*
- **Sexual Stage (Ascospore)**
  - Subdivision: Ascomycotina
  - Class: Pyrenomycetes
  - Order: Sphaeriales
  - Family: Hypocreaceae
  - Genus: Hypocrea

13.3.2 Biology of *Trichoderma*

Trichoderma spp. is a cosmopolitan of fungi known for its high genetic variability and adaptability to various environments (Grinyer et al., 2004). This is recognized as a widespread, soil-dwelling Ascomycetes that reproduce asexually and rarely sexually, commonly occurring in diverse soil types as well as environments such as decaying plant matter, manure, and wood (Singh et al., 2014). *Trichoderma* spp. is categorized as imperfect fungi within the order Hypocreales of the phylum Ascomycota (Howell, 2003). They exhibit rapid growth on various culture media such as malt agar, Czapek Dox Agar, corn meal agar, and potato dextrose agar, while initially producing white mycelium that later develops into green-colored colonies, and form conidia or spores. Some species also form thick-walled chlamydospores for survival (Chaverri et al., 2003; Lu et al., 2004).

They typically form distinct colonies with conidia in shades of green or white, while the reverse side of the Petri dish may show colors like unpigmented, yellow, amber, buff, or yellow-green. Certain species, such as *Trichoderma atroviride, Trichoderma lignorum,* and *Trichoderma fertile*, can be identified by their unique scents, which include aromas resembling coconut or mold (Gams & Bissett, 2002).

13.3.3 Morphology of *Trichoderma*

Morphological identification is a common but not highly accurate method for distinguishing *Trichoderma* species due to the high degree of similarity among them (Zhang et al., 2005). These fungi grow quickly at 25–30 °C (Latifian et al., 2007). They utilize a wide range of nutrients for their growth and sporulation (Gao et al., 2007). The asexual spore (conidia) of *Trichoderma* is typically green and produced in huge amounts, forming powdery masses that serve as a diagnostic feature, although this characteristic is shared with genera such as *Myrothecium, Clonostachys, Aspergillus*, and *Penicillium* (Chaudhari et al., 2011; Alvindia & Hirooka, 2011).

Conidiophores are extensively branched, making it difficult to classify them as loosely or densely tufted. The main branches give rise to lateral branches, which may occur in pairs or individually, with the longest typically located farther from the tip. Auxiliary branches are often paired, with the longest ones positioned closer to the main axis. Both main and auxiliary branches usually emerge at or near a 90° angle to the main axis. Conidia are generally smooth and ellipsoidal in shape. Synamorphs are identified by their solitary, vertically branched conidiophores, each bearing conidia at the tips of phialides. These structures are not easily observed in cultures younger than a week (Lu et al., 2004). Conidial shapes range from ellipsoidal to oblong, and, in some species, they may be globose or nearly spherical, with a length-to-width ratio of about 1.4 to 1.3. *Trichoderma* spp. produce chlamydospores and are typically single-celled, sub-globose, and associated with short hyphal segments, forming within the hyphal cells. Colony growth and conidiophores of *Trichoderma* spp. are shown in Figure 13.1.

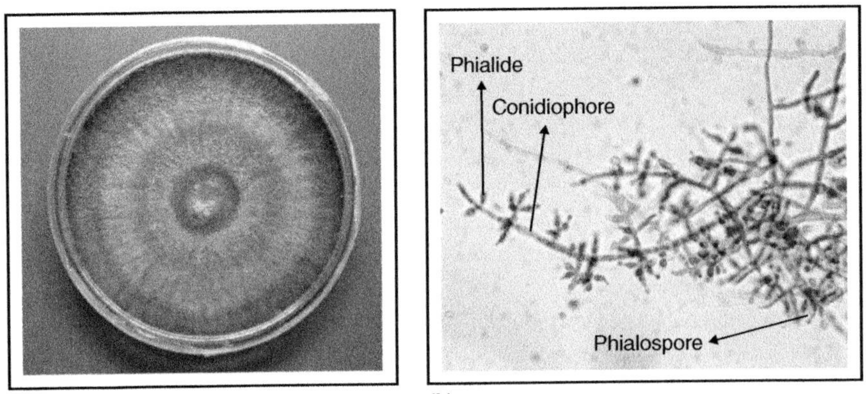

Figure 13.1 (a) Colony growth of *Trichoderma* spp. on potato dextrose agar medium, (b) conidia and conidiphore under light microscope at 40×. Maurya et al. (2024)/ARCC JOURNALS/CC BY 4.0.

13.3.4 *Trichoderma* spp. as Biocontrol Agents

Trichoderma spp. serve as potent biological control agents against various soilborne phytopathogens that cause significant agricultural losses (Kushwaha & Verma, 2014; Shahid et al., 2014). These fungi not only suppress harmful microbes but also stimulate the plant defense system and root development. Their biocontrol effectiveness stems from several mechanisms, including antibiosis, mycoparasitism, competition, and induction of systemic resistance in plants. Beyond soilborne pathogens, *Trichoderma* also plays an antagonistic role against certain postharvest fungal, bacterial, and viral pathogens (Harman, 2006; Yedidia et al., 2003). Although general nutritional requirements of *Trichoderma* are well-documented, there remains limited insight into the specific carbon and nitrogen sources that most effectively support the mass production of these beneficial strains (Rajput et al., 2014).

13.3.5 Major *Trichoderma* spp.

- *T. harzianum*
- *T. viride*
- *T. atroviride*
- *T. asperellum*
- *T. koningii*
- *T. virens*
- *T. polysporum*
- *T. longibrachiatum*

13.4 Mode of Action/Mechanisms of *Trichoderma* spp.

13.4.1 Mycoparasitism

Trichoderma exhibits mycoparasitic behavior by identifying and coiling around the hyphae of target fungi, penetrating their cells, breaking them down using enzymes like chitinases and glucanases (Benítez et al., 2004), and hampering the growth of pathogens. Various mode of action of *Trichoderma* spp. are given below in Table 13.2 and

Table 13.2 Different mechanisms of action of *Trichoderma* spp.

Different mechanism	Brief description
Mycoparasitism	*Trichoderma* hyphae coils around pathogen hyphae and penetrates and degrades them using enzymes like chitinase.
Antibiosis	Production of secondary metabolites (e.g. gliotoxin, viridin) that inhibit pathogen growth.
Competition	Competes for nutrients and space in the rhizosphere, excluding pathogens.
Induced resistance	Triggers systemic resistance in plants via jasmonic acid and ethylene signaling pathways.

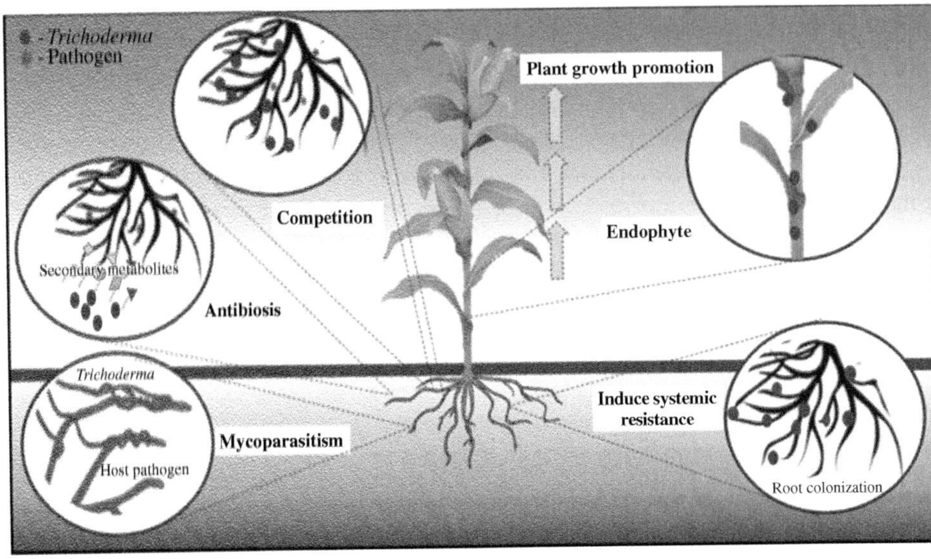

Figure 13.2 Showing different mechanisms of *Trichoderma* spp. *Source*: Hariharan et al. (2022)/ Springer Nature.

shown in Figure 13.2. This mechanism plays a crucial role in managing soilborne fungi and nematodes, involving stages such as host recognition, attachment, coiling, penetration, and parasitism (Li et al., 2020). *Trichoderma* hyphae are capable of invading pathogen cells, extracting nutrients, and eventually causing their death (Marraschi et al., 2019). This process is accompanied by the secretion of various degradative enzymes (Moo-Koh et al., 2022). The activities of enzymes like β-1,3-glucanase, chitinase, and protease significantly increase during this interaction, contributing to enhanced plant resistance against pathogens (Poveda et al., 2020).

13.4.2 Antibiosis

Trichoderma produces various secondary metabolites such as gliotoxin, viridin, and peptaibols, which effectively inhibit the growth of soilborne pathogens (Howell, 2003). It also releases compounds that retarded the growth and development of pathogens (Moo-Koh et al., 2022). These compounds are antifungals and nematocidal, including trichomycin, gliotoxin, viridin, antimicrobial peptides, β-1,3-glucanase, chitinase, polypeptides, polyketones, butyrolactones, sesquiterpenes, and a variety of volatile organic compounds like alcohols, hydrocarbons, aldehydes, esters, furans, terpenes, and aromatic compounds (Kappel et al., 2020).

13.4.3 Competition

Trichoderma rapidly colonizes in the root zone of plants, competing with pathogenic organisms for space and essential nutrients, thereby restricting their ability to thrive.

13.4.4 Induced Systemic Resistance

Certain strains of *Trichoderma* trigger systemic resistance in plants via signaling molecules such as jasmonic acid and ethylene (Shoresh et al., 2010). The plant defense response, activated by environmental stimuli, leads to physiological and metabolic changes. When *Trichoderma* species colonizes roots, it encourages the production of plant-derived secondary metabolites that act as resistance-inducing signals (Al-Hazmi & Tariq Javeed, 2016). Presently, more than 20 types of elicitors from *Trichoderma* are known to activate the plant defense system, including antitoxins, peptides, lipopeptides, cellulases, hydrophobic proteins, gene-related proteins, terpenoids, flavonoids, glycosides, and phenolic compounds (Pocurull et al., 2020). These compounds enhance defense mechanisms and support plant growth. For example, *T. hamatum* has been shown to significantly elevate the activity of defense-related enzymes in tobacco, such as phenylalanine ammonia-lyase, polyphenol oxidase, and peroxidase.

13.5 Effect of *Trichoderma* spp. Against Soil-borne Phytopathogens

Trichoderma species have shown great potential in controlling various soilborne fungal pathogens, proving to be effective BCAs. Research indicates that species like *T. harzianum* and *T. viride* possess different levels of antagonistic activity against pathogens such as *Rhizoctonia solani*, *Fusarium oxysporum*, and *Sclerotium rolfsii*. Particularly *T. harzianum* and *T. viride* have demonstrated significant inhibition of the mycelium growth of *R. solani*. Native strains of *Trichoderma* have also exhibited high levels of inhibition against *F. oxysporum*, with some achieving complete (100%) suppression. Similarly, several *Trichoderma* isolates have effectively retarded the growth of *S. rolfsii*, with some showing inhibition rates as high as 82%. Effect of *Trichoderrma* spp. against targeted pathogens are given in Table 13.3.

Table 13.3 Effect of various *Trichoderma* species against fungal soilborne phytopathogens.

S. no.	*Trichoderma* Species	Target pathogen(s)	Crop/host	Mode of action	References
1.	*T. harzianum*	*Fusarium oxysporum*, *Rhizoctonia solani*	Tomato, chickpea	Mycoparasitism, antibiosis, competition	Harman et al. (2004), Elad and Chet (1983)
2.	*T. viride*	*F. oxysporum* f. sp. *ciceri*	Chickpea	Mycoparasitism, enzyme production	Singh et al. (2004), Gaur et al. (2010)
3.	*T. asperellum*	*Sclerotium rolfsii*, *Pythium ultimum*	Groundnut, cucumber	Antibiosis, induced resistance	Yedidia et al. (1999), Mukherjee et al. (2012)

(Continued)

Table 13.3 (Continued)

S. no.	*Trichoderma* Species	Target pathogen(s)	Crop/host	Mode of action	References
4.	*T. atroviride*	*R. solani, Phytophthora capsici*	Pepper, tomato	Lytic enzyme production, competition	Benítez et al. (2004), Howell (2003)
5.	*T. koningii*	*Fusarium* spp., *R. solani*	Tomato, maize	Degradation of the cell wall, competition	Kubicek et al. (2001), Saba et al. (2012)
6.	*T. longibrachiatum*	*Macrophomina phaseolina*	Soybean, maize	Antifungal metabolites, mycoparasitism	Monte (2001), Aneja et al. (2014)
7.	*T. hamatum*	*Pythium aphanidermatum*	Cucurbits	Mycoparasitism, induced systemic resistance	Shoresh et al. (2010)

13.6 Formulation and Application Techniques

13.6.1 Formulation Types

There are various types of formulation applied for the delivery of *Trichoderma* spp. given in (Table 13.4).

13.7 Application Methods of *Trichoderma* spp.

Trichoderma species can be applied using a variety of techniques based on the crop type, targeted pathogen, and local environmental conditions. The various delivery methods are described here, such as the following:

Seed Treatment: The seeds are treated with Talc-based *Trichoderma* formulations (4–5 g/kg seed) using adhesives such as carboxymethyl cellulose (CMC). This approach protects emerging seedlings from seed- and soilborne pathogens during germination.

Soil Application: *Trichoderma* spores or formulations are mixed @ 2.5 kg/100 kg Farm yard manure (FYM) for 1 ha of field, either before sowing or during crop establishment. This encourages root colonization and helps suppress soilborne diseases.

Root Dipping: Seedling roots are dipped in a *Trichoderma* formulation suspension at 5–10 g/l of water before transplanting, ensuring early colonization of the rhizosphere.

Nursery Treatment: In nurseries, *Trichoderma* is applied directly to the growing media or trays to protect young plants from damping-off and wilt diseases.

Soil Drenching: A liquid suspension of *Trichoderma* is applied around the root zone to manage root and collar rots.

Table 13.4 Different types of formulation of *Trichoderma* spp.

S. no.	Formulation type	Description/carrier	Advantages	References
1.	Talc-based	*Trichoderma harzianum*/ *Trichoderma viride* + talc powder (1–2% carboxy methyl cellulose [CMC])	Long shelf life, easy application to soil/seeds	Jeyarajan et al. (1994), Vinale et al. (2008)
2.	Granular	*T. viride* mixed with bentonite or kaolin granules	Suitable for broadcasting in field	Fravel (2005), Harman (2006)
3.	Wettable powder (WP)	Spores in a powder base that can be suspended in water	Ease of spraying, good for foliar and seed treatment	Papavizas et al. (1984), Monte (2001)
4.	Alginate pellets	Immobilized spores in calcium alginate beads	Slow release, protection of spores in harsh soil	Lewis and Papavizas (1987), Mukherjee et al. (2012)
5.	Oil-based	Spores suspended in mineral or vegetable oil	Good adhesion to seed/foliage, improved shelf life	Sultana et al. (2006)
6.	Biopriming (seed coating)	Coating seeds with *Trichoderma* + sticker/ binder	Early root colonization, disease prevention	Callan et al. (1990), Prasad et al. (2002)
7.	Liquid formulation	Conidial suspension in sterile water + preservatives	Suitable for foliar spray, drip irrigation	Harman et al. (2004), Kumar et al. (2011)
8.	Compost-based	*Trichoderma* grown on organic compost substrates	Dual role: nutrient supply and biocontrol	Elad and Chet (1983), Shanmugam and Kanoujia (2011)

Furrow or Row Application: Formulations are placed in planting furrows or rows before covering with soil, allowing the biocontrol agent to come in close contact with emerging roots.

Use of Enriched Compost/FYM: Incorporating compost or farmyard manure enriched with *Trichoderma* enhances microbial activity and disease resistance in the soil.

Foliar Spraying: Liquid or wettable powder formulations are sprayed on foliage to control leaf diseases and induce systemic resistance in the plant.

Alginate Pellet Formulation: Spores are encapsulated in biodegradable beads, providing controlled release and improved shelf life.

Drip Irrigation: *Trichoderma* formulations can be efficiently distributed via drip systems, ensuring even root-zone colonization with minimal labor.

Broadcasting: Powdered or granular formulations 4–5 kg/ha can be spread over large fields to enhance soil health and manage a broad range of pathogens.

13.8 Merit and Demerit of *Trichoderma* spp. as Biocontrol Agents

13.8.1 Merit (Advantages)

1. **Broad-Spectrum Activity:** Effective against various soilborne phytopathogens such as *Fusarium* spp., *R. solani, S. rolfsii*, and *Pythium* spp.
2. **Diverse Mechanisms of Action:** Suppresses pathogens through mycoparasitism, competition, antifungal compound production, and induction of plant resistance.
3. **Environmentally Friendly:** Nontoxic to humans, animals, and beneficial microbes, making it a sustainable alternative to chemical fungicides.
4. **Growth Promotion:** Some strains enhance plant growth by boosting nutrient absorption, hormone production, and root development.
5. **Versatile Formulation Options:** Available in powders, liquids, granules, or alginate beads and compatible with multiple application methods.
6. **Organic Farming Compatibility:** Widely accepted in organic agriculture as a natural fungicide and soil health enhancer.
7. **Long-term Soil Presence:** Certain strains can persist in the rhizosphere, providing extended protection.

13.8.2 Demerit (Limitations)

1. **Inconsistent Field Results:** Performance may vary depending on soil conditions, climate, and microbial competition.
2. **Slower Action:** *Trichoderma* may take longer to control pathogens compared to chemical fungicides, especially under high disease pressure.
3. **Strain Specificity:** Not all strains are effective against all pathogens; selection of the right strain is crucial.
4. **Storage and Shelf Life Constraints:** Some formulations, especially liquids and powders, may degrade if not stored properly.
5. **Compatibility Challenges:** May not be compatible with certain agrochemicals or fungicides, which could affect its performance.
6. **Formulation Quality Dependence:** Efficacy relies heavily on the quality and viability of the formulation.
7. **Limited Commercial Reach:** In some regions, regulatory hurdles, limited awareness, or lack of infrastructure may restrict its use.

13.9 Challenges and Future Prospects of *Trichoderma* spp.

Although *Trichoderma* spp. is a highly effective biocontrol agent against soilborne phytopathogens, the various challenges still limit their broader adoption in agricultural systems. One of the primary constraints is the inconsistent field performance due to diverse environmental conditions. The animate and inanimate factors can significantly influence the survival, colonization, and efficacy of *Trichoderma* strains. Moreover, the specificity of strain–pathogen interactions necessitates the identification and use of well-adapted strains for each target crop and region, which can be labor-intensive and time-consuming.

Another challenge is the shorter shelf life and stability of the formulation of certain *Trichoderma* products. Improper storage or formulation can result in reduced spore viability and efficiency, especially at high temperatures or high humidity. Additionally, lack of awareness among farmers, coupled with insufficient training on the correct usage and application methods, poses a hurdle to effective deployment in the field. In some cases, compatibility issues with chemical pesticides further restrict their integration into conventional farming practices. From a regulatory and commercial point of view, a lack of uniform quality standards and region-specific biopesticide policies can limit the availability and widespread adoption of *Trichoderma* products. Moreover, intellectual property and commercialization barriers hinder the development and dissemination of indigenous strains with high biocontrol potential.

The future prospects of *Trichoderma* spp. are promising, with advancements in genomics, molecular biology, and biotechnology enabling the development of improved strains with enhanced biocontrol and plant growth-promoting traits. Additionally, consortia-based bioinoculants combining *Trichoderma* with other beneficial microbes offer a sustainable approach to disease management and crop production. Integrating *Trichoderma* into precision agriculture, seed coating, and soil health management is set to boost its use in both conventional and organic farming. Enhancing extension services, farmer education, and public–private partnerships will be key to overcoming challenges and unlocking their full potential in sustainable agriculture.

13.10 Conclusion

Soilborne diseases pose a serious hindrance to sustainable agriculture. The versatility of *Trichoderma* spp., through seed treatment, soil amendment, and foliar application, makes them well-suited for diverse cropping systems. Moreover, their compatibility with organic farming and minimal environmental impact aligns well with the goals of sustainable agriculture. However, despite their potential, several challenges such as inconsistent field performance, formulation stability, and limited farmer awareness need to be addressed to enhance their adoption. Future research focusing on strain improvement, advanced formulation technologies, and integration with other beneficial microbes will play a crucial role in overcoming these hurdles. *Trichoderma* spp. represent biologically robust and environmentally sound microbes for soilborne disease management. With continued research, innovation, and extension efforts, they can significantly contribute to resilient and sustainable agricultural systems worldwide.

References

Abbas, A., Mubeen, M., Zheng, H., Sohail, M.A., Shakeel, Q., Solanki, M.K., Iftikhar, Y., Sharma, S., Kashyap, B.K., Hussain, S., del Carmen Zuñiga Romano, M., Moya-Elizondo, E.A., & Zhou, L. 2022. *Trichoderma* spp. Genes involved in the biocontrol activity against *Rhizoctonia solani. Frontiers in Microbiology* 13, 884469. https://doi.org/10.3389/fmicb.2022.884469

Al-Hazmi, A.S., & Tariq Javeed, M. 2016. Effects of different inoculum densities of *Trichoderma harzianum* and *Trichoderma viride* against *Meloidogyne javanica* on tomato. *Saudi Journal of Biological Sciences* 23, 288–292. https://doi.org/10.1016/j.sjbs.2015.04.007

Alvindia, D.G., & Hirooka, Y. 2011. Identification of *Clonostachys* and *Trichoderma* spp. from banana fruit surfaces by cultural, morphological and molecular methods. *Mycology* 2, 109–115.

Ampt, E.A., van Ruijven, J., Raaijmakers, J.M., Termorshuizen, A.J., Mommer, L. 2019. Linking ecology and plant pathology to unravel the importance of soil-borne fungal pathogens in species-rich grasslands. *European Journal of Plant Pathology* 154(1), 141–56.

Aneja, K.R., Sharma, C., & Joshi, R. 2014. Evaluation of *Trichoderma longibrachiatum* for the management of soil-borne plant pathogens. *Indian Journal of Microbiology* 54(4), 413–418. https://doi.org/10.1007/s12088-014-0465-6

Baligar, V., Fageria, N.K., & He, Z.L. 2001. Nutrient use efficiency in plants. *Communications in Soil Science and Plant Analysis* 32(7–8), 921–50.

Bardin, M., Ajouz, S., Comby, M., Lopez-Ferber, M., Graillot, B., Siegwart, M., & Nicot, P.C. 2015. Is the efficacy of biological control against plant diseases likely to be more durable than that of chemical pesticides? *Frontiers in Plant Science* 6, 566. https://doi.org/10.3389/fpls.2015.00566

Benítez, T., Rincón, A.M., Limón, M.C., & Codón, A.C. 2004. Biocontrol mechanisms of *Trichoderma* strains. *International Microbiology* 7(4), 249–260.

Callan, N.W., Mathre, D.E., & Miller, J.B. 1990. Bio-priming seed treatment for biological control of *Pythium ultimum* pre-emergence damping-off in sh2 sweet corn. *Plant Disease* 74(6), 532–535. https://doi.org/10.1094/PD-74-0532

Chaudhari, P.J., Shrivastava, P., & Khadse, A.C. 2011. Substrate evaluation for mass cultivation of *Trichoderma viride. Asiatic Journal of Biotechnology Resources* 2, 441–446.

Chaverri, P., Castlebury, L.A., Overton, B.E., & Samuels, G.J. 2003. *Hypocrea/Trichoderma*: species with conidiophore elongations and green conidia. *Mycologia* 95, 1100–1140.

Di Marco, S., Metruccio, E.G., Moretti, S., Nocentini, M., Carella, G., Pacetti, A., Battiston, E., Osti, F., & Mugnai, L. 2022. Activity of *Trichoderma asperellum* strain ICC 012 and *Trichoderma gamsii* strain ICC 080 toward diseases of esca complex and associated pathogens. *Front Microbiology* 12, 813410. https://doi.org/10.3389/fmicb.2021.813410

Elad, Y., & Chet, I. (1983). Improved selective media for isolation of *Trichoderma* spp. or *Fusarium* spp. *Phytoparasitica* 11, 55–58. https://doi.org/10.1007/BF02980712

Fontana, D.C., de Paula, S., Torres, A.G., de Souza, V., Pascholati, S.F., & Schmidt, D. 2021. Endophytic fungi: biological control and induced resistance to phytopathogens and abiotic stresses. *Pathogens* 10, 570. https://doi.org/10.3390/pathogens10050570

Fravel, D.R. 2005. Commercialization and implementation of biocontrol. *Annual Review of Phytopathology* 43, 337–359. https://doi.org/10.1146/annurev.phyto.43.032904.092924

Gams, W., & Bissett, J., 2002. Morphology and identification of *Trichoderma*. In: C.P. Kubicek, G.E. Harman (Eds.), *Trichoderma and Gliocladium* (pp. 3–34). Gunpowder Square, London, UK: Taylor & Francis.

Gao, L., Sun, M.H., Liu, X.Z., & Che, Y.S. 2007. Effects of carbon concentration and carbon to nitrogen ratio on the growth and sporulation of several biocontrol fungi. *Mycological Research* 111, 87–92.

Gaur, A., Sharma, P., & Sharma, A.K. 2010. Antagonistic effect of *Trichoderma* species on *Fusarium oxysporum* f. sp. *ciceri*. *Legume Research* 33(3), 216–219.

Grinyer, J., McKay, M., Nevalainen, H., & Herbert, B.R. 2004. Fungal proteomics: initial mapping of biological control strain *Trichoderma harzianum*. *Current Genetics* 45, 163–169.

Harman, G.E. (2006). Overview of mechanisms and uses of *Trichoderma* spp. *Phytopathology* 96(2), 190–194. https://doi.org/10.1094/PHYTO-96-0190

Harman, G.E., Howell, C.R., Viterbo, A., Chet, I., & Lorito, M. 2004. *Trichoderma* species—opportunistic, avirulent plant symbionts. *Nature Reviews Microbiology* 2(1), 43–56.

Howell, C.R. 2003. Mechanisms employed by *Trichoderma* species in the biological control of plant diseases: the history and evolution of current concepts. *Plant Disease* 87(1), 4–10.

Janvier, C., Villeneuve, F., Alabouvette, C., Edel-Hermann, V., Mateille, T., & Steinberg, C. 2007. Soil health through soil disease suppression: which strategy from descriptors to indicators? *Soil Biology and Biochemistry* 39(1), 1–23.

Jeyarajan, R., Ramakrishnan, G., Dinakaran, D., & Sridhar, R. 1994. Development of products of *Trichoderma viride* and *Bacillus subtilis* for biological control of root diseases. *Pesticide Information* 20(2), 15–20.

Kappel, L., Münster Kotter, M., Sipos, G., Escobar Rodriguez, C., & Gruber, S. 2020. Chitin and chitosan remodeling defines vegetative development and *Trichoderma* biocontrol. *PLoS Pathogens* 16(2), e1008320.

Kubicek, C.P., Mach, R.L., Peterbauer, C.K., & Lorito, M. 2001. *Trichoderma*: from genes to biocontrol. *Journal of Plant Pathology* 83(2), 11–23.

Kubicek, C.P., Steindorff, A.S., Chenthamara, K., Manganiello, G., Henrissat, B., Zhang, J., Cai, F., Kopchinskiy, A.G., Kubicek, E.M., Kuo, A., Baroncelli, R., Sarrocco, S., Noronha, E.F., Vannacci, G., Shen, Q., Grigoriev, I.V., & Druzhinina, I.S. 2019. Evolution and comparative genomics of the most common *Trichoderma* species. *BMC Genomics* 20, 485. https://doi.org/10.1186/s12864-019-5680-7

Kumar, P., Dubey, R.C., & Maheshwari, D.K. 2011. Effect of plant growth promoting *Rhizobacteria* and *Trichoderma* spp. on development of *Macrophomina phaseolina* induced charcoal rot of soybean. *Biocontrol Science and Technology* 21(1), 21–36. https://doi.org/10.1080/09583157.2010.505637

Kushwaha, M., & Verma, A.K. 2014. Antagonistic activity of *Trichoderma* spp (a bio-control agent) against isolated and identified plant pathogens. *International Journal of Chemical and Biological Sciences* 1, 1–6.

Latifian, M., Hamidi-Esfahani, Z., & Barzegar, M. 2007. Evaluation of culture conditions for cellulase production by two *Trichoderma reesei* mutants under solid-state fermentation conditions. *Bioresource Technology* 98, 3634–3637.

Lewis, J.A., & Papavizas, G.C. 1987. Formulation and delivery systems of biocontrol fungi in agriculture. In: *Innovative Approaches to Plant Disease Control* (pp. 247–271). Wiley.

Li, J., Philp, J., Li, J., Wei, Y., Li, H., Yang, K., Ryder, M., Toh, R., Zhou, Y., Denton, M.D., Hu, J., & Wang, Y. 2020. *Trichoderma harzianum* inoculation reduces the incidence of Clubroot disease in Chinese cabbage by regulating the rhizosphere microbial community. *Microorganisms* 8, 1325. https://doi.org/10.3390/microorganisms8091325

Lu, B., Druzhinina, I.S., Fallah, P., Chaverri, P., Gradinger, C., Kubicek, C.P., & Samuels, G.J. 2004. *Hypocrea/Trichoderma* species with pachybasium-like conidiophores: teleomorphs for *T. minutisporum* and *T. polysporum* and their newly discovered relatives. *Mycologia* 96, 310–342.

Marraschi, R., Ferreira, A.B.M., da Silva Bueno, R.N., Leite, J.A., Lucon, C.M.M., Harakava, R., Leite, L.G., Padovani, C.R., & Bueno, C.J. 2019. A protocol for selection of *Trichoderma* spp. to protect grapevine pruning wounds against *Lasiodiplodia theobromae*. *Brazilian Journal of Microbiology* 50, 213–221. https://doi.org/10.1007/s42770-018-0029-y

Monte, E. 2001. Understanding *Trichoderma*: between biotechnology and microbial ecology. *International Microbiology* 4(1), 1–4.

Moo-Koh, F.A., Cristóbal-Alejo, J., Andrés, M.F., Martín, J., Reyes, F., & Tun-Suárez, J.M., & Gamboa-Angulo, M. 2022. In vitro assessment of organic and residual fractions of nematicidal culture filtrates from thirteen tropical *Trichoderma* strains and metabolic profiles of most-active. *Journal of Fungi* 8, 82. https://doi.org/10.3390/jof8010082

Mukherjee, P.K., Horwitz, B.A., Herrera-Estrella, A., Schmoll, M., & Kenerley, C.M. 2012. *Trichoderma* research in the genome era. *Annual Review of Phytopathology* 50, 291–314. https://doi.org/10.1146/annurev-phyto-081211-173728

Papavizas, G.C., Lewis, J.A., & Lumsden, R.D. 1984. Role of *Trichoderma* and *Gliocladium* in biological control. In: *Biological Control in Crop Production* (pp. 93–108). CRC Press.

Pocurull, M., Fullana, A.M., Ferro, M., Valero, P., Escudero, N., & Saus, E. 2020. Commercial formulates of *Trichoderma* induce systemic plant resistance to *Meloidogyne incognita* in tomato and the effect is additive to that of the Mi-1.2 resistance gene. *Frontiers in Microbiology* 10, 3042. https://doi.org/10.3389/fmicb.2019.03042

Poveda, J., Abril-Urias, P., & Escobar, C. 2020. Biological control of plant-parasitic nematodes by filamentous fungi inducers of resistance: *Trichoderma*, Mycorrhizal and Endophytic Fungi. *Frontiers in Microbiology* 11, 992. https://doi.org/10.3389/fmicb.2020.00992

Prasad, R.D., Rangeshwaran, R., Anuroop, C.P., & Srinivas, C. 2002. Biological control of dry root rot of chickpea with antagonistic fungi. *Indian Phytopathology* 55(2), 235–237.

Rajput, A., Khanzada, M., & Shahzad, S. 2014. Effect of different organic substrates and carbon and nitrogen sources on growth and shelf life of *Trichoderma harzianum*. *Journal of Agricultural Science and Technology* 16, 731–745.

Saba, H., Vibhash, D.R., Prashant, K., Yadav, A., & Khan, M.S. 2012. Trichoderma—a promising plant growth stimulator and biocontrol agent. *Mycosphere* 3(4), 524–531.

Shahid, M., Srivastava, M., Singh, A., Kumar, V., Rastogi, S., Pathak, N., & Srivastava, A. 2014. Comparative study of biological agents, *Trichoderma harzianum* (Th-Azad) and *Trichoderma viride* (01PP) for controlling wilt disease in pigeon pea. *Journal of Microbial and Biochemical Technology* 6, 110–115.

Shanmugam, V., & Kanoujia, N. 2011. Biological management of vascular wilt of tomato caused by *Fusarium oxysporum* f. sp. *lycopersici* by *Pseudomonas fluorescens* and *Trichoderma harzianum*. *Biological Control* 57(2), 85–93. https://doi.org/10.1016/j.biocontrol.2011.02.008

Shoresh, M., Harman, G.E., & Mastouri, F. 2010. Induced systemic resistance and plant responses to fungal biocontrol agents. *Annual Review of Phytopathology* 48, 21–43.

Singh, A., Shahid, M., Srivastava, M., Pandey, S., Sharma, A., & Kumar, V. 2014. Optimal physical parameters for growth of *Trichoderma* species at varying pH, temperature and agitation. *Virology and Mycology* 3, 1–7.

Singh, S.D., Singh, G., & Reddy, M.V. 2004. Biological control of *Fusarium* wilt of chickpea. *Indian Phytopathology* 57(2), 187–190.

Sultana, N., Ghaffar, A., & Ghaffar, A. 2006. Effect of seed treatment with *Trichoderma* species and *Gliocladium virens* on root-rot disease of mung bean (*Vigna radiata* L.). *Pakistan Journal of Botany* 38(2), 417–424.

Tilocca, B., Cao, A., & Migheli, Q. 2020. Scent of a killer: microbial volatilome and its role in the biological control of plant pathogens. *Frontiers in Microbiology* 11, 41. https://doi.org/10.3389/fmicb.2020.00041

Vinale, F., Sivasithamparam, K., Ghisalberti, E.L., Marra, R., Barbetti, M.J., Li, H., Woo, S.L., & Lorito, M. 2008. A novel role for *Trichoderma* secondary metabolites in the interactions with plants. *Physiological and Molecular Plant Pathology* 72(1–3), 80–86. https://doi.org/10.1016/j.pmpp.2008.05.005

Wang, H., Zhang, R., Mao, Y., Jiang, W., Chen, X., Shen, X., Yin, C., & Mao, Z. 2022. Effects of *Trichoderma asperellum* 6S-2 on apple tree growth and replanted soil microbial environment. *Journal of Fungi* 8, 63. https://doi.org/10.3390/jof8010063

Yedidia, I., Benhamou, N., & Chet, I. 1999. Induction of systemic resistance in cucumber plants (*Cucumis sativus* L.) by the biocontrol agent *Trichoderma harzianum*. *Applied and Environmental Microbiology* 65(3), 1061–1070.

Yedidia, I., Shoresh, M., Kerem, Z., Benhamou, N., Kapulnik, Y., & Chet, I., 2003. Concomitant induction of systemic resistance to *Pseudomonas syringae* pv. *lachrymans* in cucumber by *Trichoderma asperellum* (T-203) and accumulation of phytoalexins. *Applied and Environmental Microbiology* 69, 7343–7353.

Zhang, C.-L., Druzhinina, I.S., Kubicek, C.P., & Xu, T. 2005. *Trichoderma* biodiversity in China: evidence for a North to South distribution of species in East Asia. *FEMS Microbiology Letters* 251, 251–257.

14

Significance of Microbial Genome Protective Enzymes in Tolerating Synthetic Dyes

Riya Raghav[1]*, Raj Kishor Kapardar*[2]*, and Rajpal Srivastav*[1]*

[1] *Amity Institute of Biotechnology, Amity University Uttar Pradesh, Noida, Uttar Pradesh, India*
[2] *The Energy and Resources Institute, New Delhi, India*

* Corresponding author: rsrivastav2@amity.edu

14.1 Introduction

Dyes are the chemical compounds used to impart color to substances, and they have an affinity for the fabric. They are used in dyeing various substrates like cosmetics, pharmaceuticals, textiles, leather, and paper. They can color because of certain properties like the presence of a chromophore, absorption of light in the visible spectrum, a conjugated system of alternate double and single bonds, and resonance of electrons. They also contain a structure called auxochrome, which is not helpful in coloring but increases the solubility of the dye. Dyes are designed specifically for the substrate to increase the affinity between the substrate and the dye.

They contain high affinity for the substrate and a high degree of stability under diverse industrial conditions (Sobczyk et al., 2025). Earlier, the powder extracted from natural sources like plants and microbes was used as colorants, but due to their limitation of poor stability more focus has shifted to producing synthetic dyes. The synthetic dyes are highly stable in diverse conditions because of the presence of conjugated bonds. This high stability makes them the best choice for use in industrial applications. However, the wastewater and contaminants released by these industries pollute water bodies and the environment if discharged untreated.

The treatment of wastewater is a very crucial step, and biological processes are used for the treatment, like microbial discoloration by laccases, bioremediation, and biodegradation. The microbial treatment is now preferred to reduce the contaminants of the discharged water and to improve water quality. The microbial species, including bacteria and fungi, are useful bioresources for biodegrading the wastewater. Further, advancements are being made to produce dyes with more stable properties that do not cause any harmful impact on the environment and humans. For example, to increase the dyeing characteristics,

Microbial Consortia in Plant Science and Sustainable Agriculture, First Edition. Edited by Ajay Tomar, et al.

liposome-mediated dyeing has been introduced in which dyes are encapsulated in the liposome and then applied on the cellulosic fiber (Bilal et al., 2022).

It improves the fiber quality, fastness property, and smoothness of the fiber. Further, it's important to find ways to improve the dyeing process as it has vast applications in different industries (Sulthana et al., 2025). There are many challenges to the industries, including better quality and stability, particularly in the textile industry. Dyes that are mainly used in the textile industry are being produced with antimicrobial and related properties. This causes serious synthetic dyes, particularly azo dyes, as these are toxic, and their end products may be mutagenic. As they are highly stable and contain aromatic structures, azo dyes are hard to degrade by microbial action and persist in the environment for a long time (Srivastav & Suneja, 2019; Ngo & Tischler, 2022). The impact of microbial consortium activities on the bioremediation of synthetic dyes is an important area of ecological research, and still there are many unexplored aspects associated with it. The effects of these synthetic dyes on health and the environment are harmful, and there is a need to understand the details of the homeostasis mechanisms of the host to tackle these problems and hazards.

14.2 Variation in Chemical Nature

Natural dyes are those that are extracted from natural resources, including plants and minerals. However, natural dyes have poor reproducibility and require a mordant to increase fastness properties. Archaeological research investigated the composition and the source of dye stuff used in the old textile industry. It was seen that the powder from the various natural sources is extracted and used in textiles and cosmetics. Egyptians used biocolorants including Madder and Murex. In ancient India, dyers used natural colorants including turmeric, indigo, henna, Patnag aluminum, chromium, iron, and tin as mordants to increase the fastness property (Zhou et al., 2023). Natural biocolorants include madder (*Rubia tinctoria*), cochineal (*Dactylopius coccus*), and Tyrian purple (Sigurdson et al., 2017). It is seen that leather dyed with annatto extract has better coloring and fastness properties. It is used in food commodities like dairy products, butter, and dyeing synthetic (nylon and polyester), as well as natural fibers (wool, silk, cotton) (Sathiya Mala et al., 2015). Research has been going on to use annatto as an alternative to bromophenol blue, as bromophenol blue is a synthetic dye and has a carcinogenic effect.

Synthetic dyes are produced from organic molecules or are man-made from petrochemicals. Accidentally it was found that aniline could color silk. A reddish-purple dye from coal tar, named mauve, was found to have interesting coloring properties. This discovery revolutionized the dye industry, and a new industry of synthetic dye was developed. Synthetic dyes have better fastness properties than natural dyes. Synthetic dyes are widely preferred in diverse industries. Over all the advantages it has, it is a fact that synthetic dyes are harmful to the environment. The industrial wastewater dye discharge pollutes the rivers and water bodies.

Mordant is a chemical used to increase the affinity of dye for the fiber. They are applied when there is less or no affinity for the material. It includes all the dyes that can form complexes with a metal mordant. Vat dyes are called so because the dyeing process takes place in the bucket. The colored forms are insoluble, while their reduction can generate soluble colorless forms. Insoluble form can be regained by the re-oxidation of the dye. Examples of vat dyes are indigo and Tyrian purple. Direct dyes are water-soluble organic molecules that can be applied to cellulosic fibers directly. Because of the difference in chemical nature, direct dyes have poor fastness. But it can be improved after treatment. Examples of direct dyes are annatto, turmeric, harda, safflower, and pomegranate. Acid dyes are types of direct dyes for polyamide fibers, like silk, nylon, and wool. Acid dyes have either sulfonic acid or carboxylic acid groups, and they function in acidic conditions. Saffron is the best example of acid dyes and has two carboxylic groups. Basic dyes are also known as cationic dyes and are applied in near-neutral medium conditions. The ionization yields colored ions, forming bonds with the carboxyl groups of fibers, for example, berberine. Nitroso dyes are the polynitro derivative of phenol and contain at least one nitro (NO_2) or nitroso (N—O) group at ortho or para position. They have a nitro group as a chromophore and a hydroxyl group as an auxochrome. Nitro dyes include picric acid, Manchester yellow, sulfonic acid, naphthol yellow S, and amido yellow E, and nitroso dyes include Gambine-Y. Amido yellow E is made by reaction between 1-chloro-2,4-dinitroenzene and 4-aminodiphenylamine-2-sulfonic acid. Martius yellow or Manchester yellow is used to dye silk or wool.

They are the largest class of synthetic dyes. They contain azo (—N=N—) group as the primary chromophore. Additional groups are also present in the dyes, which increase the solubility and different shades of the dye like sulfonic acid, hydroxyl, or amino group. Azo dye synthesis is a two-step process: diazotization and diazo coupling. Conversion of aromatic amine to diazo compound is called diazotization; it is followed by the reaction between the diazo group and phenol, naphthol, or a compound with an active methyl group. This process is called diazo coupling. Based on the number of azo groups, it can be divided into mono-azo-, bisazo, triazo-, and so on, and subdivided into acid, basic, direct, and ingrain according to the mode of application. They are widely used in textile, paper, pharmaceutical, cosmetic, and various other industries.

These dyes are highly stable in diverse conditions because of the presence of resonance and π-conjugated azo bond characteristics. This high stability makes them a suitable candidate for industrial applications. Azo dyes are extensively used in the dye industry, and the wastewater produced by these industries contains a significant proportion of azo dyes and derivatives. This causes a serious concern because these azo dyes and derivatives are toxic and carcinogenic. As they are highly stable and contain aromatic structures, azo dyes are hard to degrade by microbial action and remain in the environment for a long period of time. Methyl orange and acid orange are the mono-azo dyes used in dyeing wool and silk. Methyl orange is fleeting and used as an acid–base indicator in titration as it gives an orange color in an alkaline solution and a red color in an acidic solution.

Synthetic dyes are produced from organic molecules or are man-made from petrochemicals. Synthetic dyes are superior to natural dyes in terms of fastness property, cost, and coloring capacity (Vyas & Krishnaswamy, 2006). Based on the number of azo groups, it can be divided into mono-azo-, bisazo, triazo-, and so on and subdivided into acid, basic, direct, and ingrain, according to mode of application. They are widely used in textile, paper, pharmaceuticals, cosmetic, and various other industries. These dyes are highly stable in diverse

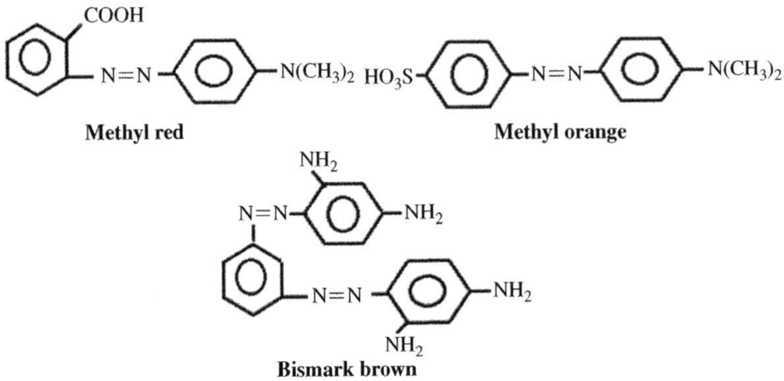

Figure 14.1 Representation of the typical structure of some important azo dyes.

conditions because of the presence of resonance and π-conjugated azo bond characteristics. Azo dyes are extensively used in the dye industry, and the wastewater is discharged into water bodies. This causes a serious concern because azo dyes are toxic, and their products, that is, aromatic amines, are mutagenic and carcinogenic. As they are highly stable and contain aromatic structures, azo dyes are persistent and difficult substrates for biodegradation (Ngo & Tischler, 2022).

Methyl orange and acid orange are the mono-azo dyes used in dyeing wool and silk. Methyl orange is fleeting and is used as an acid–base indicator in titration as it gives an orange color in an alkaline solution and a red color in an acidic solution. The complexity of the representative azo dye is shown in Figure 14.1.

14.3 Microbial Pigments

Pigments extracted from the microorganisms are the potential source of biocolorants due to their advantages over plants. Microorganisms can be grown rapidly and easily in low-cost medium, which can be easily processed, and growth is independent of weather conditions, whereas plants are available seasonally and less stable; cost efficiency is high, but they are labor intensive, and so on. Slowly, natural and synthetic dyes are replaced by microbial pigments. The carotenoids, melanins, flavins, quinines, monascins, and violancein are some of the microbial pigments produced by microorganisms (Banerjee et al., 2025). These coloring agents are used in a variety of industries like cosmetic, food, textile, and pharmaceutical and also possess antibacterial, antifungal, antioxidant, and anti-inflammatory properties (Varjani et al., 2020).

Microorganisms can be cultivated by either solid-state fermentation or submerged fermentation (Roukas, 2016). In solid-state fermentation, microbial biomass occurs on the soil surface, and in submerged fermentation, growth of microorganisms occurs in the liquid medium with proper aeration and agitation. Fermentation can be done using raw material or industrial organic waste, like corn steep liquor, as a substrate. The economy of the industry depends on factors like input cost and output. Here, to reduce the cost, industrial waste is used as a substrate, like grape waste and corn steep liquor. During the production process, consideration should be given that the microorganism should not produce any toxin and can tolerate a wide range of pH changes. For controlling toxin production, important measures should be taken, including a nonpathogenic strain, control of the biosynthesis of metabolites, and media selection (Pswarayi & Gänzle, 2022).

Many fungal and bacterial strains are used to produce biocolorants (Srivastav et al., 2013). Filamentous fungi produce pigments like carotenoids, melanins, indigo, and quinones (Zomer et al., 2024). Monascus produce can yield different color pigments like ankaflavine, monascine, rubropunctatine, and purple pigment, which have antimicrobial, anticancer, antimutagenic, and antiobesity properties (Chaudhary et al., 2022). *Eurotium, Fusarium, Curvularia*, and *Drechslera* produce anthraquinone, naphthoquinone, melanin, flavin, and cynodontin (Sandmann, 2022). Marine organisms live in different environments like low temperature, absence of light, and high salinity. Indeed, these organisms can produce unique substances. Marine endophytic fungi produce a variety of pigments, and their association with algae and coral imparts various colors (Kim et al., 2025).

Bacterial strains are more advantageous than fungal strains. They are available everywhere, like soil, desert, sand, and fresh water and can tolerate a wide range of temperatures. These microorganisms have a short life span and can be modified genetically, like *Streptomyces, Nocardia, Micromonospora, Thermomonospora, Actinoplanes*, and *Microbispora*. Researchers are interested in finding a new source of pigment and increasing the yield by optimizing the present strain in a process called strain improvement (Nodwell, 2019). Strain improvement by UV mutation increases the prodigiosin production. By genetic engineering, microorganisms can be modified to produce the pigment of interest (Srivastav & Suneja, 2019). *Streptomyces coelicolor* originally produces blue pigment, and by modification it can produce bright yellow (Kalafungin), orange, or yellow pigment (anthraquinones) (Hoskisson & van Wezel, 2019). Carotenoids are produced by zygomycetes (*Phycomyces, Blakeslea*), Basidiomycetes (*Rhodosporidium, Sclerotium*), and Ascomycetes (*Aspergillus, Penicillium, Aschersonia*).

14.4 Environmental Toxicity

The wastewater released into the environment without treatment causes harmful impacts on the environment and water bodies. They have a negative impact on land and aquatic life, and their products are even carcinogenic to several life forms (Tripathi et al., 2023; Radhakrishnan et al., 2023) (Table 14.1). They also interrupt the photosynthetic process in the water system by blocking the penetration of light into deeper layers. Physiological methods that are used to treat wastewater have various limitations like high cost and a large amount of sludge production. As extremophiles, they can survive in harsh environmental conditions where normal bacteria cannot survive. Studies show that halotolerant and halophilic bacteria have the potential to decolorize dyes in the presence of nitrate, sulfate, chloride, and carbonate salts. Therefore, halophilic bacteria are now exploited to treat the wastewater released from the dye industry.

Table 14.1 The effects of the dyes on the host and their consequences on cellular integrity.

Dye	Microbe(s)	Effect(s)	Target genes	Pathways involved	References
Melanin	*Streptomyces, Aspergillus*	Antioxidant, radio protective. Generally nonmutagenic	—	It inhibits lipid peroxidation.	Tomoshige et al. (2017)
Violacein	*Chromobacterium violaceum*	Anticancerous, antimicrobial, induces DNA fragmentation	BAX, BCL2, CASP3	p53 pathway; intrinsic apoptosis	Kanelli et al. (2018)
Riboflavin	*Ashbya gossypii, Bacillus subtilis*	Vitamin B2, cofactor. Under UV, produces reactive oxygen species (ROS) (possible DNA damage)	DNA repair genes	DNA repair, redox regulation, oxidative stress	Liang et al. (2017)
Indigoidine	*Streptomyces*	Antioxidant, anti-inflammatory, nonmutagenic	Nrf2, IL-6 (modulatory)	Nrf2/ARE, anti-inflammatory	Huang et al. (2024), Xu et al. (2015)
Prodigiosin	*Serratia marcescens*	Antitumor, immune suppressive. ROS	MYC, BCL2, P21, TP53	Mitochondrial apoptosis, p53, ROS pathway	Abdel-Rafei et al. (2022)
Pyocyanin	*Pseudomonas aeruginosa*	Cytotoxic, pro-inflammatory. Oxidative DNA damage	Nrf2, HO-1, IL-8	Nrf2/ARE, NF-kB, ROS pathway	Saunders et al. (2020), Kovacic and Somanathan (2014)
Actinorhodin	*Streptomyces coelicolor*	Antibacterial, inhibits DNA/RNA synthesis	Binds DNA → inhibits polymerases	DNA replication stress, SOS response	Nayyara et al. (2024), Wang et al. (2020)
Flexirubin	*Flavobacterium* spp., *Chryseobacterium* sp.	Antimicrobial, antibiofilm. Possible membrane disruption	Not characterized	Membrane permeability, bacterial signaling	Bertolini et al. (2021)
Pulcherrimin	*Metschnikowia pulcherrima* (yeast)	Antifungal, iron-chelating (stress-induced mutation leads to growth inhibition)	Iron-dependent genes (e.g. fepA)	Iron metabolism, oxidative stress	Chen et al. (2025)
Phycocyanin	*Spirulina platensis* (cyanobacteria)	Antioxidant, anti-inflammatory DNA protection from oxidative damage	COX-2, NF-κB, Nrf2	Nrf2/ARE, NF-κB, MAPK	Kissoudi et al. (2018)

Haloarchaea are also used in the degradation of toxic materials, liposome construction, and food fermentation (Metwally et al., 2023). Halophilic archaea are suitable for the decoloration of azo dyes in industries that involve harsh conditions like high temperature and salt concentration (El Awady et al., 2024). Different strains of haloarchaea are examined under different environmental conditions. Halophilic bacteria that have decolorization ability are *Halogeometricum, Haloferax, Haloarcula, Halococcus, Haloarchaeobius, Halovivax, Halopenitus*, and *Halorientalis*.

Laccase or benzenediol oxygen oxidoreductase is an enzyme produced by many bacterial or fungal strains. It catalyzes the oxidation of phenolic, nonphenolic, and aromatic substrates present in the mixture. Laccases are well known for their dye decolorization ability (Janusz et al., 2020). Laccases used so far are from fungi. So, our interest is to explore laccases, which are thermostable or alkali-stable. The enzyme laccases, when thermostable, have wide application in industries. Various microorganisms, including *Aquifex aeolicus, Bacillus subtilis*, and *Thermus thermophilus*, have the ability to produce thermostable laccases (Bertrand et al., 2017).

In a typical process of recombinant DNA technology, the target gene of thermostable laccase was cloned into a suitable vector for protein expression. The heterologous expression of thermostable laccase TthLAC in an expression system is performed (Mustafa et al., 2024). The biochemical assays can be performed to assess the biological activity of the protein, produced by recombinant methods (Srivastav et al., 2014). These preparations of the thermostable enzymes are useful when they are produced at low cost. The production cost defines the applicability of the method on an industrial scale. The thermostable characteristic of laccases is due to their kinetic stability, that is, slow unfolding. It can be used in redox and oxidation applications involving decolorization and delignification. This thermostable laccase can decolorize different dyes as per the recent observations. It can also decolorize Congo red or other dyes in the presence of a substrate acting as a redox mediator.

Synthetic dyes like azo dyes are widely used in different industries like food, pharmaceuticals, cosmetics, and textiles. There are many challenges to the textile industry. The clothes provide the best environment for the microorganisms to grow, which may cause bad odors, skin infections, and a decrease in fiber quality. It is important to find ways to prevent the growth of microorganisms. Moreover, hybrid pigments, surface activation of fiber by plasma, and many other advancements are taking place in this field.

14.5 Role of Microbes and Genome Protection

Textile industries are showing themselves as one of the major contributors to discharging dye-laden effluents, increasing the environmental pollution. These synthetic dyes are toxic, carcinogenic in nature, and even resistant to the conventional wastewater treatments (Suneja & Srivastav, 2021). This increases demand to achieve a sustainable method of dye removal due to their eco-friendliness, cost-effectiveness, and efficiency. There are many microorganisms that contribute to dye decolorization primarily through two mechanisms: adsorption onto microbial biomass and biodegradation via enzymatic or molecular pathways. In adsorption, dye molecules physically bind to the surface of microbial cells. In case of toxic dye-laden effluents, they inhibit the support to microbial metabolism. They do not affect the chemical composition of the dye such as degrading the dye into fragments, but cause its physical removal. Microbial absorbents may include bacteria, fungi, and microalgae, possessing unique surface properties that facilitate dye binding on the microbial cell's surface.

In biodegradation, it involves the enzymatic breakdown of dye molecules into smaller fragments, frequently less harmful fragments. Many microbial enzymes such as laccase, lignin peroxidase, azoreductase, and tyrosinase have been identified for their role in dye degradation. Complex dye structures, such as azo bonds and aromatic rings, can be broken down by these enzymes, converting them into simpler organic compounds or even achieving complete mineralization. Fungal species such as *Phanerochaete chrysosporium* and *Neurospora crassa* are known for their high laccase production and are well known for their capacity to extract dye pigments and phenolic compounds from industrial effluents. Significant dye-degrading properties have also been shown by several fungal species, including *Trametes versicolor, Bjerkandera adusta, Aspergillus ochraceus, Pleurotus* spp., and *Phlebia* spp. Techniques like gel entrapment and matrix adherence can further enhance the performance of these fungal species by stabilizing the biomass and facilitating sustained enzyme activity.

Saccharomyces cerevisiae MTCC 463 has demonstrated encouraging enzymatic activity among yeasts, producing laccase, lignin peroxidase, and several reductases. Complex dyes are broken down by these enzymes into simpler substances like aliphatic amines, which are subsequently broken down into even simpler forms. Remarkably, when grown in nutrient-rich environments like molasses, *S. cerevisiae* has also demonstrated the capacity to bioaccumulate colors.

Other ascomycete yeasts, including *Candida tropicalis, Debaryomyces polymorphus*, and *Issatchenkia occidentalis*, have been studied for their ability to decolorize azo dyes. *Trichosporon beigelii* has shown complete degradation of dyes such as Navy Blue HER, while *Galactomyces geotrichum* MTCC 1360 is efficient against both azo and reactive dyes. Mixed microbial cultures often outperform individual strains due to synergistic interactions that enhance dye breakdown and detoxification. Aromatic amines and other dye byproducts can be effectively broken down by these consortia. Many dye-degrading bacteria, however, need additional nutrients to be active in aerobic environments since they are unable to use colors as their only carbon source. Rhodamine B, a xanthene-based compound, is of particular concern as it was widely used. Although it has genotoxic, carcinogenic, and neurotoxic qualities, it is utilized in many different industries, including food coloring, paints, and textiles. Its presence in wastewater presents serious threats to human health and the environment, underscoring the need for efficient biological treatment techniques.

In the past, the 1970s saw the first reports of pure bacterial cultures that could break down colors, including *Bacillus cereus, Aeromonas hydrophila*, and *B. subtilis*. Since then, many strains of bacteria that degrade dye have been identified from various habitats; some of these strains have demonstrated higher decolorization effectiveness than commercial alternatives. However, enzyme activity and microbiological growth can still be inhibited by high dye concentrations, highlighting the significance of adjusting treatment settings. The microbes are attracted to their surface through noncovalent interactions like hydrophilic, electrostatic, or Van der Waals interactions (Srivastav et al., 2019). These interactions are mediated by the functional groups present in the polymers. The degradation of these synthetic polymers happens at a very slow rate in some cases.

Many pathogenic strains developed drug resistance, and it made it necessary for researchers to find other antimicrobial agents. So, microbial pigment acts as an alternative as they exhibit antimicrobial, antioxidant, anticancer, and anti-inflammatory activities. If free radicals are present in the body, they can cause various diseases like cancer, diabetes, and autoimmune disorders. The free radical scavenging activity of antioxidants neutralizes the free radicals by donating electrons to the radicals. Anthraquinone from fungi *Stemphylium lucopersici* and melanin from *S. glaucescens* have antioxidant properties.

The use of microbial pigments has created a new treatment strategy for different diseases and conditions. Many microbial pigments have anticancer properties. Prodiginines, pigments isolated from *Pseudoalteromonas* spp., have a cytotoxic effect on U937 Leukemia cells. *Streptomyces glaucescens* NEAE-H has an anticancer effect on skin cancer cell lines. Some microbial pigments also have the property to induce apoptosis and cell cycle inhibition. Prodigiosin from *Serratia marcescens* has an apoptotic effect on human cervical cancerous cells in a dose-dependent manner. Prodiginine pigment produced from *Vibro* spp. has antimicrobial activity for *Staphylococcus aureus* and *E. coli*. Pyocyanin and pyorubin pigments isolated from *Pseudomonas aeruginosa* have antimicrobial activity against *Citrobacter* spp., which cause wound infection and urinary tract infection. Pigments from *Streptomyces hygroscopicus* also show antimicrobial activity against *S. aureus*, which is resistant to methicillin and vancomycin.

With an emphasis on the molecular mechanisms causing genomic instability and possible therapeutic vulnerabilities, there are implications of synthetic dyes on host cells, including humans and microbes. The genome integrity is governed by various molecular determinants (Radhakrishnan et al., 2024b). Sustaining genome stability throughout cell division depends on accurate DNA replication, and errors can result in replication stress (RS), which is a major contributing factor to early-stage malignant growth. Both external (like UV light and chemical mutagens) and intrinsic (such as nucleotide imbalances, oncogene activation, transcription replication conflicts) mechanisms cause RS. The cells react to RS, especially DNA repair mechanisms, fork remodeling, and checkpoint activation, regulated by ATR under cellular stress. The roles of key regulators like ATR kinase, CHK1, and WEE1 kinase, as well as MRN complex and PARP1 have significant roles in tolerating mutagens. Several RS response-related replication and repair proteins of DNA, including RPA, RAD51, and FANCD2, create discrete focal regions within the nucleus. These focal points may indicate the renovation or fixing occurrences in the replication forks that have stopped or collapsed.

Additionally, double-strand breaks (DSBs) caused by extended stalling and nucleolytic cleavage of replication forks could be tracked using nuclear foci and levels of proteins of KAP1 p-S824, ATM p-S1981, Chk2 p-T68, and RPA2 p-T21. There are precise ways to track these changes within the cell when external chemicals or mutagens are introduced. Pulsed-field gel electrophoresis can be used to directly detect the DSBs caused by RS.

RS refers to the condition that cells enter when exposed to genotoxic factors, which hinder DNA replication and stall replication forks (Kapil et al., 2024). This initiates the stress response, which refers to a temporary halt of the progression of the cell cycle resulting from the activation of DNA damage response (DDR) pathways. DDR's main job is to slow down DNA replication and synthesis so that DNA repair has enough time to complete the process. Cellular senescence or apoptosis may occur if RSR is not effectively managed. Single-stranded breaks result in the activation of kinases and DNA-PK or DNA-dependent protein kinase. These kinases further activate ATR kinase as well as its downstream effectors, WEE1-like protein kinase (WEE1) and checkpoint kinase 1 (CHK1). These kinases then cause cell cycle arrest and initiate the recovery process of the stalled replication forks. Further, MCM helicases cause constant unwinding of the DNA helix (Radhakrishnan et al., 2024a). This unwinding results in the exposure of the ssDNA in halted replication forks. This exposed ssDNA is rapidly coated by RPA protein to stop degradation and promote repair. Timeless and Tipin heterodimers form a fork protection complex that is drawn to these stalled forks by the displacement of proliferating nuclear antigen. This complex binds to RPA to improve fork stability and stop replication-associated DNA damage. ATR kinase can be captured and activated more easily due to its connection with ATR-interacting protein, also known as ATRIP, which binds to ssDNA coated with replication protein A at stalled replication forks. ATR activation is further increased by its association with DNA topoisomerase 2-binding protein 1 (TOPBP1) or ETAA1 activator of ATR kinase. ATR phosphorylates several important substrates, such as checkpoint kinase 1 (CHK1) and histone H2AX (γH2AX), to start a chain of signals that control the RS response.

Recently, a method known as the neutral comet test has also found relevance. Additionally, the duration and magnitude of RS have an impact on the degree to which DSBs associated with replication occur. While mild, prolonged RS does not result in a significant DSB reaction, it does raise DSBs at RFSs. The functioning and progression of RS, which ultimately results in SDBs, needs further in-depth analysis (Grover et al., 2019).

ATM doesn't seem to directly affect the repair of DNA, which is under stress during replication, while being crucial for repairing DSBs. Tel1's introduction to the MRX complex is one of its essential roles; it keeps Mre11 attached to DNA, stabilizes stalled replication forks, promotes replisome stability, and stops extensive chromosomal reorganization during RS (González-Acosta & Lopes, 2024). ATMTel1 also has a vital part in connecting RS with modifications in cell metabolism at a larger cellular level. Senescence, which stops defective DNA from being transmitted on to subsequent generations and lowers the chance of cancer, can occur when cells undergo RS. Senescent cells can, however, avoid senescence if ATMTel1 is deactivated in response to RS brought on by nucleotide deficiencies. This is because ATMTel1 promotes the metabolism of glutamine and glucose as well as the activity of glucose-6-phosphate dehydrogenase, which helps restore the levels of dNTPs (Mognato et al., 2021). These metabolic alterations are caused by p53 suppression and c-MYC overexpression brought on by ATM inactivation. This suggests that ATM's activation of p53 is essential for linking the senescence phenomena to DNA and cellular damage (Williams & Zhang, 2021).

14.6 Control of DNA Damage

Checkpoint activation is another crucial reaction to RS. By responding to DNA damage and offering a chance for repair prior to duplication, checkpoint activation controls the course of the cell cycle when there is external stress or mutagen (Grover et al., 2019). The complexity of these systems is demonstrated by the involvement of checkpoint kinases such as

WEE1 and the key regulatory protein SOG1 (Suppressor of Gamma Responses 1). To coordinate the response to DDR, for example, SOG1 controls a group of genes linked to stress response and repair. According to Herbst et al. (2024), SOG1 is a transcription activator that regulates the expression of genes involved in DNA repair and cell cycle regulation (Herbst et al., 2024). Cells may stop in the S phase or the G2 phase, depending on when DNA damage occurs during the cell cycle. Protein kinases SOG1 and WEE1 regulate an intra-S checkpoint that is set off by RS.

Through different ways, SOG1 can also result in a G2 cell cycle halt. First, ATR and SOG1 work together to control the proteasome breakdown of mitotic CDKB2;1. Afterward, SOG1 promotes endoreduplication by controlling the amount of expression of genes that generate SMR5 and SMR7, which inhibits the action of CDK (Chen et al., 2025). The WEE1 kinase also prevents CDK activity by blocking the G2/M switch. Additionally, SOG1 promotes the production of G2-specific CYCLINB1, which may be related to CYCB1,1's role in DNA repair and is believed to postpone mitosis. According to a recent study, SOG1-dependent transcriptome alterations brought on by DNA damage are largely mediated by MYB3R repressor activation, which prevents the G2/M cell cycle-related genes from being expressed. MYB3R transcription factors, namely MYB3R4, MYB3R3, and MYB3R5, are known to govern the G2/M switch. Additionally, according to the proteins it engages with, MYB3R1 can function either as an activator and as a repressor (Pedroza-Garcia et al., 2022a). According to Chen et al. (2025), rep-MYB3R is necessary for growth inhibition caused by DNA damage. After zeocin administration, MYB3R3 accumulates in root meristems, causing G2 arrest and stopping cell division. In this study, researchers demonstrated that CDKs phosphorylate MYB3R3, which causes the proteasome to degrade it. Therefore, the decrease in CDK activity caused by CDK inhibitor stimulation is probably a factor in the buildup of Rep-MYB3Rs as an aftermath of DNA damage. These findings provide new insights into the processes behind SOG1's inhibition of CDKB2,1 accumulation.

SOG1 helps control activators of anaphase-promoting complex. The decrease in CDKB2;1 caused by DNA damage may be due to the simultaneous repression of CDKB2;1 gene by Rep-MYB3Rs and the destruction of the protein molecule by APC/C. According to recent reports, the transcription factors ANAC044 and ANAC085 are both targets of SOG1 and intricately connected to help promote the accumulation of rep-MYB3R after DNA damage (Pedroza-Garcia et al., 2022). According to a genetic study, SOG1, in cooperation with ANAC044 and ANAC085, regulates the cell cycle arrest by increasing rep-MYB3R levels rather than by activating SMR or DNA repair genes. Since ANAC044 and ANAC085 do not directly affect Rep-MYB genes, the precise mechanism by which they influence rep-MYB3R protein levels is yet unknown. There is a need for more investigation to understand the intricacies of these mechanisms.

14.7 Conclusive Remarks

Understanding the microbial dynamics and synergistic activities of microbial consortia at the molecular level has broader applications in environmental toxicity management of synthetic dyes and chemicals. There are many unknown biomarkers that need to be explored to fully understand the effects of these chemicals on the host systems and the ecosystem.

Acknowledgment

All authors acknowledge Amity University Uttar Pradesh for infrastructural support during the preparation of this manuscript.

References

Abdel-Rafei, M.K., Thabet, N.M., & Amin, M.M. 2022. Concerted regulation of OPG/ RANKL/ NF-κB/MMP-13 trajectories contribute to ameliorative capability of prodigiosin and/or low dose γ-radiation against adjuvant- induced arthritis in rats. *International Immunopharmacology* 111, 109068. https://doi.org/10.1016/j.intimp.2022.109068

Banerjee, S., Sarkar, A., & Rao, K.V.B. 2025. Extraction and characterization of carotenoid pigments with antioxidant and antibacterial potential from marine yeast *Rhodotorula* sp. KSB1. *International Microbiology: The Official Journal of the Spanish Society for Microbiology* 28(1), 137–156. https://doi.org/10.1007/s10123-024-00529-3

Bertolini, D., Jiménez, M.E.P., Dos Santos, C., Corrêa, A.P.F., & Brandelli, A. 2021. Microbial bioconversion of feathers into antioxidant peptides and pigments and their liposome encapsulation. *Biotechnology Letters* 43(4), 835–844. https://doi.org/10.1007/s10529-020-03067-w

Bertrand, B., Martínez-Morales, F., & Trejo-Hernández, M.R. 2017. Upgrading laccase production and biochemical properties: strategies and challenges. *Biotechnology Progress* 33(4), 1015–1034. https://doi.org/10.1002/btpr.2482

Bilal, M., Ihsanullah, I., Hassan Shah, M.U., Bhaskar Reddy, A.V., & Aminabhavi, T.M. 2022. Recent advances in the removal of dyes from wastewater using low-cost adsorbents. *Journal of Environmental Management* 321, 115981. https://doi.org/10.1016/j.jenvman.2022.115981

Chaudhary, V., Katyal, P., Poonia, A.K., Kaur, J., Puniya, A.K., & Panwar, H. 2022. Natural pigment from Monascus: the production and therapeutic significance. *Journal of Applied Microbiology* 133(1), 18–38. https://doi.org/10.1111/jam.15308

Chen, T., Uzunovic, H., Brul, S., & Hugenholtz, J. 2025. Developing *Bacillus subtilis* as cell factory for the production of the natural biocontrol compound pulcherrimin. *Bioresource Technology* 427, 132433. https://doi.org/10.1016/j.biortech.2025.132433

El Awady, M.E., El-Shall, F.N., Mohamed, G.E., Abd-Elaziz, A.M., Abdel-Monem, M.O., & Hassan, M.G. 2024. Exploring the decolorization efficiency and biodegradation mechanisms of different functional textile azo dyes by *Streptomyces albidoflavus* 3MGH. *BMC Microbiology* 24(1), 210. https://doi.org/10.1186/s12866-024-03347-9

Grover, R., Burse, S., Shankrit, S., Aggarwal, A., Kirty, K., Narta, K., Srivastav, R., Ray, A.K., Malik, G., Vats, A., Motiani, R.K., Thukral, L., Roy, S.S., Bhattacharya, S., Sharma, R., Natarajan, K., Mukerji, M., Pandey, R., Gokhale, R.S., & Natarajan, V.T. 2019. Mygl exonuclease couples the nuclear and mitochondrial translational programs through RNA processing, *Nucleic Acids Research* 47(11), 5852–5866.

Herbst, J., Li, Q.-Q., & De Veylder, L. (2024). Mechanistic insights into DNA damage recognition and checkpoint control in plants. *Nature Plants* 10(4), 539–550. https://doi.org/10.1038/s41477-024-01652-9

Hoskisson, P.A., & van Wezel, G.P. 2019. *Streptomyces coelicolor. Trends in Microbiology* 27(5), 468–469. https://doi.org/10.1016/j.tim.2018.12.008

Huang, X., Gan, L., He, Z., Jiang, G., & He, T. 2024. Bacterial pigments as a promising alternative to synthetic colorants: from fundamentals to applications. *Journal of Microbiology and Biotechnology* 34(11), 2153–2165. https://doi.org/10.4014/jmb.2404.04018

Janusz, G., Pawlik, A., Świderska-Burek, U., Polak, J., Sulej, J., Jarosz-Wilkołazka, A., & Paszczyński, A. 2020. Laccase properties, physiological functions, and evolution. *International Journal of Molecular Sciences* 21(3), 966. https://doi.org/10.3390/ijms21030966

Kanelli, M., Mandic, M., Kalakona, M., Vasilakos, S., Kekos, D., Nikodinovic-Runic, J., & Topakas, E. 2018. Microbial production of violacein and process optimization for dyeing polyamide fabrics with acquired antimicrobial properties. *Frontiers in Microbiology* 9, 1495. https://doi.org/10.3389/fmicb.2018.01495

Kapil, T., Gupta, P., Radhakrishnan, A., & Srivastav, R. 2024. Multifunctional nanocomposites enhance biotherapeutics sensitivity in replication-associated diseases. In: A. Kumar, M.K. Solanki (Eds.), *Microbial Biocontrol Techniques. Microorganisms for Sustainability* (Vol. 54, pp. 311–321). Springer, Singapore. https://doi.org/10.1007/978-981-97-8739-5_16

Kim, H.J., Kim, S., Lee, Y., Shin, Y., Choi, S., Oh, J., Jeong, J., Park, H., Ahn, J., Joo, J.C., Choi, K.Y., Bhatia, S.K., & Yang, Y.H. 2025. Production of bio-indigo from engineered *Pseudomonas putida* KT2440 harboring tryptophanase and flavin-containing monooxygenase. *Enzyme and Microbial Technology* 182, 110529. https://doi.org/10.1016/j.enzmictec.2024.110529

Kissoudi, M., Sarakatsianos, I., & Samanidou, V. 2018. Isolation and purification of food-grade C-phycocyanin from *Arthrospira platensis* and its determination in confectionery by HPLC with diode array detection. *Journal of Separation Science* 41(4), 975–981. https://doi.org/10.1002/jssc.201701151

Kovacic, P., & Somanathan, R. 2014. Toxicity of imine-iminium dyes and pigments: electron transfer, radicals, oxidative stress and other physiological effects. *Journal of Applied Toxicology: JAT* 34(8), 825–834. https://doi.org/10.1002/jat.3005

Liang, J.Y., Yuann, J.P., Hsie, Z.J., Huang, S.T., & Chen, C.C. 2017. Blue light induced free radicals from riboflavin in degradation of crystal violet by microbial viability evaluation. *Journal of Photochemistry and Photobiology. B, Biology* 174, 355–363. https://doi.org/10.1016/j.jphotobiol.2017.08.018

Metwally, R.A., El Nady, J., Ebrahim, S., El Sikaily, A., El-Sersy, N.A., Sabry, S.A., & Ghozlan, H.A. 2023. Biosynthesis, characterization and optimization of TiO_2 nanoparticles by novel marine halophilic *Halomonas* sp. RAM2: application of natural dye-sensitized solar cells. *Microbial Cell Factories* 22(1), 78. https://doi.org/10.1186/s12934-023-02093-3

Mognato, M., Burdak-Rothkamm, S., & Rothkamm, K. 2021. Interplay between DNA replication stress, chromatin dynamics and DNA-damage response for the maintenance of genome stability. *Mutation Research/Reviews in Mutation Research* 787, 108346. https://doi.org/10.1016/j.mrrev.2020.108346

Mustafa, Z., Haq, I.U., Nawaz, A., Alessa, A.H., Aftab, M.N., Alsaigh, A.A., & Rehman, A.U. 2024. Laccase production optimization from recombinant *E. coli* BL21 codon plus containing novel laccase gene from *Bacillus megaterium* for removal of wastewater textile dye. *Molecules (Basel, Switzerland)* 29(23), 5514. https://doi.org/10.3390/molecules29235514

Nayyara, P., Permana, D., Ermawar, R.A., & Fahayana, R. 2024. Computational analysis into the potential of azo dyes as a feedstock for actinorhodin biosynthesis in *Pseudomonas putida*. *PLoS One* 19(3), e0299128. https://doi.org/10.1371/journal.pone.0299128

Ngo, A.C.R., & Tischler, D. 2022. Microbial degradation of azo dyes: approaches and prospects for a hazard-free conversion by microorganisms. *International Journal of Environmental Research and Public Health* 19(8), 4740. https://doi.org/10.3390/ijerph19084740

Nodwell, J.R. 2019. Microbe profile: *Streptomyces coelicolor*: a burlesque of pigments and phenotypes. *Microbiology (Reading, England)* 165(9), 953–955. https://doi.org/10.1099/mic.0.000821

Pswarayi, F., & Gänzle, M. 2022. African cereal fermentations: a review on fermentation processes and microbial composition of non-alcoholic fermented cereal foods and beverages. *International Journal of Food Microbiology* 378, 109815. https://doi.org/10.1016/j.ijfoodmicro.2022.109815

Radhakrishnan, A., Gangopadhyay, R., Sharma, C., Kapardar, R.K., Sharma, N.K., & Srivastav, R. 2024a. Unwinding helicase MCM functionality for diagnosis and therapeutics of replication abnormalities associated with cancer: a review. *Molecular Diagnosis & Therapy* 28, 249–264. https://doi.org/10.1007/s40291-024-00701-5

Radhakrishnan, A., Sharma, C., Malviya, V.N., & Srivastav, R. 2024b. Deciphering the molecular functionality of Cdc45 in replisomal complex. *Biochemistry and Biophysics Reports* 37, 101643.

Radhakrishnan, A., Balaganesh, P., Vasudevan, M., Natarajan, N., Chauhan, A., Arora, J., Ranjan, A., Rajput, V.D., Sushkova, S., Minkina, T., Basniwal, R.K., Kapardar, R., & Srivastav, R. 2023. Bioremediation of hydrocarbon pollutants: recent promising sustainable approaches, scope, and challenges. *Sustainability* 15(7), 5847

Roukas, T. 2016. The role of oxidative stress on carotene production by *Blakeslea trispora* in submerged fermentation. *Critical Reviews in Biotechnology* 36(3), 424–433. https://doi.org/10.3109/07388551.2014.989424

Sandmann, G. 2022. Carotenoids and their biosynthesis in fungi. *Molecules (Basel, Switzerland)* 27(4), 1431. https://doi.org/10.3390/molecules27041431

Sathiya Mala, K., Prabhakara Rao, P., Prabhavathy, M.B., & Satyanarayana, A. 2015. Studies on application of annatto (*Bixa orellena* L.) dye formulations in dairy products. *Journal of Food Science and Technology* 52(2), 912–919. https://doi.org/10.1007/s13197-013-1038-3

Saunders, S.H., Tse, E.C.M., Yates, M.D., Otero, F.J., Trammell, S.A., Stemp, E.D.A., Barton, J.K., Tender, L.M., & Newman, D.K. 2020. Extracellular DNA promotes efficient extracellular electron transfer by pyocyanin in *Pseudomonas aeruginosa* Biofilms. *Cell* 182(4), 919–932.e19. https://doi.org/10.1016/j.cell.2020.07.006

Sigurdson, G.T., Tang, P., & Giusti, M.M. 2017. Natural colorants: food colorants from natural sources. *Annual Review of Food Science and Technology* 8, 261–280. https://doi.org/10.1146/annurev-food-030216-025923

Sobczyk, R., Galmiche, L., Mongin, C., Djendli, M., Leray, I., & Méallet, R. 2025. Cyanine dye-embedded fluorescent film for ratiometric pH measurement. *Photochemical & Photobiological Sciences: Official Journal of the European Photochemistry Association and the European Society for Photobiology* 24(1), 165–179. https://doi.org/10.1007/s43630-025-00681-3

Srivastav, R., Kumar, D., Grover, A., Singh, A., Manjasetty, B., Sharma, R., & Taneja, B. 2014. Unique subunit packing in mycobacterial nanoRNase leads to alternate substrate recognitions in DHH phosphodiesterases. *Nucleic Acids Research* 42(12), 7894–7910.

Srivastav, R., Sharma, R., Tandon, S., & Tandon, C. 2019. Role of DHH superfamily proteins in nucleic acids metabolism and stress tolerance in prokaryotes and eukaryotes. *International Journal of Biological Macromolecules* 127, 66–75.

Srivastav, R., Singh, A., Jangir, PK., Kumari, C., Muduli, S., & Sharma, R. 2013. Genome sequence of *Staphylococcus massiliensis* strain S46 isolated from surface of healthy human skin. *Genome Announcements* 1(4), e00553-13.

Srivastav, R., & Suneja, G. 2019. Recent advances in microbial genome sequencing. In: V. Tripathi, P. Kumar, P. Tripathi, A. Kishore, M. Kamle (Eds.), *Microbial Genomics in Sustainable Agroecosystems* (1st ed., Vol. 2). Springer. https://doi.org/10.1007/978-981-32-9860-6

Sulthana, R., Taqui, S.N., Deepa Kumari, H.N., Mir, R.A., Syed, A.A., Saad, H.M., Bashir, M.N., Fouad, Y., Jathar, L., & Shelare, S. (2025). Bioremediation of brilliant green cationic dye from water using Nutraceutical Industrial Coriander Seed Spent as an adsorbent: adsorption isotherms, kinetic models, and thermodynamic studies. *International Journal of Phytoremediation* 27(1), 1–12. https://doi.org/10.1080/15226514.2024.2391949

Suneja, G., & Srivastav, R. 2021. Impact of microbial genome sequencing advancements in understanding extremophiles. *Extreme Environments, Unique Ecosystems—Amazing Microbes* (1st ed., p. 13). CRC Press.

Tomoshige, S., Kobayashi, Y., Hosoba, K., Hamamoto, A., Miyamoto, T., & Saito, Y. 2017. Cytoskeleton-related regulation of primary cilia shortening mediated by melanin-concentrating hormone receptor 1. *General and Comparative Endocrinology* 253, 44–52. https://doi.org/10.1016/j.ygcen.2017.08.021

Tripathi, M., Singh, S., Pathak, S., Kasaudhan, J., Mishra, A., Bala, S., Garg, D., Singh, R., Singh, P., Singh, P.K., Shukla, A.K., & Pathak, N. 2023. Recent strategies for the remediation of textile dyes from wastewater: a systematic review. *Toxics* 11(11), 940. https://doi.org/10.3390/toxics11110940

Varjani, S., Rakholiya, P., Ng, H.Y., You, S., & Teixeira, J.A. 2020. Microbial degradation of dyes: an overview. *Bioresource Technology* 314, 123728. https://doi.org/10.1016/j.biortech.2020.123728

Vyas, H., & Krishnaswamy, G. 2006. Paul Ehrlich's "Mastzellen"--from aniline dyes to DNA chip arrays: a historical review of developments in mast cell research. *Methods in Molecular Biology (Clifton, N.J.)* 315, 3–11.

Wang, Q., Xie, F., Tong, Y., Habisch, R., Yang, B., Zhang, L., Müller, R., & Fu, C. 2020. Dual-function chromogenic screening-based CRISPR/Cas9 genome editing system for actinomycetes. *Applied Microbiology and Biotechnology* 104(1), 225–239. https://doi.org/10.1007/s00253-019-10223-4

Williams, R. M., & Zhang, X. (2021). Roles of ATM and ATR in DNA double strand breaks and replication stress. *Progress in Biophysics and Molecular Biology* 161, 27–38. https://doi.org/10.1016/j.pbiomolbio.2020.11.005

Xu, F., Gage, D., & Zhan, J. (2015). Efficient production of indigoidine in Escherichia coli. *Journal of Industrial Microbiology & Biotechnology* 42(8), 1149–1155. https://doi.org/10.1007/s10295-015-1642-5

Zhou, S.M., Wang, F., Yan, S.Y., Zhu, Z.M., Gao, X.F., & Zhao, X.L. 2023. Phylogenomics and plastome evolution of Indigofera (Fabaceae). *Frontiers in Plant Science* 14, 1186598. https://doi.org/10.3389/fpls.2023.1186598

Zomer, A., Ingham, C.J., von Meijenfeldt, F.A.B., Escobar Doncel, Á., van de Kerkhof, G. T., Hamidjaja, R., Schouten, S., Schertel, L., Müller, K.H., Catón, L., Hahnke, R.L., Bolhuis, H., Vignolini, S., & Dutilh, B.E. 2024. Structural color in the bacterial domain: the ecogenomics of a 2-dimensional optical phenotype. *Proceedings of the National Academy of Sciences of the United States of America* 121(29), e2309757121. https://doi.org/10.1073/pnas.2309757121

15

Microbial Consortia: Enhanced Availability of Micronutrients and Elimination of Biotic Stresses in Plants

Vikash Kumar[1], Lucky Gupta[2], Saurabh Singh[2], Nikhil Raghuvanshi[3], and Lokesh Sisodiya[2]*

[1] *School of Agriculture Science and Technology, Kargil Campus, University of Ladakh, Ladakh, India*
[2] *Faculty of Agricultural Sciences, GLA University, Mathura, Uttar Pradesh, India*
[3] *Department of Agronomy, DDU Gorakhpur University, Gorakhpur, Uttar Pradesh, India*

* Corresponding author: vikasagronomy@gmail.com

15.1 Introduction

Sustainable and resilient agricultural techniques are more important than ever before in light of the increasing worldwide need for food, which is being driven by factors such as climate uncertainty, soil degradation, and escalating insect threats (Saleem et al., 2024). Microbial consortia are one of the many new farming ideas that are showing promise since they help plants deal with biotic challenges while also increasing their availability of micronutrients. Structured communities of two or more microbial species that coexist, frequently in shared settings, are known as microbial consortia (Behera et al., 2021). Consortia like these are more accurate representations of the complex, coevolved interactions between plants and the wide variety of microbes found in the soil and roots than single-strain inoculants. In most cases, these connections are not simple or one-sided but rather represent a symbiotic network of interdependent interconnections that affect plant vitality and yield (Bhatt et al., 2021). For their capacity to provide a wide range of agroecological functions, multi-strain microbial formulations are gaining popularity in the agricultural environment. Rajpal et al. (2022) list soil bioremediation as one of these, while Chen et al. (2023) and Kaur et al. (2022a) list plant growth enhancement, and Upamanya et al. (2020) include pest and disease suppression. The ecological robustness of microbial consortia is one of their main advantages. Combining species with different metabolic features, dietary requirements, and action mechanisms allows consortia to operate better in a wide range of environments (Qian et al., 2021; Trivedi et al., 2021). Farmers now have a more robust biological tool thanks to its stability, which lessens the biological restrictions seen with single-microbe formulations. Many studies have shown that microbial consortia are useful in practice; thus, their advantages aren't simply theoretical. For instance, in chickpeas, a combination of nitrogen-fixing *Rhizobium* species and *Purpureocillium lilacinum*

Microbial Consortia in Plant Science and Sustainable Agriculture, First Edition. Edited by Ajay Tomar, et al.

effectively controlled *Meloidogyne javanica*, a root-knot nematode, while simultaneously promoting plant growth. This effect was not as noticeable when only one organism was used (Maciag et al., 2023).

Curiously, plant extracts are also included in some of the most efficient microbial consortia, which adds a level of biocontrol efficacy. One study found that a bacterial consortium supplemented with extracts of neem (*Azadirachta indica*), *Ipomoea carnea*, and mustard (*Brassica juncea*) effectively suppressed *Meloidogyne incognita* in cucumbers (Mahmoud et al., 2024). Microbial consortia provide an effective means of micronutrient mobilization in addition to pathogen control. Iron, zinc, manganese, and copper are only a few of the vital micronutrients that can be soluble or bound by rhizospheric bacteria and fungi. These minerals are frequently present in soils in inaccessible forms. This helps plants absorb nutrients better and makes them stronger in the long run. According to Sharma et al. (2023), biotic stress is still a major obstacle to agricultural output and food stability, and it is caused by diseases, insects, and parasitic weeds. The overuse of chemical pesticides has caused harm to the environment, the development of insect resistance, and worries about human health, even though they have historically been the first line of defense. According to Syed-Ab-Rahman et al. (2018), this has led to an increase in the popularity of biological alternatives such as biofertilizers, biopesticides, and integrated pest management (IPM) methods, which rely on microbial consortia.

Enzymatic breakdown of pathogen structures, induced systemic resistance (ISR), nutritional competition, and antibacterial chemical synthesis are common biological control methods utilized by these consortia (Chaudhary et al., 2024). These microbial solutions provide sustainable and environmentally safe substitutes for synthetic inputs when used properly. However, there are several obstacles to overcome to put microbial consortia into practice. Persistent success in the field is still elusive because of factors including environmental adaptation, native soil microbiota suppressing competition, and microbial incompatibility (Xu & Yu, 2021). However, microbial consortia have a bright future in agriculture thanks to developments in omics-based research, improvements in microbial formulation technology, and better information regarding interactions in plants and microbes at the system level. In this chapter, we look at how microbial consortia might help crops in two ways: by increasing micronutrient availability and by protecting them from biotic stressors. For academics, agronomists, and policymakers seeking sustainable agricultural solutions, it offers an in-depth examination of these biological tools, exploring their composition, mechanisms, field applications, and prospects.

15.2 Source of Microbial Consortia

Microbial consortia are living, breathing communities of multiple microbial taxa, including bacteria (e.g. *Bacillus subtilis, Pseudomonas fluorescens, Rhizobium leguminosarum*), fungi (e.g. *Trichoderma harzianum, Beauveria bassiana, Metarhizium anisopliae*), actinomycetes (e.g. *Streptomyces* spp.), and others that work synergistically to enhance plant growth, nutrient acquisition, and resistance against a broad spectrum of phytopathogens and insect pests (Kaur et al., 2022a). These consortia are primarily sourced from diverse ecological niches such as the rhizosphere, phyllosphere, endosphere, and even composted organic matter, where microbes naturally coexist and coevolve, exhibiting functional synergy and ecological compatibility (Berg et al., 2020).

In agricultural systems, microbial consortia are isolated from healthy soils, plant roots, decaying plant residues, and insect cadavers. For example, *B. subtilis* and *P. fluorescens* strains are commonly sourced from the rhizosphere for their plant growth-promoting and ISR-inducing capabilities (Pieterse et al., 2014a). Entomopathogenic fungi such as *M. anisopliae* and *B. bassiana* are isolated from infected insect hosts in the soil and are key components in biopesticide formulations (de Faria & Wraight, 2007). Microbes in consortia are selected not just for individual efficacy but for complementary traits such as nutrient solubilization, pathogen suppression, and abiotic stress tolerance. For instance, *Bacillus megaterium* solubilizes phosphorus, *Frateuria aurantia* mobilizes potassium, and *Pseudomonas* spp. produce siderophores that chelate iron-making micronutrients more bioavailable to plants (Rajendran et al., 2012; Vassilev et al., 2006). Actinomycetes like *Streptomyces* spp. are also included for their antibiotic production and disease-suppressive roles (Khan et al., 2023).

Some consortia are designed by artificially combining microbes from different functional groups or phyla. Spescha et al. (2023) demonstrated the compatibility of a tripartite consortium consisting of *Pseudomonas chlororaphis*, the entomopathogenic nematode *Steinernema feltiae*, and the fungus *Metarhizium brunneum*, which enhanced suppression of the cabbage maggot (*Delia radicum*). Native microbial consortia can also be enriched from local soils using selective media and bioassays. Iwanicki et al. (2019) reported that indigenous strains of *M. anisopliae* played a significant role in pest control, even after the application of a commercial isolate, underscoring the value of native sources. These microbial partnerships are naturally evolved systems that provide stability and resilience, outperforming single-strain inoculants in terms of persistence and efficacy. Identifying and employing microbial consortia from various habitats provides an effective alternative for chemical pesticides and fertilizers, which is consistent with eco-friendly and sustainable agriculture practices.

15.2.1 Bacteria

Bacteria are among the most crucial and extensively examined components of microbial communities, significantly contributing to soil fertility, nutrient cycling, and the health of plants. Bacterial strains from various genera such as *Bacillus, Pseudomonas, Rhizobium, Azospirillum, Azotobacter, Bradyrhizobium, Streptomyces, Paenibacillus, Agrobacterium*, and *Burkholderia* have garnered significant attention for their potential roles as plant growth-promoting rhizobacteria and biocontrol agents (Ahemad and Kibert, 2013). Consortia of bacteria significantly help in the availability of micronutrients like zinc, iron, and other micronutrients to be soluble and mobilized. For example, by secreting organic acids, strains such as *B. megaterium* and *P. fluorescens* can solubilize inorganic phosphates, hence boosting Zn, Mn, and P availability to plants (Elhaissoufi et al., 2022).

Species of *Bacillus* and *Azotobacter* improve iron absorption by generating siderophores – low-molecular-weight molecules that chelate Fe^{3+} from the soil and make it available to plants. Particularly in legumes, symbiotic nitrogen fixation and micronutrient availability depend critically on nitrogen-fixing bacteria including *Rhizobium* and *Bradyrhizobium* (Zhang et al., 2025). Free-living nitrogen-fixers such as *Azospirillum* and *Azotobacter* help supply nitrogen for nonlegumes. Some strains of *Paenibacillus polymyxa* have been documented to release K and B from mineral sources, hence improving K and B nutrition. Apart from nutrient solubilization, these bacteria promote plant growth by synthesizing phytohormones that include gibberellins, indole-3-acetic acid (IAA), 1-aminocyclopropane-1-carboxylic acid (ACC) deaminase, and cytokinin, reducing stress-induced ethylene (ET) levels in plants and so promoting root elongation and general biomass accumulation (Aasfar et al., 2021).

Bacterial consortia demonstrate significant biocontrol capabilities by inhibiting a range of phytopathogens, including fungi, bacteria, and nematodes. *Bacillus* and *Pseudomonas* species are commonly employed as biocontrol agents because of their capacity to generate a diverse range of antimicrobial metabolites, including lipopeptides (surfactin, iturin, fengycin), siderophores, volatile organic compounds (VOCs), and hydrolytic enzymes (chitinases, glucanases, cellulases, and proteases) (Dimkić et al., 2022). *Bacillus velezensis* has garnered considerable interest due to its broad-spectrum antagonistic effects against fungal pathogens including *Rhizoctonia solani, Fusarium oxysporum*, and *Sclerotinia sclerotiorum*. The effectiveness of biocontrol is linked to the generation of a variety of bioactive secondary metabolites and the capacity to successfully colonize plant roots. Recent findings indicate that *B. velezensis* QST713 is used commercially to manage green mold disease in mushrooms (*Agaricus bisporus*) caused by *Trichoderma aggressivum* (Cheng et al., 2024). *Pseudomonas* spp. are recognized for their capacity to trigger systemic resistance in host plants. This process enhances the plant's immune response, enabling it to react more efficiently to pathogen threats. The role of enzymes like polyphenol oxidase (PPO), peroxidase (PO), and phenylalanine ammonia-lyase (PAL) in the defense mechanisms of plants has been extensively studied in crops inoculated with *Pseudomonas*. In the realm of nematode management, bacterial strains like *B. subtilis* and *Burkholderia cepacia* have demonstrated efficacy against root-knot nematodes (*Meloidogyne* spp.), leading to a notable decrease in gall formation and nematode population density in both roots and soil. These bacteria are capable of producing compounds that are toxic to nematodes or disrupting the hatching and movement of these organisms (Mehmood et al., 2023).

15.2.2 Fungus

Fungal species are a crucial part of microbial communities, known for their remarkable ability to inhabit various ecological environments, generate bioactive compounds, and establish intricate mutualistic relationships with plant hosts. The fungi in question, especially the endophytic and rhizospheric strains, perform essential functions in the solubilization of micronutrients, the biological control of pests and pathogens, and the enhancement of systemic resistance in plants (Saini et al., 2025). Their presence in microbial consortia significantly boosts plant resilience and productivity. Among the fungal genera, *Trichoderma, Aspergillus, Penicillium, Beauveria, Metarhizium*, and *Verticillium* have attracted significant interest for their capabilities in biocontrol and promoting plant growth. This group of fungi demonstrates various strategies such as competing for nutrients and space, engaging in mycoparasitism, producing antibiotic substances, and boosting plant immune responses. The genus *Trichoderma* is notably one of the most extensively examined and utilized fungi within the realm of agricultural biotechnology (Adedayo et al., 2023). Fungi, especially phosphate and zinc-solubilizing species such as *Aspergillus niger* and *Penicillium citrinum*, facilitate the transformation of insoluble forms of micronutrients into plant-available forms through the secretion of organic acids (e.g. citric, gluconic, and oxalic acids) (Saxena et al., 2013). These acids chelate metallic ions and release them into the soil solution, thereby improving nutrient uptake efficiency. Additionally, fungal hyphae significantly extend the root surface area, enabling access to otherwise inaccessible nutrient zones (Wahab et al., 2023).

Through their extraradical hyphal networks, some arbuscular mycorrhizal fungi (AMF), particularly *Glomus* spp., establish symbiotic relationships with plant roots and improve absorption of copper, zinc, and phosphorus. AMF also improves soil structure, enhances water retention, and helps in alleviating abiotic stresses that include salinity and drought.

Fungal members of microbial consortia perform essential functions in the suppression of plant diseases and insect pests, acting through both direct antagonism and indirect plant defense modulation. Among these, *Trichoderma* spp. are extensively documented for their mycoparasitic activity against several economically important phytopathogens, including *R. solani, S. sclerotiorum, Botrytis cinerea*, and *F. oxysporum* (Harman et al., 2004; Mukherjee et al., 2013). *Trichoderma* species deploy an arsenal of hydrolytic enzymes including β-1,3-glucanases, proteases, and chitinases that degrade fungal cell walls, facilitating parasitism and pathogen lysis (Benítez et al., 2004).

In addition to enzymatic degradation, *Trichoderma* spp. produce secondary metabolites including gliotoxin, viridin, and peptaibols, which exhibit antifungal and antibiotic activity (Vinale et al., 2008). Moreover, they can elicit ISR in host plants by activating salicylic acid and jasmonic acid (JA) pathways, thereby priming the plant immune system against a broad range of pathogens (Shoresh et al., 2010). Entomopathogenic fungi such as *B. bassiana, M. anisopliae*, and *Lecanicillium lecanii* are integral components of insect pest biocontrol strategies. These fungi infect various insect pests including aphids, whiteflies, thrips, and lepidopteran larvae, by adhering to and penetrating the insect cuticle using enzymes such as lipases, proteases, and chitinases (Zimmermann, 2007). Once inside, they colonize the hemocoel, leading to insect mortality followed by external sporulation, which facilitates epizootic transmission to other hosts (de Faria & Wraight, 2007). Notably, *B. bassiana* has been effectively used to control *Helicoverpa armigera* and *Spodoptera litura* in various cropping systems (Yao et al., 2023). These fungal agents are eco-friendly, are highly selective, and pose minimal risk to nontarget organisms, making them a sustainable alternative to chemical pesticides. Their integration into microbial consortia enhances the overall efficacy of biological pest suppression while promoting agroecological balance.

Fungal strains generate a diverse array of bioactive metabolites including terpenoids, lipopeptides, phenolics, alkaloids, and polyketides. *Trichoderma* spp. synthesize compounds such as peptaibols, gliotoxin, viridin, and harzianic acid, which demonstrate significant antifungal and antibacterial properties. These metabolites play a dual role by not only inhibiting pathogens but also acting as triggers for plant immunity (Khan et al., 2020). Endophytic fungi, residing within plant tissues without inducing disease, represent a promising source of novel antimicrobial and growth-promoting compounds. These entities establish close relationships with the host plant and perform pivotal functions in enhancing stress tolerance through modulation of hormonal levels, antioxidant enzyme activities, and uptake of nutrients. *Trichoderma asperellum, T. harzianum*, and *T. virens* exhibit significant potential in enhancing systemic acquired resistance (SAR) and ISR across various crops (Jha et al., 2023).

Fungal incorporation into microbial consortia enhances disease and pest resistance while also promoting microbial stability and survival within the rhizosphere. Their filamentous growth supports physical root colonization and promotes the simultaneous establishment of bacterial partners (Chen et al., 2024). Fungal species serve as structural frameworks for the formation of bacterial biofilms, thereby improving the resilience and longevity of the entire microbial community. The co-inoculation of *Trichoderma,* along with *B. subtilis, P. fluorescens*, or *Azotobacter,* demonstrates enhanced root colonization, increased plant biomass, and improved protection against pathogens when compared to the use of single inoculants. These collaborative groups can be strategically developed to address various agricultural issues at once, such as nutrient shortages, pathogen threats, and environmental stressors.

15.2.3 Actinomycetes

Actinomycetes, particularly those from the genus *Streptomyces*, are filamentous, gram-positive bacteria known for their ability to promote plant growth and provide biocontrol benefits. Their involvement in phosphate solubilization is crucial and frequently associated with their metabolic flexibility. While the majority of actinomycetes do not possess the pyrroloquinoline quinone biosynthetic gene cluster and the related dehydrogenase enzymes required for gluconic acid production through the Entner–Doudoroff pathway, several high-efficiency strains demonstrate alternative mechanisms, including the glyoxylate bypass, for the solubilization of mineral phosphate. *Streptomyces griseoviridis*, originally isolated from sphagnum peat, stands out as one of the most extensively researched biocontrol actinomycetes (Vurukonda et al., 2016). It demonstrates antagonistic properties toward various phytopathogens, such as *Alternaria brassicola, B. cinerea, Fusarium avenaceum*, and *Fusarium culmorum*. Commercial biofungicides such as *Mycostop* utilize this species and are effectively employed in crop protection via root-dip or growth nutrient applications in potted plants, greenhouse vegetables, and cut flowers. Actinomycetes are known for their significant production of phytohormones, especially IAA, which plays a vital role in regulating essential plant developmental processes, including cell elongation, apical dominance, and root hair formation. Various species of *Streptomyces* produce IAA via the indole-3-acetamide (IAM) pathway, processing intermediates such as IAM, indole-3-lactic acid, indole-3-ethanol, and indole-3-acetaldehyde (Brazhnikova et al., 2025). Another significant aspect of actinomycetes is their ability to generate siderophores, which are iron-chelating compounds that improve iron absorption in plants when iron is scarce. For instance, certain species of *Streptomyces* are known to secrete enterobactin and heterobactin, whereas other genera such as *Nocardia* are responsible for the production of nocardichelins, bioactive siderophores that may possess anticancer properties. The synthesis of siderophores such as desferrioxamine B (produced by *Streptomyces pilosus*) has shown significant pharmacological and agronomic importance. Gene knockout studies, including those focused on the desD-like gene in *Streptomyces* sp. GMKU 3100, provides evidence that siderophore biosynthesis is directly associated with enhanced root and shoot biomass in rice and mungbean (Miethke & Marahiel, 2007).

15.2.4 Viruses

Entomopathogenic viruses are gaining attention as effective and environmentally safe biocontrol agents, although their potential in agriculture has only recently been acknowledged. The classification of these viruses is mainly into two categories: inclusion viruses (IVs) and noninclusion viruses (NIVs). Inclusion viruses, including granulosis and polyhedrosis viruses, generate distinctive inclusion bodies within host cells. They are further categorized into nuclear polyhedrosis viruses and cytoplasmic polyhedrosis viruses, based on their replication sites (Chaudhary et al., 2024). Prominent viral genera utilized in biocontrol consist of Baculovirus, Cytoplasmic polyhedrosis virus, Granulovirus, Iridovirus, Parvovirus, Poxvirus, and Polydnavirus, along with RNA viruses including Reovirus, Nodavirus, Picorna-like viruses, and Tetravirus. Baculoviruses stand out as the most widely employed, representing more than 10% of viruses that affect insects. These viruses are characterized by their enveloped structure and double-stranded DNA, possessing a genome size ranging from 80 to 180 kb. Their particle dimensions vary, measuring between 230 and 385 nm in length and 40–60 nm in diameter. Their host specificity and nonpathogenicity to vertebrates render them excellent candidates for IPM (Chaudhari et al., 2021). While viral biopesticides offer certain benefits, they encounter obstacles including elevated costs, a more gradual mechanism of action, and sensitivity to UV exposure. Nonetheless, the effectiveness of these methods in controlling chewing insect pests in vegetable and field crops is thoroughly established (Ahmad et al., 2024).

15.2.5 Algae

Algae, particularly cyanobacteria (blue-green algae), serve as both biofertilizers and components of microbial consortia due to their nitrogen-fixing ability, phytohormone production, and soil-conditioning effects. Genera such as *Anabaena, Nostoc, Aulosira*, and *Tolypothrix* are widely recognized for their symbiotic and free-living roles in agriculture. Cyanobacteria improve soil fertility through biological nitrogen fixation, exopolysaccharide secretion that enhances soil aggregation, and release of organic compounds that promote beneficial microbial colonization (Singh et al., 2016). Moreover, microalgae such as *Chlorella* and *Scenedesmus* are increasingly explored for their plant biostimulant properties. They secrete bioactive molecules like amino acids, vitamins, and hormones (e.g. cytokinins and auxins), which promote seed germination, root elongation, and stress resistance. Algal biomass or extracts have also shown promise in suppressing soilborne pathogens through antimicrobial metabolites, positioning them as an eco-friendly tool for sustainable agriculture (Moon et al., 2024).

15.3 Microbial Consortia for Biotic Stress

Plants encounter a diverse array of phytopathogens throughout their life cycle, which can significantly affect their growth, development, and overall productivity. To endure such biotic stress conditions, plants have developed complex defense mechanisms that operate at both local and systemic levels. The responses are meticulously controlled via intricate signaling networks that assimilate environmental signals and those derived from pathogens, leading to considerable reprogramming at the transcriptomic and metabolomic levels. In response to environmental changes, plants increase the production of primary and secondary metabolites, regulate phytohormone signaling pathways, produce reactive oxygen species (ROS), and modify osmotic and hydraulic balances (Compant et al., 2010; Sahu et al., 2022). A highly effective and environmentally responsible approach to addressing biotic stress is to utilize the advantageous characteristics of microbial consortia. Microbial consortia represent collaborative groups of various advantageous microorganisms that engage with plants, offering enhanced defense against pathogens in contrast to single-strain inoculations. The establishment of pathogens is hindered by these consortia through competition for space and nutrients, the production of antimicrobial metabolites, and the activation of plant immune responses. Plant immune responses can be classified into two primary categories: SAR and ISR. ISR is mainly initiated by beneficial microbes associated with roots, including plant growth-promoting rhizobacteria (PGPR), and is governed by JA and ET signaling pathways (Kaur et al., 2022b; Pršić & Ongena, 2020). The identification of pathogen-associated molecular patterns (PAMPs) results in PAMP-triggered immunity, whereas effector-triggered immunity is characterized by the recognition of specific pathogen effectors by plant resistance (R) proteins (Ding et al., 2022). The presence of microbial consortia notably enhances these layers of immunity, incorporating a diverse array of beneficial bacteria and fungi, including *Bacillus, Pseudomonas, Trichoderma, Azospirillum, Azotobacter, Streptomyces*, and mycorrhizal fungi. Microbial consortia have shown significant efficacy in mitigating biotic stress. For example, the pairing of *Aspergillus spinulosporus* and *Bacillus amyloliquefaciens* has demonstrated promise in alleviating biotic stress and enhancing growth in *Oryza sativa* (Behera et al., 2021). A synergistic effect was noted when *Streptomyces africanus* KAI-32 and *Streptomyces setonii* CAI-127 were co-inoculated, resulting in the suppression of *F. oxysporum* f. sp. ciceri via the activation of the PAL pathway. A recent investigation demonstrated that a tetra-combination of *Trichoderma viride, B. subtilis*, and *P. fluorescens* markedly improved resistance in *Solanum tuberosum* L. by stimulating phenylpropanoid derivatives and redox homeostasis pathways. Furthermore, the simultaneous inoculation of *T. atrobruneum, Pseudomonas* sp., *B. amyloliquefaciens*, and *B. velezensis* in cumin showed significant improvements in disease suppression, plant biomass, and levels of chlorophyll and carotenoids.

The utilization of microbial consortia is essential in the development of seed treatment strategies. A consortium of *Bacillus* spp. coated on pea seeds demonstrated effective control of wilt disease and enhanced plant growth parameters, providing a biobased alternative to chemical seed treatments (Ankati et al., 2021). A consortium consisting of *Pseudomonas putida, B. subtilis*, and *Trichoderma* spp. demonstrated dual suppression of bacterial leaf blight (*Xanthomonas oryzae*) and rice blast (*P. oryzae*), thus contributing to sustainable disease management in rice cultivation (Bestha et al., 2025). The collective findings emphasize that microbial consortia present a strong, environmentally friendly, and sustainable approach for managing biotic stress. Through the enhancement of plant immunity and the promotion of growth, these microbial assemblies play a crucial role in decreasing reliance on chemical pesticides and fostering resilient crop production systems (Jambhulkar et al., 2018).

15.3.1 Microbial Consortia for Plant Diseases

Microbial consortia are being utilized more frequently in agriculture to manage plant diseases effectively. Multiple studies demonstrate that utilizing various microbial strains with different mechanisms of action provides enhanced management of phytopathogens (Table 15.1). For example, Akhtar and Siddiqui (2009) showed that a combined application of the filamentous fungus *P. lilacinum* and nitrogen-fixing *Rhizobium* species

Table 15.1 Microbes involved in disease management.

Microbe	Pathogen controlled	Host	Control mechanism	References
Trichoderma spp.	*Fusarium*, *Rhizoctonia*, *Pythium* spp.	Plants (e.g. tomato, cucumber)	Mycoparasitism, competition, enzyme secretion, ISR	Harman et al. (2004)
Bacillus subtilis	*Xanthomonas*, *Fusarium*, *Botrytis* spp.	Plants (e.g. rice, grapes)	Antibiosis (lipopeptides), biofilm formation, ISR	Ongena and Jacques (2008)
Pseudomonas fluorescens	Root rot, damping-off fungi	Plants (e.g. wheat, maize)	Antibiotic production (phenazines), siderophores, ISR	Weller (2007)
Saccharomyces boulardii	*Clostridium difficile*, *E. coli*	Humans	Toxin neutralization, restoration of gut flora, anti-inflammatory effects	McFarland (2010)
Bacillus thuringiensis	Insect pests (e.g. *Helicoverpa*, mosquitoes)	Crops/animals	Production of crystal (Cry) proteins toxic to insects	Schnepf et al. (1998)
Rhizobium spp.	Indirect control of soilborne pathogens	Legumes	Nitrogen fixation, competition for root colonization	Zahran (2001)
Azospirillum spp.	Soil pathogens	Cereals, grasses	Phytohormone production, root stimulation, pathogen resistance	Bashan and de-Bashan (2010)
Bacteriophages	*Staphylococcus aureus*, *Pseudomonas aeruginosa*	Humans, animals	Targeted bacterial lysis	Abedon et al. (2011)
Mycorrhizal fungi	*Phytophthora*, *Fusarium* spp.	Plants	Nutrient uptake, physical pathogen barrier, ISR	Smith and Read (2008)

improved chickpea resistance to the root-knot nematode *M. javanica* more effectively than when each was applied separately. The enhanced resistance was linked to the parasitism of nematode eggs and females by *P. lilacinum*, along with the antimicrobial compounds and phytoalexins produced by *Rhizobium*. In a similar vein, Guetsky et al. (2001) found that the combination of the yeast *Pichia guilliermondii* and *Bacillus mycoides* effectively suppressed grey mold (*B. cinerea*) in strawberries through various mechanisms, such as nutrient competition and the inhibition of conidial germination. Panpatte et al. (2021) integrated botanical extracts like *A. indica, I. carnea*, and *B. juncea* into a bacterial consortium that included *Providencia vermicola* and two strains of *Pseudomonas*, resulting in the most significant reduction of *M. incognita* eggs in cucumber. Kaewsalong et al. (2019) demonstrated that a combination of *T. harzianum, T. asperellum*, and *C. fenestratum* extract has been efficacious in managing diseases associated with *Fusarium moniliforme, Curvularia lunata, Alternaria padwickii*, and *Bipolaris oryzae* in rice.

15.3.1.1 Induced Systemic Response

Plants possess the remarkable ability to enhance their defensive capacity upon stimulation by both pathogenic and nonpathogenic microorganisms. This enhanced immunity typically manifests as a restriction in pathogen proliferation and a reduction in disease severity (Nishad et al., 2020). Induced resistance can be triggered by various factors, including chemical agents, nonpathogenic microbes, attenuated forms of pathogens, or avirulent strains, particularly when environmental conditions delay infection onset (Table 15.2) (Vallad & Goodman, 2004). Plants employ several types of resistance strategies in response to these environmental stimuli, which primarily fall into ISR and SAR (Durrant & Dong, 2004; Conrath, 2006).

ISR is typically activated by beneficial microbes, such as PGPRs and mycorrhizal fungi, without causing any visible disease symptoms. In contrast, SAR is usually initiated following a pathogen attack, leading to the activation of defense pathways throughout the plant (Beceiro et al., 2013; Harman et al., 2004). A critical molecular distinction between ISR and SAR lies in their signaling molecules: while SAR predominantly relies on salicylic acid (SA) as a signal molecule, ISR is generally mediated by JA and ET pathways (Foyer & Noctor, 2005; Mittler, 2002). The role of SA in SAR is well established, and inhibition of SA biosynthesis significantly compromises SAR efficacy (Vlot et al., 2009). Interestingly, rhizobacteria that produce SA at the root surface have been shown to trigger SAR-like responses in host plants (Nishad et al., 2020). Molecular studies, especially those on *Arabidopsis thaliana*, have identified NPR1 (Nonexpressor of Pathogenesis-Related Genes 1) as a central regulatory node connecting the SAR and ISR pathways, although ISR signaling networks are generally more complex and multifaceted than those of SAR (Pieterse et al., 2014a). Furthermore, ISR is uniquely associated with beneficial soil microbes rather than pathogenic ones. Numerous bacterial genera, which include *Bacillus, Pseudomonas*, and *Streptomyces*, along with beneficial fungi like *Trichoderma* spp. and *Serendipita indica*, have been reported for inducing systemic resistance in host plants (Vurukonda et al., 2016; Azcón-Aguilar & Barea, 1997). These beneficial microbes can reinforce plant defense by strengthening cell wall structures, enhancing mechanical barriers, or altering physiological and metabolic processes that lead to synthesizing antimicrobial compounds (Sharma et al., 2012; Upadhyay et al., 2012).

Table 15.2 Key differences between systemic acquired resistance (SAR) and induced systemic resistance (ISR). Adapted from Durrant and Dong (2004).

Feature	ISR	Systemic acquired resistance (SAR)
Trigger	Beneficial nonpathogenic microbes such as PGPRs and mycorrhizal fungi (Pieterse et al., 2014a; Van Wees et al., 2008)	Pathogenic attack causing localized necrosis (Durrant & Dong, 2004)
Signaling molecules	Jasmonic acid (JA) and ethylene (ET) (Pieterse et al., 2014b; Conrath, 2006)	Salicylic acid (SA) (Durrant & Dong, 2004; Vlot et al., 2009)
Key regulatory gene	NPR1 mediates ISR response (Pieterse et al., 2014b)	NPR1 also governs SAR (Durrant & Dong, 2004; Vlot et al., 2009)
Primary defense response	Priming of defense genes for rapid response upon attack (Conrath, 2006)	Systemic expression of pathogenesis-related (PR) proteins (Vallad & Goodman, 2004)
Induction method	Root colonization by beneficial microbes (Van Wees et al., 2008)	Infection at one site triggers systemic resistance (Durrant & Dong, 2004)
Characteristic feature	No direct defense activation; primed state achieved (Conrath, 2006)	Activation of systemic defense responses including PR proteins (Vlot et al., 2009)
Examples of inducers	*Pseudomonas fluorescens, Bacillus subtilis, Trichoderma* spp., *Serendipita indica* (Pieterse et al., 2014a)	Tobacco mosaic virus, *Pseudomonas syringae* (Durrant & Dong, 2004)
Pathogens controlled	Broad-spectrum including fungi, bacteria, viruses, and nematodes (Pieterse et al., 2014a)	Mainly biotrophic pathogens like fungi and viruses (Vallad & Goodman, 2004)
Type of immunity	Primed and faster activation of defense (Van Wees et al., 2008)	Preactivated defense across the entire plant (Durrant & Dong, 2004)
Energy cost to plant	Low, only activated when needed (Pieterse et al., 2014b)	High, continuous production of PR proteins (Conrath, 2006)
Duration	Moderate to long-lasting, depending on microbe persistence (Van Wees et al., 2008)	Long-lasting but may diminish over time (Vallad & Goodman, 2004)
Pathway cross-talk	Interaction with SAR via NPR1 and partial overlap of SA–JA–ET signaling (Pieterse et al., 2014b)	Limited cross talk with JA–ET pathways (Vlot et al., 2009)
Main defense compounds	Strengthened cell wall, secondary metabolites, and defense enzymes (Van Wees et al., 2008)	PR proteins that include β-1,3-glucanases and chitinases (Durrant & Dong, 2004)
Discovery	Observed during plant–microbe beneficial interaction studies (Pieterse et al., 2014a)	Discovered in response to pathogen infection studies (Durrant & Dong, 2004)

The activation of ISR often involves the upregulation of defense-related enzymes, including chitinases, POs, and proteinase inhibitors, contributing to both direct pathogen inhibition and fortification of plant tissues (Kumawat et al., 2022). Additionally, exposure to methyl jasmonate and ET precursor ACC significantly enhances plant resistance to a broad range of pathogens, highlighting the importance of JA–ET pathways in ISR activation (Tufail et al., 2022). One prominent example is *P. fluorescens* WCS417r, which can induce ISR in *Arabidopsis* by influencing NPR1 activity and modulating JA–ET-mediated signaling cascades (Pieterse et al., 2014a; Dimkpa et al., 2009). Furthermore, by increasing the activity of defense enzymes including PAL, PPO, PO, β-1,3-glucanases, and chitinases, PGPR strains have been demonstrated to generate systemic resistance (Beneduzi et al., 2012). Interestingly, there is evidence of cross talk between SA-dependent SAR pathways and JA–ET-mediated ISR pathways, suggesting a degree of overlap in their downstream responses (Van Wees et al., 2008). For instance, *P. fluorescens* CHA0 has been shown to trigger both SAR and ISR-like resistance in tobacco plants, where bacterial production of SA acts not only as a siderophore under iron-limited conditions but also contributes to host defense activation (Beneduzi et al., 2012; Asada, 2006). Thus, the ability of plants to integrate multiple signaling pathways in response to both beneficial and pathogenic microorganisms highlights the sophistication of their immune systems and opens new avenues for the development of sustainable disease management strategies in agriculture.

Two critically important processes by which plants protect themselves from various biotic stressors are ISR and SAR. ISR is mostly induced by beneficial rhizosphere bacteria via ET and JA communication, as opposed to SAR, which is primarily caused by pathogenic assault and depends on salicylic acid accumulation. Importantly, although both resistance mechanisms use different upstream signaling pathways, they both culminate at the NPR1 regulatory node (Durrant & Dong, 2004; Vlot et al., 2009). In comparison to SAR's high energetic demands, ISR's cheap energy cost makes it an interesting option for sustainable crop protection measures. Research into the molecular mechanisms and interactions between these pathways has the potential to lead to new plant kinds that are resistant to diseases (Van Wees et al., 2008; Conrath, 2006).

15.3.1.2 ROS and Antioxidant Activity

Reactive species are a broad category of extremely reactive molecules that include free radicals and their nonradical derivatives. They are primarily divided into three groups: reactive sulfur species, reactive nitrogen species, and ROS. These entities are continuously produced in plant cells as natural by-products of various metabolic pathways. However, when their production exceeds the plant's scavenging capacity, it results in oxidative stress, leading to potential cellular damage (Compant et al., 2010). Among these, ROS forms a particularly critical class, encompassing both radical species that includes hydroxyl radical (OH), organic peroxyl (ROO), and superoxide anion (O_2^-), in addition to nonradical molecules including hypochlorous acid (HOCl), molecular oxygen (O_2), ozone (O_3), and hydrogen peroxide (H_2O_2) (Sahu et al., 2022; Sharma et al., 2012).

ROS constitutes significant signaling molecules that control plant growth, development, and defense responses, despite their reputation to be damaging agents (Foyer & Noctor, 2005). However, when produced in excess, ROS can oxidize lipids, proteins, and nucleic acids, ultimately compromising cellular integrity and viability. Therefore, plants rely heavily on a sophisticated antioxidant defense system to maintain redox homeostasis. Antioxidants, either enzymatic or nonenzymatic, act by neutralizing ROS, thereby protecting cellular components from oxidative injuries. Low-molecular-weight substances including ascorbic acid, glutathione, phenolic compounds, carotenoids, tocopherols, and tocotrienols are examples of nonenzymatic antioxidants. These compounds function in redox buffering in plant cells and directly scavenge ROS. As the first line of defense against oxidative stress, enzymatic antioxidants, on the other hand, comprise various essential enzymes, including glutathione PO (GPx; EC 1.11.1.9), ascorbate PO (APX; EC 1.11.1.11), catalase (CAT; EC 1.11.1.6), and superoxide dismutase (SOD) (Mittler, 2002). These enzymes exhibit remarkable efficiency in detoxifying superoxide radicals and hydrogen peroxide, thereby mitigating oxidative damage.

The importance of antioxidant mechanisms is further demonstrated through plant–microbe interactions. For instance, the root endophytic fungus *Piriformospora indica* has been shown to elevate antioxidant enzyme levels in host plants like Chinese cabbage and barley, enhancing their drought and salt stress tolerance (Jomova et al., 2023). Under optimal conditions, ROSs are produced at low, manageable levels primarily as by-products of photosynthesis and respiration (Asada, 2006). However, the photosynthetic electron transport chain is significantly disrupted by stressors, including drought, salinity, chilling, and intense light, resulting in ROS generation. The oxidative burst triggered during such stress conditions can cause lipid peroxidation, membrane destabilization, and ultimately cell death if not controlled appropriately (Sahu et al., 2022). Microbial inoculation has emerged as an effective strategy to enhance plant antioxidant capacity. Numerous studies have demonstrated that PGPRs prime plants to upregulate antioxidant defenses, thus conferring improved resilience against environmental stresses (Vurukonda et al., 2016). For example, a hexa-microbial consortium consisting of *T. harzianum, P. polymyxa, P. putida, B. megaterium, P. fluorescens*, and *P. aeruginosa* had been employed to inoculate tomato plants. This resulted in a considerable increase in crop yield, soil fertility, drought tolerance, and antioxidant enzyme activities.

T. harzianum inoculation under chilling stress conditions also improved photosynthetic performance and growth rates in tomato plants by decreasing lipid peroxidation, ROS accumulation, and electrolyte leakage while simultaneously increasing proline and leaf water content (Mukherjee et al., 2013; Harman et al., 2004). In rice (*O. sativa*), *Bacillus pumilus* inoculation resulted in improved plant growth and elevated antioxidant enzyme synthesis, along with reduced sodium ion accumulation under salt stress (Upadhyay et al., 2012). Further research has emphasized the potential of *B. subtilis* in alleviating drought stress in potato plants. Upon bacterial inoculation, research discovered that total soluble sugar content increased, PO, SOD, and catalase activity increased, and ROS and mono-dehydroascorbate concentrations decreased (Benítez et al., 2004; Dimkpa et al., 2009).

The combined inoculation of maize plants with *S. indica, A. chroococcum*, and *R. intraradices* improved various physiological and biochemical traits compared to uninoculated controls. These plants exhibited significantly higher levels of antioxidant enzymes that include catalase, PO, superoxide dismutase, and PPO, ultimately contributing to better growth and stress tolerance (Jomova et al., 2023; Azcón-Aguilar & Barea, 1997). Collectively, these findings emphasize the crucial role of ROS and antioxidant machinery in plant stress responses and highlight the promising potential of microbial inoculants as bioenhancers to bolster antioxidative defenses in crops under adverse environmental conditions.

15.3.1.3 Induction of Plant Osmolyte Synthesis

Drought, salinity, and severe temperatures impair plant water potential and ion balance. To counteract these adverse effects, plants have developed two fundamental adaptive strategies: osmoregulation and osmoprotection. Osmoregulation involves the active adjustment of internal solute concentrations to maintain water uptake, while osmoprotection focuses on safeguarding cellular structures through the accumulation of compatible solutes, known as osmolytes. The synthesis of osmolytes leads to a reduction in osmotic potential (Ψ), facilitating the maintenance of cellular turgor and overall plant vigor. These osmolytes include a broad array of compounds including polyamines (e.g. spermidine), sugar alcohols (e.g. mannitol, sorbitol), amino acids (e.g. proline, glutamate), soluble sugars (e.g. sucrose, trehalose), and quaternary ammonium compounds like glycine betaine. Osmolytes not only stabilize proteins and membranes but also scavenge ROS, thereby enhancing the plant's defense mechanisms against oxidative damage (Ahmad et al., 2020).

Given its complex function in stress tolerance, proline is perhaps the most extensively investigated osmolyte. It acts as an osmoprotectant, antioxidant, signaling molecule, and stabilizer of subcellular structures. Under conditions of osmotic stress, a significant accumulation of proline has been observed in several plant species, contributing to enhanced stress resilience. Studies have shown that beneficial microorganisms such as *Pseudomonas* spp. and *Rhizobium* spp. can promote the endogenous synthesis of proline and other osmoprotectants. In *Zea mays* L., for instance, microbial inoculation led to improved water and ion homeostasis, thereby boosting biomass production, nutrient uptake, and overall plant development (Liang et al., 2013). Munir et al. (2022), who worked on *Arthrobacter* strains EZB4, EZB18, and EZB20, demonstrated that these beneficial microbes enhance proline accumulation in *Capsicum annuum* L. under abiotic stress, suggesting their potential use as bioinoculants for stress management in agriculture. In addition to proline, trehalose serves as an important osmoprotectant. During a period of stress, trehalose, a nonreducing disaccharide composed of two glucose units, serves as a reserve of energy. Trehalose production in microorganisms is primarily mediated by the trehalose-6-phosphate synthase (TPS)/trehalose-6-phosphate phosphatase (TPP) pathway. Trehalose enables plant cells to maintain osmotic equilibrium, stabilize membranes, and protect enzymes from stress (Munir et al., 2022). Application of *Pseudomonas* strains has been linked to enhanced trehalose accumulation, along with the production of ACC deaminase, leading to improved biomass and chlorophyll content in plants.

Microbial–plant interactions often induce the release of organic acids such as citric, oxalic, and malic acids from root systems. These organic acids mobilize essential minerals like potassium and magnesium, enhancing nutrient uptake and sustaining osmoregulation. Recent studies also highlight the synergistic potential of microbial consortia. For instance, the co-application of *Enterobacter* potassium solubilizing bacteria (KSB) with native symbiont *Bradyrhizobium japonicum* significantly enhanced soybean growth under salt stress by improving osmotic adjustment mechanisms and promoting better physiological functioning (Feng et al., 2023). Overall, the induction of osmolyte synthesis, facilitated by microbial inoculation, represents a promising and sustainable strategy to enhance plant tolerance to abiotic stresses, thereby contributing to stable crop productivity in challenging environments.

15.3.2 Microbial Consortia for Pest

The potential of microbial consortia extends beyond phytopathogen control and into the domain of entomopathogenic pest management. Application of microbial agents that include bacteria, fungi, and nematodes to manage insect pests is a long-standing practice, initially demonstrated with a control. While single-strain applications are common, research increasingly supports the superior efficacy of microbial consortia, especially due to the enhanced virulence and production of immunosuppressive and antioxidative metabolites. Soth et al. (2022) found that although four *Beauveria* species exhibited low virulence individually, their combination resulted in 100% mortality of the diamondback moth (*Plutella xylostella*), evidencing a strong synergistic impact. Similarly, Stuart et al. (2022) reported increased mortality of *Duponchelia fovealis* larvae when treated with a consortium of *B. bassiana* strains Bov 2 and Bov 3. This enhanced effect was linked to elevated levels of chitinase and lipase, which compromise the insect cuticle and suppress immune responses.

Field applications reveal additional complexities. Iwanicki et al. (2019), using microsatellite molecular markers, showed that native *M. anisopliae* strains significantly contributed to the control of *Mahanarva* spp. in sugarcane, even after the application of a commercial strain (ESALQ1604). This highlights the importance of understanding interactions between introduced and resident entomopathogens for optimizing field performance. Multitrophic microbial consortia, including members from different phyla, also demonstrate potential in IPM. Spescha et al. (2023) investigated a tripartite consortium comprising the bacterium *P. chlororaphis*, entomopathogenic nematode *S. feltiae*, and fungus *M. brunneum* against cabbage maggot (*D. radicum*). Their study confirmed the compatibility of these agents and suggested synergistic effects, thereby broadening the scope for designing complex microbial consortia in sustainable pest control strategies.

15.3.2.1 Mechanisms of Insect Pest Suppression by Microbial Consortia

Microbial consortia have emerged as promising tools in sustainable agriculture for the suppression of insect pests, offering an ecologically sound alternative to synthetic chemical pesticides. These consortia, composed of multiple microbial strains or species, interact synergistically to enhance pest suppression through various mechanisms, including the production of insecticidal metabolites, enzyme secretion, immune modulation, and competition for ecological niches. One of the primary mechanisms by which microbial consortia suppress insect pests is through the production of insecticidal secondary metabolites. These include toxins, antibiotics, and VOCs that interfere with the physiology and behavior of insects. For example, species of *Beauveria* and *Metarhizium* produce destruxins and beauvericin, which act as neurotoxins disrupting insect nervous systems (de Faria & Wraight, 2007). When used in consortia, the combined metabolite profiles can exhibit additive or synergistic toxic effects. Stuart et al. (2022) reported that *B. bassiana* strains (Bov 2 and Bov 3) consortium exhibited elevated production of chitinase and lipase, enhancing virulence against *D. fovealis*. Enzymatic degradation of the insect cuticle represents another crucial pest suppression mechanism. Insect cuticles are composed primarily of chitin and proteins, forming a protective barrier. Microbes such as entomopathogenic fungi secrete hydrolytic enzymes like chitinases, proteases, and lipases that degrade these components, facilitating fungal penetration and colonization (Sharma et al., 2020). Combined microbial strains increase enzyme synthesis and cuticle degradation. Soth et al. (2022) showed that a four-species *Beauveria* consortium achieved 100% mortality of *P. xylostella* due to elevated enzymatic activity. Another essential mechanism involves immune system modulation in the insect host. Entomopathogens produce compounds that suppress the insect immune response, facilitating microbial colonization and reproduction. For instance, *M. anisopliae* produces destruxins that interfere with hemocyte function and reduce the phenoloxidase activity involved in melanization, a key insect immune defense (Stuart et al., 2022). The use of microbial consortia often amplifies this immunosuppressive effect, increasing pest mortality rates. Competition for ecological niches and resources is also a critical indirect mechanism. Microbial consortia can colonize the insect cuticle or gut, thereby outcompeting pathogenic or symbiotic microbes essential for insect survival. For example, *Bacillus thuringiensis* can inhibit the growth of midgut microbiota, impairing nutrient assimilation in larvae (Broderick et al., 2009). When combined with fungi or nematodes, this competitive exclusion can significantly compromise pest health and development. Recent studies have also emphasized the role of multitrophic microbial interactions in enhancing pest suppression. Spescha et al. (2023) indicated that a tripartite consortium consisting of *M. brunneum, S. feltiae*, and *P. chlororaphis*, showed increased effectiveness against *D. radicum* due to synergistic interactions across microbial kingdoms. Such combinations enable the targeting of pests at multiple biological levels (e.g. cuticle penetration, gut colonization, systemic infection), thereby increasing the robustness and reliability of pest control (Mikiciuk et al., 2024).

Microbial consortia may influence insect behavior through the emission of VOCs that act as repellents or disrupt host-seeking and feeding patterns. For instance, *P. fluorescens* produces 2,4-diacetyl phloroglucinol, a VOC known to affect insect behavior and survival (Haas & Défago, 2005). When included in microbial consortia, such effects can be enhanced or diversified. These results demonstrate that microbial consortia may reduce insect pests in various methods. Group-based biocontrol in IPM techniques reduces chemical input and pest resistance, making pest control sustainable and resilient. Future research should focus on optimizing microbial combinations, understanding their ecological interactions, and enhancing field-level performance through advanced formulation and delivery systems.

15.4 Microbial Consortia for Micronutrient Availability

15.4.1 Siderophore Production and Micronutrient Availability

Under iron-limiting conditions, microorganisms synthesize low-molecular-weight, high-affinity iron-chelating siderophores (Table 15.3). These molecules bind ferric ions (Fe^{3+}) and facilitate their transport into microbial cells. The term "siderophore," introduced

Table 15.3 Microorganisms involved in micronutrient availability.

Microorganism	Micronutrients	Role in plant growth	Mechanism of action	References
Rhizobium spp.	Fe, Mo	Nitrogen fixation; micronutrient mobilization	Siderophore production; Mo needed for nitrogenase	Zahran (2001)
Azotobacter spp.	Zn, Fe	Nitrogen fixation; Zn solubilization	Organic acids and siderophores	Vessey (2003)
Azospirillum spp.	Fe, Zn, Cu	Root development, growth promotion	Siderophores, indole-3-acetic acid production	Bashan and de-Bashan (2010)
Penicillium spp.	Zn, Fe	Mineral solubilization	Secretion of citric and oxalic acids	Whitelaw (2000)
Frankia spp.	Fe, Mo	Nitrogen fixation in nonlegumes	Enhances Fe/Mo uptake for nitrogenase	Benson and Silvester (1993)
Mycorrhizal fungi (AMF)	Zn, Cu, Mn, Fe, P	Nutrient uptake, drought tolerance	Hyphal extensions, metal mobilization	Smith and Read (2008)
Thiobacillus spp.	Fe, Mn, Cu	Sulfur oxidation; micronutrient release	Produces sulfuric acid; pH reduction	Ghosh and Dam (2009)
Gluconacetobacter diazotrophicus	Fe, Zn, Cu	Nitrogen fixation; growth promotion	Produces gluconic acid and siderophores	Cavalcante and Döbereiner (1988)
Burkholderia spp.	Fe, Zn	Nutrient mobilization	Siderophore and organic acid production	Compant et al. (2005)
Streptomyces spp.	Zn, Cu, Fe	Soil health, nutrient cycling	Siderophore and secondary metabolite production	Jog et al. (2014)

in 1973, encompasses a wide range of structurally diverse molecules classified primarily based on the nature of their iron-chelating ligands: catecholates, hydroxamates, carboxylates, and mixed-type siderophores (Chand et al., 2025). Microbes and plants require iron for respiration, chlorophyll production, and enzymatic activities. Due to insoluble ferric hydroxide forms, iron is scarce in soils, especially under aerobic and high pH conditions, despite its abundance in Earth's crust. Siderophores generated by PGPRs and other bioinoculants help plants and related microbial communities absorb iron from these unavailable forms (Dhankhar et al., 2022).

Rhizosphere micronutrient management is enhanced by siderophore-producing microbial consortia. These consortia not only improve iron acquisition but also enhance the bioavailability of other essential micronutrients such as zinc, manganese, cobalt, molybdenum, and nickel through similar chelation mechanisms. For example, siderophores possess electron-rich donor atoms, such as oxygen or nitrogen, which facilitate metal binding and transport. Nonessential and potentially toxic heavy metals, including Al, Cd, Cu, Pb, and Zn may be chelated to reduce their impact on plant development (Khoso et al., 2024). In plant–microbe interactions, siderophores function as key tools for microbial competitiveness in the rhizosphere. By sequestering iron from the surrounding environment, beneficial microbes not only fulfill their own nutritional needs but also limit iron availability to phytopathogens, thus exerting a biocontrol effect. This iron deprivation inhibits pathogen growth, sporulation, nucleic acid synthesis, and cell wall development, contributing to overall disease suppression.

In addition, metal stress increases siderophores biosynthesis. Soils contaminated with heavy metals typically exhibit iron deficiency symptoms in plants, yet microbial siderophores can alleviate this stress by enhancing iron solubilization and uptake. PGPRs, including *Streptomyces, Bacillus, Enterobacter, Pseudomonas, Bradyrhizobium, Rhizobium,* and *Serratia,* have been widely recognized for their siderophore production and ability to modulate iron dynamics in the rhizosphere (Garg et al., 2021).

Plants, too, exhibit mechanisms for iron acquisition, such as the release of phytosiderophores in dicots and the activity of root membrane reductases in monocots, which convert Fe^{3+} to the more soluble Fe^{2+} form. However, microbial siderophores often act synergistically with plant mechanisms to optimize iron availability. Additionally, siderophore–Fe complexes are reduced and taken up by plant and microbial cells, ensuring a continuous supply of bioavailable iron (Crowley, 2006). Microbial consortia that include multiple siderophore-producing organisms offer a robust strategy for enhancing micronutrient availability under diverse soil conditions. Such consortia can adapt to varying pH levels, compete effectively against native soil microbes, and collectively contribute to improved plant nutrition, growth, and resilience against environmental stresses. Overall, siderophore-mediated micronutrient mobilization is a critical component of sustainable agricultural practices and biofertilizer formulations (Ahmed et al., 2014).

15.4.2 Zinc-solubilizing Bacteria for Zinc Solubilization and Micronutrient Availability in Crops

Zinc-solubilizing bacteria (ZSB) have garnered significant attention in recent years for their potential to enhance zinc (Zn) bioavailability in soils and improve plant nutrition, thereby addressing a major micronutrient deficiency that affects crop yield and human health. These PGPRs colonize plant roots and enhance development via direct and indirect mechanisms (Sethi et al., 2025). Zinc deficiency is prevalent in many agricultural soils worldwide, particularly in calcareous and alkaline soils, where Zn is often present in insoluble forms such as zinc carbonate, zinc oxide, and zinc phosphate. ZSB has the remarkable ability to convert these unavailable forms into bioavailable ones through mechanisms such as organic acid secretion (citric acid, gluconic acid, oxalic acid, etc.), proton extrusion, and the release of chelating agents that lower the pH and dissolve mineral-bound Zn (Masood et al., 2022). These biochemical processes not only improve Zn uptake by plants but also contribute to improved plant metabolism, chlorophyll synthesis, and stress resistance. Several PGPR genera that include *Azospirillum, Pseudomonas, Burkholderia, Bacillus,* and *Enterobacter* have presented promising Zn-solubilizing potential across various agro-ecological zones

(Khoso et al., 2024). The integration of such bioinoculants into crop production systems has been shown to significantly enhance biomass, grain Zn content, and overall productivity. For instance, *Pseudomonas* species were found to enhance Zn availability in wheat, leading to better growth and yield attributes (Kumari et al., 2023), while *B. cepacia* improved Zn uptake in rice and increased dry matter accumulation (Upadhayay et al., 2022). In chickpeas, *Bacillus altitudinis* not only mobilized Zn but also improved nodulation and nitrogen fixation, indicating multifunctional benefits (Upadhayay et al., 2025). The application of endophytic strains such as *Pseudomonas plecoglossicida* in pigeon pea demonstrated that these microbes could efficiently colonize internal plant tissues and directly transfer solubilized Zn to developing tissues (Tripathi et al., 2022). Notably, the use of microbial consortia-combinations of compatible microbial strains has been shown to outperform single-strain inoculants in both laboratory and field conditions. Synergistic microbe interactions enhance Zn solubilization, plant hormone synthesis (e.g. indole-3-acetic acid, gibberellins), and pathogen control in these consortia (Liu et al., 2023). A well-documented example includes the co-inoculation of *Burkholderia* and *Acinetobacter* species, which significantly increased Zn accumulation in wheat grains (Upadhayay et al., 2022). Similarly, a microbial mixture involving *Enterobacter* and *Serratia marcescens* resulted in a 32% increase in Zn uptake under pot trials and a 23% increase in field trials, suggesting potential for large-scale application (Bhatta et al., 2023). Moreover, co-application of ZSB with inorganic Zn fertilizers like $ZnSO_4$ or ZnO nanoparticles has been reported to further improve Zn use efficiency and plant uptake. This integrated approach has been found effective in crops such as rice, maize, and wheat, as it combines the immediate availability of mineral Zn with the prolonged solubilization and mobilization capabilities of microbes (Verma et al., 2022). *Cyanobacteria*-based biofilm formulations, such as the *Anabaena-Azotobacter* consortium, have also shown enhanced Zn biofortification in maize, illustrating the scope of microbial innovations in micronutrient management (Bidyarani et al., 2016). Another important consideration is the formulation and delivery mechanism of these microbial biofertilizers. Carrier-based formulations, seed-coating techniques, and biochar-supported inoculants are being explored to enhance the shelf life, survivability, and field performance of ZSB. Overall, the integration of ZSB into modern nutrient management practices represents a sustainable and eco-friendly strategy to tackle Zn deficiency in agriculture, reduce dependence on chemical fertilizers, and achieve nutrient-rich crop production. Future research should emphasize the development of region-specific consortia, molecular characterization of efficient ZSB strains, and field-scale validation under diverse agroclimatic conditions to ensure consistent performance and scalability (Saif et al., 2021).

15.4.3 Other Micronutrients

Micronutrients like molybdenum (Mo), nickel (Ni), boron (B), and manganese (Mn) are vital for plant physiological functions, although required in trace amounts. Their availability is often limited in soils due to various abiotic stresses or imbalanced fertilization, making microbial mediation a promising strategy for enhancing their uptake and maintaining homeostasis. Molybdenum is taken up by plants in the form of molybdate ions (MoO_4^{2-}), and although it is among the least abundant essential elements in plant tissues apart from copper, its deficiency is relatively rare in most cropping systems (Lilay et al., 2024). Nevertheless, when deficiencies do occur, they impact the functioning of Mo-dependent enzymes such as nitrate reductase. Supplementation through foliar sprays can restore Mo levels, but microbial assistance offers a more sustainable solution. Studies have demonstrated that AM fungi like *Claroideoglomus etunicatum* significantly enhance Mo uptake. This fungus strain's inoculation increased concentrations of Mo in roots and shoots of sweet sorghum cultivated in soil spiked with MoS_2 (Kovács et al., 2015). Similarly, in maize grown in soil treated with varying concentrations of ammonium molybdate ($(NH_4)_2\ MoO_4$), *C. etunicatum* boosted Mo content while also modulating the shoot-to-root Mo ratio under moderate to severe contamination scenarios, indicating a buffering role of AM fungi under excessive metal exposure. Nickel (Ni), although only required in trace amounts for the functioning of the enzyme urease in higher plants, often enters the plant-soil system through fertilizer additives such as rock phosphate, which naturally contains Ni ranging from 16.8 to 50.4 mg/kg (Prasad et al., 2011). While mechanisms governing Ni uptake remain poorly understood, Ni contamination due to industrial and agricultural sources poses phytotoxic threats. Therefore, microbial research in this context focuses more on stress mitigation rather than direct nutrient bioavailability. Certain PGPRs and AM fungi can induce tolerance to Ni stress by reducing its uptake or enhancing antioxidant defense mechanisms in plants (Ahmed et al., 2020). Such microbe-assisted phytostabilization strategies are gaining importance for their dual role in nutrient management and environmental detoxification.

Boron (B), absorbed by plants in its uncharged boric acid form, is another micronutrient with a narrow window between deficiency and toxicity. Though B deficiencies are widespread and can be managed through soil or foliar applications, the risks of overapplication necessitate alternative methods to regulate their uptake. Beneficial microbes, especially AM fungi like *Rhizophagus irregularis*, have shown potential in modulating B homeostasis. For instance, in maize subjected to varying B levels (25 and 100 μM) under drought and well-watered conditions, inoculation with *R. irregularis* improved plant dry weight irrespective of water regime, suggesting an improved balance of B uptake and utilization (García-Sánchez et al., 2020). This effect demonstrates that microbial associations can enhance the efficacy of micronutrient use by plants, even under stress conditions.

Manganese (Mn), an essential micronutrient and a critical component in industries like steel, batteries, and glass, is often mined from high-grade ores. With growing demand surpassing the supply of quality Mn ores, and improper management of mining tailings causing widespread environmental degradation, microbial recovery of Mn from low-grade ores has emerged as a sustainable alternative (Khoshru et al., 2023). Laboratory-scale experiments have shown that certain bacterial strains like *Acinetobacter* MSB5 and *Lysinibacillus* MSB11 are effective bioleachers of Mn. Among them, *Acinetobacter* MSB5 exhibited superior bioleaching potential when used alone, but when combined with *Lysinibacillus* MSB11, the mixed culture resulted in enhanced bioleaching efficiency with Mn recovery reaching 74%. These microbial systems not only contribute to the recycling of industrially important elements but also help in restoring ecological balance by preventing heavy metals from leaching into the environment (Ghosh et al., 2017). Thus, microbial consortia perform an essential function in the sustainable extraction and mobilization of Mn while offering avenues to reduce dependency on high-grade ore reserves.

Acknowledgment

The authors would like to thank the reviewers and editors for contributing to improving the manuscript.

References

Aasfar, A., Bargaz, A., Yaakoubi, K., Hilali, A., Bennis, I., Zeroual, Y., & Meftah Kadmiri, I. 2021. Nitrogen fixing *Azotobacter* species as potential soil biological enhancers for crop nutrition and yield stability. *Frontiers in Microbiology* 12, 628379. https://doi.org/10.3389/fmicb.2021.628379

Abedon, S.T., Kuhl, S.J., Blasdel, B.G., & Kutter, E.M. 2011. Phage treatment of human infections. *Bacteriophage* 1(2), 66–85. https://doi.org/10.4161/bact.1.2.15845

Adedayo, A.A., & Babalola, O.O. 2023. Fungi that promote plant growth in the rhizosphere boost crop growth. *Journal of Fungi* 9(2), 239. https://doi.org/10.3390/jof9020239

Ahemad, M., & Kibret, M. 2013. Recent trends in microbial biosorption of heavy metals: a review. *Biochemistry & Molecular Biology* 1(1), 19–26. https://doi.org/10.12966/bmb.06.02.2013

Ahmad, F., Singh, A., & Kamal, A. 2020. Osmoprotective role of sugar in mitigating abiotic stress in plants. In: M.A. Hossain, V. Kumar, D.J. Burritt, M. Fujita, & P.S.A. Mäkelä (Eds.), *Osmoprotectant-Mediated Abiotic Stress Tolerance in Plants* (pp. 39–59). Wiley. https://doi.org/10.1002/9781119552154.ch3

Ahmad, M.F., Ahmad, F.A., Alsayegh, A.A., Zeyaullah, M., AlShahrani, A.M., Muzammil, K., Saati, A.A., Wahab, S., Elbendary, E.Y., Kambal, N., Abdelrahman, M.H., & Hussain, S. 2024. Pesticides impacts on human health and the environment with their mechanisms of action and possible countermeasures. *Heliyon* 10(7), e29128. https://doi.org/10.1016/j.heliyon.2024.e29128

Ahmed, E., & Holmström, S.J. 2014. Siderophores in environmental research: roles and applications. *Microbial Biotechnology* 7(3), 196–208.

Ahmed, M.F., Talaat, A., Vattakandy Jasin, R., & Abdul Kareem, K. 2020. Modulation of plant defense system in response to microbial interactions. *Frontiers in Microbiology* 11, 1298. https://doi.org/10.3389/fmicb.2020.01298

Akhtar, M.S., & Siddiqui, Z.A. 2009. Effects of phosphate solubilizing microorganisms and *Rhizobium* sp. on the growth, nodulation, yield and root-rot disease complex of chickpea under field condition. *African Journal of Biotechnology* 8(15), 3489–3496.

Ankati, S., Srinivas, V., Pratyusha, S., & Gopalakrishnan, S. 2021. *Streptomyces* consortia – mediated plant defense against *Fusarium* wilt and plant growth-promotion in chickpea. *Microbial Pathogenesis* 157, 104961. https://doi.org/10.1016/j.micpath.2021.104961

Asada, K. 2006. Production and scavenging of reactive oxygen species in chloroplasts and their functions. *Plant Physiology* 141(2), 391–396. https://doi.org/10.1104/pp.106.082040

Azcón-Aguilar, C., & Barea, J. M. 1997. Arbuscular mycorrhizas and biological control of soil-borne plant pathogens – an overview of the mechanisms involved. *Mycorrhiza* 6(6), 457–464. https://doi.org/10.1007/s005720050147

Bashan, Y., & de-Bashan, L.E. 2010. How the plant growth-promoting bacterium *Azospirillum* promotes plant growth – a critical assessment. *Advances in Agronomy* 108, 77–136. https://doi.org/10.1016/S0065-2113(10)08002-8

Beceiro, A., Tomás, M., & Bou, G. 2013. Antimicrobial resistance and virulence: a successful or deleterious association in the bacterial world? *Clinical Microbiology Reviews* 26(2), 185–230. https://doi.org/10.1128/CMR.00059-12

Behera, B., Das, T.K., Raj, R., & Adak, A. 2021. Microbial consortia for sustaining productivity of non-legume crops: prospects and challenges. *Agricultural Research* 10, 1–14. https://doi.org/10.1007/s40003-020-00482-3

Beneduzi, A., Ambrosini, A., & Passaglia, L.M. 2012. Plant growth-promoting rhizobacteria (PGPR): their potential as antagonists and biocontrol agents. *Genetics and Molecular Biology* 35(4 Suppl), 1044–1051. https://doi.org/10.1590/s1415-47572012000600020

Benítez, T., Rincón, A.M., Limón, M.C., & Codón, A.C. 2004. Mecanismos de biocontrol de cepas de *Trichoderma*. *International Microbiology* 7(4), 249–260. https://doi.org/10.2436/im.v7i4.9531

Benson, D.R., & Silvester, W.B. 1993. Biology of *Frankia* strains, actinomycete symbionts of actinorhizal plants. *Microbiological Reviews* 57(2), 293–319. https://doi.org/10.1128/mmbr.57.2.293-319.1993

Berg, J.S., Duverger, A., Cordier, L., & Jézéquel, D. 2020. Rapid pyritization in the presence of a sulfur/sulfate-reducing bacterial consortium. *Scientific Reports* 10, 8264. https://doi.org/10.1038/s41598-020-64990-6

Bestha, S.T., Jamwal, G., Gupta, V., Verma, M., Sharma, A., Sharma, A., & Pandit, V. 2025. Biological control of bacterial leaf blight (BLB) in rice–a sustainable approach. *Heliyon* 11(2), e41769. https://doi.org/10.1016/j.heliyon.2025.e41769

Bhatt, P., Bhatt, K., Sharma, A., Zhang, W., Mishra, S., & Chen, S. 2021. Biotechnological basis of microbial consortia for the removal of pesticides from the environment. *Critical Reviews in Biotechnology* 41(3), 317–338. https://doi.org/10.1080/07388551.2020.1853032

Bhatta, D., Adhikari, A., Kang, S.-M., Kwon, E.-H., Jan, R., Kim, K.-M., & Lee, I.-J. 2023. Hormones and the antioxidant transduction pathway and gene expression, mediated by *Serratia marcescens* DB1, lessen the lethality of heavy metals (As, Ni, and Cr) in *Oryza sativa* L. *Ecotoxicology and Environmental Safety* 263, 115377. https://doi.org/10.1016/j.ecoenv.2023.115377

Bidyarani, N., Prasanna, R., Babu, S., Hossain, F., & Saxena, A.K. 2016. Enhancement of plant growth and yields in chickpea (*Cicer arietinum* L.) through novel cyanobacterial and biofilmed inoculants. *Microbiological Research* 188–189, 97–105. https://doi.org/10.1016/j.micres.2016.04.007

Brazhnikova, Y., Belimov, A., Ignatova, L., Mukasheva, T., Karpenyuk, T., & Goncharova, A. 2025. The antagonistic activity of beneficial fungi and mechanisms underlying their protective effects on plants against phytopathogens. *Sustainability* 17(2), 450. https://doi.org/10.3390/su17020450

Broderick, N.A., Robinson, C.J., McMahon, M.D., Holt, J., Handelsman, J., & Raffa, K.F. 2009. Contributions of gut bacteria to *Bacillus thuringiensis*-induced mortality vary across a range of Lepidoptera. *BMC Biology* 7, 11. https://doi.org/10.1186/1741-7007-7-11

Cavalcante, V.A., & Dobereiner, J. 1988. A new acid-tolerant nitrogen-fixing bacterium associated with sugarcane. *Plant and Soil* 108, 23–31. https://doi.org/10.1007/BF02370096

Chand, U., Hegde, G., & Mesta, R. 2025. The potential of siderophores in biological control in plant diseases. *Asia Pacific Biofertilizers and Biopesticides Information Platform* 1(1), 1–25. https://doi.org/10.56669/HGRS2291

Chaudhari, H.V., Inamdar, M.M., & Kondabagil, K. 2021. Scaling relation between genome length and particle size of viruses provides insights into viral life history. *iScience* 24(5), 102452. https://doi.org/10.1016/j.isci.2021.102452

Chaudhary, R., Nawaz, A., Khattak, Z., Butt, M.A., Fouillaud, M., Dufossé, L., Munir, M., Haq, I.U., & Mukhtar, H. 2024. Microbial bio-control agents: a comprehensive analysis on sustainable pest management in agriculture. *Journal of Agriculture and Food Research* 18, 101421. https://doi.org/10.1016/j.jafr.2024.101421

Chen, J., Ke, Y., Zhu, Y., Xu, M., Chen, C., & Xie, S. 2023. Enrichment of tetracycline-degrading bacterial consortia: microbial community succession and degradation characteristics and mechanism. *Journal of Hazardous Materials* 448, 130984. https://doi.org/10.1016/j.jhazmat.2023.130984

Chen, Q., Song, Y., An, Y., Lu, Y., & Zhong, G. 2024. Mechanisms and impact of rhizosphere microbial metabolites on crop health, traits, functional components: a comprehensive review. *Molecules* 29(24), 5922. https://doi.org/10.3390/molecules29245922

Cheng, Y., Lou, H., He, H., He, X., Wang, Z., Gao, X., & Liu, J. 2024. Genomic and biological control of *Sclerotinia sclerotiorum* using an extracellular extract from *Bacillus velezensis* 20507. *Frontiers in Microbiology* 15, 1385067. https://doi.org/10.3389/fmicb.2024.1385067

Compant, S., Clément, C., & Sessitsch, A. 2010. Plant growth-promoting bacteria in the rhizo- and endosphere of plants: their role, colonization, mechanisms involved and prospects for utilization. *Soil Biology and Biochemistry* 42(5), 669–678. https://doi.org/10.1016/j.soilbio.2009.11.024

Compant, S., Duffy, B., Nowak, J., Clément, C., & Barka, E.A. 2005. Use of plant growth-promoting bacteria for biocontrol of plant diseases: principles, mechanisms of action, and future prospects. *Applied and Environmental Microbiology* 71(9), 4951–4959. https://doi.org/10.1128/AEM.71.9.4951-4959.2005

Conrath, U. 2006. Systemic acquired resistance. *Plant Signaling & Behavior* 1(4), 179–184. https://doi.org/10.4161/psb.1.4.3221

Crowley, D. 2006. Microbial siderophores in the plant rhizosphere. In: L.L. Barton, & J. Abadia (Eds.), *Iron Nutrition in Plants and Rhizospheric Microorganisms* (pp. 169–188). Springer. https://doi.org/10.1007/1-4020-4743-6_8

de Faria, M. R., & Wraight, S. P. 2007. Mycoinsecticides and mycoacaricides: a comprehensive list with worldwide coverage and international classification of formulation types. *Biological Control* 43(3), 237–256. https://doi.org/10.1016/j.biocontrol.2007.08.001

Dhankhar, R., Gupta, S., & Gulati, P. 2022. Insights on plant–microbe interactions in soil in relation to iron dynamics. *Vegetos* 36, 670–679. https://doi.org/10.1007/s42535-022-00467-3

Dimkić, I., Janakiev, T., Petrović, M., Degrassi, G., & Fira, D. 2022. Plant-associated *Bacillus* and *Pseudomonas* antimicrobial activities in plant disease suppression via biological control mechanisms – a review. *Physiological and Molecular Plant Pathology* 117, 101754. https://doi.org/10.1016/j.pmpp.2021.101754

Dimkpa, C., Weinand, T., & Asch, F. 2009. Plant–rhizobacteria interactions alleviate abiotic stress conditions. *Plant, Cell & Environment* 32(12), 1682–1694. https://doi.org/10.1111/j.1365-3040.2009.02028.x

Ding, L.-N., Li, Y.-T., Wu, Y.-Z., Li, T., Geng, R., Cao, J., Zhang, W., & Tan, X.-L. 2022. Plant disease resistance-related signaling pathways: recent progress and future prospects. *International Journal of Molecular Sciences* 23(24), 16200. https://doi.org/10.3390/ijms232416200

Elhaissoufi, W., Ghoulam, C., Barakat, A., Zeroual, Y., & Bargaz, A. 2022. Phosphate bacterial solubilization: a key rhizosphere driving force enabling higher P use efficiency and crop productivity. *Journal of Advanced Research* 38, 13–28. https://doi.org/10.1016/j.jare.2021.08.014

Feng, C., Gao, H., Zhou, Y., Jing, Y., Li, S., Yan, Z., Xu, K., Zhou, F., Zhang, W., Yang, X., Hussain, M.A., & Li, H. 2023. Unfolding molecular switches for salt stress resilience in soybean: recent advances and prospects for salt-tolerant smart plant production. *Frontiers in Plant Science* 14, 1162014. https://doi.org/10.3389/fpls.2023.1162014

Foyer, C.H., & Noctor, G. 2005. Redox homeostasis and antioxidant signaling: a metabolic interface between stress perception and physiological responses. *The Plant Cell* 17(7), 1866–1875. https://doi.org/10.1105/tpc.105.033589

García-Sánchez, F., Simón-Grao, S., Nicolás, J.J., Alfosea-Simón, M., Liu, C., Chatzissavvidis, C., Pérez-Pérez, J., & Cámara-Zapata, J.M. 2020. Multiple stresses occurring with boron toxicity and deficiency in plants. *Journal of Hazardous Materials* 397, 122713. https://doi.org/10.1016/j.jhazmat.2020.122713

Garg, G., Kumar, S., & Bhati, S. 2021. Siderophore in plant nutritional management: role of endophytic bacteria. In: A. Rana, A. Saneja, S. Kumar, & E. Lichtfouse (Eds.), *Sustainable Agriculture Reviews 56* (pp. 265–287). Springer. https://doi.org/10.1007/978-3-030-65447-4_14

Ghosh, S., Bal, B., & Das, A. 2017. Enhancing manganese recovery from low-grade ores by using mixed culture of indigenously isolated bacterial strains. *Geomicrobiology Journal* 35(7), 583–590. https://doi.org/10.1080/01490451.2017.1362080

Ghosh, W., & Dam, B. 2009. Biochemistry and molecular biology of lithotrophic sulfur oxidation by chemolithoautotrophic bacteria. *FEMS Microbiology Reviews* 33(6), 999–1043. https://doi.org/10.1111/j.1574-6976.2009.00187.x

Guetsky, R., Shtienberg, D., Elad, Y., & Dinoor, A. 2001. Combining biocontrol agents to reduce the variability of biological control. *Biological Control* 91(7), 621–627. https://doi.org/10.1094/PHYTO.2001.91.7.621

Haas, D., & Défago, G. 2005. Biological control of soil-borne pathogens by fluorescent pseudomonads. *Nature Reviews Microbiology* 3(4), 307–319. https://doi.org/10.1038/nrmicro1129

Harman, G.E., Howell, C.R., Viterbo, A., Chet, I., & Lorito, M. 2004. *Trichoderma* species – Opportunistic, avirulent plant symbionts. *Nature Reviews Microbiology* 2(1), 43–56. https://doi.org/10.1038/nrmicro797

Iwanicki, N.S., Pereira, A.A., Botelho, A.B.R.Z., Rezende, J.M., Moral, R.D.A., Zucchi, M.I., & Delalibera Júnior, I. 2019. Monitoring of the field application of *Metarhizium anisopliae* in Brazil revealed high molecular diversity of *Metarhizium* spp. in insects, soil, and sugarcane roots. *Scientific Reports* 9, 4443. https://doi.org/10.1038/s41598-019-38594-8

Jambhulkar, P., Sharma, P., Manokaran, R., Lakshman, D., Rokadia, P., & Jambhulkar, N. 2018. Assessing synergism of combined applications of *Trichoderma harzianum* and *Pseudomonas fluorescens* to control blast and bacterial leaf blight of rice. *European Journal of Plant Pathology* 152(3), 707–717. https://doi.org/10.1007/s10658-018-1519-3

Jha, P., Kaur, T., Chhabra, I., Panja, A., Paul, S., Kumar, V., & Malik, T. 2023. Endophytic fungi: Hidden treasure chest of antimicrobial metabolites – interrelationship of endophytes and metabolites. *Frontiers in Microbiology* 14, 1227830. https://doi.org/10.3389/fmicb.2023.1227830

Jog, R., Pandya, M., Nareshkumar, G., & Rajkumar, S. 2014. Mechanism of phosphate solubilization and antifungal activity of *Streptomyces* spp. isolated from wheat roots and rhizosphere and their application in improving plant growth. *Microbiology (Reading)* 160(Pt 4), 778–788. https://doi.org/10.1099/mic.0.074146-0

Jomova, K., Raptova, R., Alomar, S.Y., Alwasel, S.H., Nepovimova, E., Kuca, K., & Valko, M. 2023. Reactive oxygen species, toxicity, oxidative stress, and antioxidants: chronic diseases and aging. *Archives of Toxicology* 97(10), 2499–2574. https://doi.org/10.1007/s00204-023-03562-9

Kaewsalong, N., Songkumarn, P., Duangmal, K., & Dethoup, T. (2019). Synergistic effects of combinations of novel strains of *Trichoderma* species and *Coscinium fenestratum* extract in controlling rice dirty panicle. *Journal of Plant Pathology* 101, 367–372. https://doi.org/10.1007/s42161-018-0191-

Khan, S., Srivastava, S., Karnwal, A., & Malik, T. 2023. Streptomyces as a promising biological control agent for plant pathogens. *Frontiers in Microbiology* 14, 1285543. https://doi.org/10.3389/fmicb.2023.1285543

Kaur, T., Devi, R., Kumar, S., Sheikh, I., Kour, D., & Yadav, A.N. 2022a. Microbial consortium with nitrogen fixing and mineral solubilizing attributes for growth of barley (*Hordeum vulgare* L.). *Heliyon* 8(4), e09326. https://doi.org/10.1016/j.heliyon.2022.e09326

Kaur, T., Kour, D., Pericak, O., Olson, C., Mohan, R., Yadav, A., Mishra, S., Kumar, M., Rai, A.K., & Yadav, A.N. 2022b. Structural and functional diversity of plant growth promoting microbiomes for agricultural sustainability. *Journal of Applied Biology and Biotechnology* 10(Suppl 1), 70–89. https://doi.org/10.7324/JABB.2022.10s108

Khan, R.A.A., Najeeb, S., Hussain, S., Xie, B., & Li, Y. 2020. Bioactive secondary metabolites from *Trichoderma* spp. against phytopathogenic fungi. *Microorganisms* 8(6), 817. https://doi.org/10.3390/microorganisms8060817

Khoshru, B., Mitra, D., Nosratabad, A.F., Reyhanitabar, A., Mandal, L., Farda, B., Djebaili, R., Pellegrini, M., Guerra-Sierra, B.E., Senapati, A., Panneerselvam, P., & Mohapatra, P.K.D. 2023. Enhancing manganese availability for plants through microbial potential: a sustainable approach for improving soil health and food security. *Bacteria* 2(3), 129–141. https://doi.org/10.3390/bacteria2030010

Khoso, M.A., Wagan, S., Alam, I., Hussain, A., Ali, Q., Saha, S., Poudel, T.R., Manghwar, H., & Liu, F. 2024. Impact of plant growth-promoting rhizobacteria (PGPR) on plant nutrition and root characteristics: current perspective. *Plant Stress* 11, 100341. https://doi.org/10.1016/j.stress.2023.100341

Kovács, B., Puskás-Preszner, A., Huzsvai, L., Lévai, L., & Bódi, É. 2015. Effect of molybdenum treatment on molybdenum concentration and nitrate reduction in maize seedlings. *Plant Physiology and Biochemistry* 96, 38–44. https://doi.org/10.1016/j.plaphy.2015.07.013

Kumawat, K.C., Sharma, B., Nagpal, S., Kumar, A., Tiwari, S., & Nair, R.M. 2022. Plant growth-promoting rhizobacteria: salt stress alleviators to improve crop productivity for sustainable agriculture development. *Frontiers in Plant Science* 13, 1101862. https://doi.org/10.3389/fpls.2022.1101862

Kumari, K.S., Padma Devi, S.N., Ranjithkumar, R., Djearamane, S., Tey, L.-H., Wong, L.S., Kayarohanam, S., Arumugam, N., Almansour, A.I., & Perumal, K. 2023. Organic remobilization of zinc and phosphorus availability to plants by application of mineral solubilizing bacteria *Pseudomonas aeruginosa*. *Heliyon* 9(11), e22128. https://doi.org/10.1016/j.heliyon.2023.e22128

Liang, X., Zhang, L., Natarajan, S.K., & Becker, D.F. 2013. Proline mechanisms of stress survival. *Antioxidants & Redox Signaling* 19(9), 998–1011. https://doi.org/10.1089/ars.2012.5074

Lilay, G.H., Thiébaut, N., Mee, D.d., Assunção, A.G.L., Schjøerring, J.K., Husted, S., & Persson, D.P. 2024. Linking the key physiological functions of essential micronutrients to their deficiency symptoms in plants. *New Phytologist*, 242(3), 881–902. https://doi.org/10.1111/nph.19645

Liu, X., Mei, S., & Falcao Salles, J. 2023. Inoculated microbial consortia perform better than single strains in living soil: a meta-analysis. *Applied Soil Ecology* 190, 105011. https://doi.org/10.1016/j.apsoil.2023.105011

Maciag, T., Kozieł, E., Rusin, P., Otulak-Kozieł, K., Jafra, S., & Czajkowski, R. 2023. Microbial consortia for plant protection against diseases: more than the sum of its parts. *International Journal of Molecular Sciences* 24(15), 12227. https://doi.org/10.3390/ijms241512227

Mahmoud, G.A., Rashed, N.M., El-Ganainy, S.M., & Salem, S.H. 2024. Unveiling the neem (*Azadirachta indica*) effects on biofilm formation of food-borne bacteria and the potential mechanism using a molecular docking approach. *Plants* 13(18), 2669. https://doi.org/10.3390/plants13182669

Masood, F., Ahmad, S., Malik, A. 2022. Role of rhizobacterial bacilli in zinc solubilization. In: S.T. Khan, &A. Malik (Eds.) Microbial Biofertilizers and Micronutrient Availability. Springer, Cham. https://doi.org/10.1007/978-3-030-76609-2_15

McFarland, L.V. 2010. Systematic review and meta-analysis of *Saccharomyces boulardii* in adult patients. *World Journal of Gastroenterology* 16(18), 2202–2222. https://doi.org/10.3748/wjg.v16.i18.2202

Mehmood, N., Saeed, M., Zafarullah, S., Hyder, S., Rizvi, Z.F., Gondal, A.S., Jamil, N., Iqbal, R., Ali, B., Ercisli, S., & Kupe, M. 2023. Multifaceted impacts of plant-beneficial *Pseudomonas* spp. in managing various plant diseases and crop yield improvement. *ACS Omega* 8(25), 22296–22315. https://doi.org/10.1021/acsomega.3c00870

Miethke, M., & Marahiel, M.A. 2007. Siderophore-based iron acquisition and pathogen control. *Microbiology and Molecular Biology Reviews* 71(3), 413–451. https://doi.org/10.1128/MMBR.00012-07

Mikiciuk, G., Miller, T., Kisiel, A., Cembrowska-Lech, D., Mikiciuk, M., Łobodzińska, A., & Bokszczanin, K. 2024. Harnessing beneficial microbes for drought tolerance: a review of ecological and agricultural innovations. *Agriculture* 14(12), 2228. https://doi.org/10.3390/agriculture14122228

Mittler, R. 2002. Oxidative stress, antioxidants, and stress tolerance. *Trends in Plant Science* 7(9), 405–410. https://doi.org/10.1016/S1360-1385(02)02312-9

Moon, J., Park, Y.J., Choi, Y.B., Truong, T.Q., Huynh, P.K., Kim, Y.B., & Kim, S.M. 2024. Physiological effects and mechanisms of *Chlorella vulgaris* as a biostimulant on the growth and drought tolerance of *Arabidopsis thaliana*. *Plants* 13(21), 3012. https://doi.org/10.3390/plants13213012

Mukherjee, P.K., Horwitz, B.A., Herrera-Estrella, A., Schmoll, M., & Kenerley, C.M. 2013. *Trichoderma* research in the genome era. *Annual Review of Phytopathology* 51, 105–129. https://doi.org/10.1146/annurev-phyto-082712-102353

Munir, N., Hanif, M., Abideen, Z., Sohail, M., El-Keblawy, A., Radicetti, E., Mancinelli, R., & Haider, G. 2022. Mechanisms and strategies of plant microbiome interactions to mitigate abiotic stresses. *Agronomy* 12(9), 2069. https://doi.org/10.3390/agronomy12092069

Ongena, M., & Jacques, P. 2008. *Bacillus* lipopeptides: versatile weapons for plant disease biocontrol. *Trends in Microbiology* 16(3), 115–125. https://doi.org/10.1016/j.tim.2007.12.009

Panpatte, D.G., Shelat, H.N., Jhala, Y.K., & Vyas, R.V. 2021. Fortified bacterial consortium – A novel approach to control root knot nematode in cucumber (*Cucumis sativus*). *Biological Control 155*, 104528. https://doi.org/10.1016/j.biocontrol.2020.104528

Pieterse, C.M.J., Van der Does, D., Zamioudis, C., Leon-Reyes, A., & Van Wees, S.C.M. 2014a. Hormonal modulation of plant immunity. *Annual Review of Cell and Developmental Biology* 30, 489–521. https://doi.org/10.1146/annurev-cellbio-100913-013211

Pieterse, C.M.J., Zamioudis, C., Berendsen, R.L., Weller, D.M., Van Wees, S.C.M., & Bakker, P.A.H.M. 2014b. Induced systemic resistance by beneficial microbes. *Annual Review of Phytopathology* 52, 347–375. https://doi.org/10.1146/annurev-phyto-082712-102340

Prasad, A., Kumar, S., Khaliq, A., & Pandey, A. 2011. Heavy metals and arbuscular mycorrhizal (AM) fungi can alter the yield and chemical composition of volatile oil of sweet basil (*Ocimum basilicum* L.). *Biology and Fertility of Soils* 47(7), 853–861. https://doi.org/10.1007/s00374-011-0590-0

Pršić, J., & Ongena, M. 2020. Elicitors of plant immunity triggered by beneficial bacteria. *Frontiers in Plant Science* 11, 594530. https://doi.org/10.3389/fpls.2020.594530

Qian, Z., Lu, Z.-M., Chai, L.-J., Zhang, X.-J., Li, Q., Wang, S.-T., Shen, C.-H., Shi, J.-S., & Xu, Z.-H. 2021. Cooperation within the microbial consortia of fermented grains and pit mud drives organic acid synthesis in strong-flavor Baijiu production. *Food Research International* 147, 110449. https://doi.org/10.1016/j.foodres.2021.110449

Rajendran, G., Patel, M.H., & Joshi, S.J. 2012. Isolation and characterization of nodule-associated *Exiguobacterium* sp. from the root nodules of fenugreek (*Trigonella foenum-graecum*) and their possible role in plant growth promotion. *International Journal of Microbiology* 2012, 693982. https://doi.org/10.1155/2012/693982

Rajpal, N., Ratan, J.K., Divya, N., & Hebbani, A.V. 2022. Bioremediation of greywater using a novel bacterial–fungal consortium: optimization and validation of the operating parameters *in vitro*. *Environmental Technology*, 43(16), 2430–2442. https://doi.org/10.1080/09593330.2021.1882582

Sahu, P.K., Jayalakshmi, K., Tilgam, J., Gupta, A., Yalavarthi, N., Kumar, A., Hamid, S., Singh, H.V., Minkina, T., Rajput, V.D., & Rajawat, M.V.S. 2022. ROS generated from biotic stress: effects on plants and alleviation by endophytic microbes. *Frontiers in Plant Science* 13, 1042936. https://doi.org/10.3389/fpls.2022.1042936

Saif, S., Abid, Z., Ashiq, M., & Ashraf, R. 2021. Biofertilizer formulations. In: M. Ashraf, Z. Abid, & R. Ashraf (Eds.), *Biofertilizers (Chapter 7)*. Wiley. https://doi.org/10.1002/9781119724995.ch7

Saini, H. P., Meena, M., Sahoo, A., & Mehta, T. 2025. A review on fungal endophytes of the family Fabaceae, their metabolic diversity and biological applications. *Heliyon* 11(3), e42153. https://doi.org/10.1016/j.heliyon.2025.e42153

Saleem, A., Anwar, S., Nawaz, T., Fahad, S., Saud, S., Ur Rahman, T., Khan, M.N.R., & Nawaz, T. 2024. Securing a sustainable future: the climate change threat to agriculture, food security, and sustainable development goals. *Journal of Umm Al-Qura University for Applied Sciences*. https://doi.org/10.1007/s43994-024-00177-3

Saxena, J. 2013. Phosphate solubilization by a few fungal strains belonging to the genera *Aspergillus* and *Penicillium*. *African Journal of Microbiology Research* 7(48), 4862–4869. https://doi.org/10.5897/AJMR2013.5991

Schnepf, E., Crickmore, N., Van Rie, J., Lereclus, D., Baum, J., Feitelson, J., Zeigler, D.R., & Dean, D.H. 1998. *Bacillus thuringiensis* and its pesticidal crystal proteins. *Microbiology and Molecular Biology Reviews* 62(3), 775–806. https://doi.org/10.1128/MMBR.62.3.775-806.1998

Sethi, G., Behera, K.K., Sayyed, R., Adarsh, V., Sipra, B.S., Singh, L., Alamro, A.A. & Behera, M. 2025. Enhancing soil health and crop productivity: The role of zinc-solubilizing bacteria in sustainable agriculture. *Plant Growth Regulation* 105(3), 601–617 https://doi.org/10.1007/s10725-025-01294-7

Sharma, P., Jha, A.B., Dubey, R.S., & Pessarakli, M. 2012. Reactive oxygen species, oxidative damage, and antioxidative defense mechanism in plants under stressful conditions. *Journal of Botany* 2012, 1–26. https://doi.org/10.1155/2012/217037

Sharma, A., Srivastava, A., Shukla, A.K., Srivastava, K., Srivastava, A.K., & Saxena, A.K. 2020. Entomopathogenic fungi: a potential source for biological control of insect pests. In: M. Solanki, P. Kashyap, & B. Kumari (Eds.), *Phytobiomes: Current Insights and Future Vistas* (pp. 185–212). Springer. https://doi.org/10.1007/978-981-15-3151-4_9

Sharma, S., Rathod, Z.R., Jain, R., Goswami, D., & Saraf, M. 2023. Strategies to evaluate microbial consortia for mitigating abiotic stress in plants. In: D. K. Maheshwari, & S. Dheeman (Eds.), *Sustainable Agrobiology: Microorganisms for Sustainability* (Vol. 43, pp. 145–165). Springer. https://doi.org/10.1007/978-981-19-9570-5_9

Shoresh, M., Harman, G.E., & Mastouri, F. 2010. Induced systemic resistance and plant responses to fungal biocontrol agents. *Annual Review of Phytopathology* 48, 21–43. https://doi.org/10.1146/annurev-phyto-073009-114450

Singh, J.S., Kumar, A., Rai, A.N., & Singh, D.P. 2016. Cyanobacteria: a precious bio-resource in agriculture, ecosystem, and environmental sustainability. *Frontiers in Microbiology* 7, 529. https://doi.org/10.3389/fmicb.2016.00529

Smith, S.E., & Read, D.J. 2008. *Mycorrhizal Symbiosis* (3rd ed.). Academic Press.

Soth, S., Glare, T.R., Hampton, J.G., Card, S.D., & Brookes, J.J. 2022. Biological control of diamondback moth – increased efficacy with mixtures of *Beauveria* fungi. *Microorganisms* 10(3), 646. https://doi.org/10.3390/microorganisms10030646

Spescha, A., Weibel, J., Wyser, L., Brunner, M., Hess Hermida, M., Moix, A., Scheibler, F., Guyer, A., Campos-Herrera, R., Grabenweger, G., & Maurhofer, M. 2023. Combining entomopathogenic *Pseudomonas* bacteria, nematodes and fungi for biological control of a below-ground insect pest. *Ecosystems & Environment* 348, 108414. https://doi.org/10.1016/j.agee.2023.108414

Stuart, A.K.d.C., Furuie, J.L., Cataldi, T.R., Stuart, R.M., Zawadneak, M.A.C., Labate, C.A., & Pimentel, I.C. 2022. Fungal consortium of two *Beauveria bassiana* strains increases their virulence, growth, and resistance to stress: a metabolomic approach. *PLoS ONE* 17(7), e0271460. https://doi.org/10.1371/journal.pone.0271460

Syed-Ab-Rahman, S.F., Carvalhais, L.C., Chua, E., Xiao, Y., Wass, T.J., & Schenk, P.M. 2018. Identification of soil bacterial isolates suppressing different *Phytophthora* spp. and promoting plant growth. *Frontiers in Plant Science* 9, 1502. https://doi.org/10.3389/fpls.2018.01502

Tripathi, A., Pandey, P., Tripathi, S.N., & Kalra, A. 2022. Perspectives and potential applications of endophytic microorganisms in cultivation of medicinal and aromatic plants. *Frontiers in Plant Science* 13, 985429. https://doi.org/10.3389/fpls.2022.985429

Trivedi, S., Butani, N., Patel, H., Nath, M., & Bhatt, D. 2021. Microbial consortia and their application for the development of a sustainable environment. In: P. Bhatt, S. Gangola, D. Udayanga, & G. Kumar (Eds.), *Microbial Technology for Sustainable Environment* (pp. 145-166). Springer. https://doi.org/10.1007/978-981-16-3840-4_11

Tufail, M., Ayyub, M., Irfan, M., Shakoor, A., Chibani, C., & Schmitz, R. 2022. Endophytic bacteria perform better than endophytic fungi in improving plant growth under drought stress: a meta-comparison spanning 12 years (2010–2021). *Physiologia Plantarum* 174, e13806. https://doi.org/10.1111/ppl.13806

Upadhayay, V.K., Gangola, S., Taj, G., Gaurav, K., Rani, A., Kumar, S., Garg, S., Gupta, G., Ali, H., Siddiqui, S., Alamri, S.A.M., Mittal, A., Alrumman, S.A., & Pandey, M. 2025. Zinc-solubilizing bacterial consortia: a promising approach for zinc biofortification of crops. *Frontiers in Microbiology* 16, 1575514. https://doi.org/10.3389/fmicb.2025.1575514

Upadhayay, V.K., Singh, A.V., Khan, A., Singh, J., Pareek, N., & Raghav, A. 2022. FE-SEM/EDX based zinc mobilization analysis of *Burkholderia cepacia* and *Pantoea rodasii* and their functional annotation in crop productivity, soil quality, and zinc biofortification of paddy. *Frontiers in Microbiology* 13, 852192. https://doi.org/10.3389/fmicb.2022.852192

Upadhyay, S.K., Singh, J.S., Singh, D.P., & Singh, D.P. 2012. Exopolysaccharide-producing plant growth-promoting rhizobacteria under salinity condition. *Pedosphere* 22(5), 543–549. https://doi.org/10.1016/S1002-0160(12)60040-7

Upamanya, G.K., Bhattacharyya, A., & Dutta, P. 2020. Consortia of entomopathogenic fungi and bio-control agents improve the agro-ecological conditions for brinjal cultivation of Assam. *3 Biotech* 10(450). https://doi.org/10.1007/s13205-020-02439-3

Vallad, G.E., & Goodman, R.M. 2004. Systemic acquired resistance and induced systemic resistance in conventional agriculture. *Crop Science* 44(6), 1920–1934. https://doi.org/10.2135/cropsci2004.1920

Van Wees, S.C.M., Van der Ent, S., & Pieterse, C.M.J. 2008. Plant immune responses triggered by beneficial microbes. *Current Opinion in Plant Biology* 11(4), 443–448. https://doi.org/10.1016/j.pbi.2008.05.005

Vassilev, N., Vassileva, M., & Nikolaeva, I. 2006. Simultaneous P-solubilizing and biocontrol activity of microorganisms: potentials and future trends. *Applied Microbiology and Biotechnology* 71, 137–144. https://doi.org/10.1007/s00253-006-0501-2

Verma, D., Meena, R., Sukhwal, A., Jat, G., Meena, S., Upendra, S., & Jain, D. 2022. Effect of ZSB with graded levels of zinc fertilizer on yield and zinc uptake under maize cultivation. *Proceedings of the National Academy of Sciences, India Section B: Biological Sciences* 93, 1–7. https://doi.org/10.1007/s40011-022-01433-4

Vessey, J.K. 2003. Plant growth-promoting rhizobacteria as biofertilizers. *Plant and Soil* 255, 571–586. https://doi.org/10.1023/A:1026037216893

Vinale, F., Sivasithamparam, K., Ghisalberti, E.L., Marra, R., Woo, S.L., & Lorito, M. 2008. *Trichoderma*–plant–pathogen interactions. *Soil Biology and Biochemistry* 40(1), 1–10. https://doi.org/10.1016/j.soilbio.2007.07.002

Vlot, A.C., Dempsey, D.A., & Klessig, D.F. 2009. Salicylic acid, a multifaceted hormone to combat disease. *Annual Review of Phytopathology* 47, 177–206. https://doi.org/10.1146/annurev.phyto.050908.135202

Vurukonda, S.S.K.P., Vardharajula, S., Shrivastava, M., & SkZ, A. 2016. Enhancement of drought stress tolerance in crops by plant growth promoting rhizobacteria. *Microbiological Research* 184, 13–24. https://doi.org/10.1016/j.micres.2015.12.003

Wahab, A., Muhammad, M., Munir, A., Abdi, G., Zaman, W., Ayaz, A., Khizar, C., & Reddy, S.P.P. 2023. Role of arbuscular mycorrhizal fungi in regulating growth, enhancing productivity, and potentially influencing ecosystems under abiotic and biotic stresses. *Plants* 12(17), 3102. https://doi.org/10.3390/plants12173102

Weller, D.M. 2007. *Pseudomonas* biocontrol agents of soilborne pathogens: looking back over 30 years. *Phytopathology* 97(2), 250–256. https://doi.org/10.1094/PHYTO-97-2-0250

Whitelaw, M.A. 2000. Growth promotion of plants inoculated with phosphate-solubilizing fungi. *Advances in Agronomy* 69, 99–151. https://doi.org/10.1016/S0065-2113(08)60948-7

Xu, C., & Yu, H. 2021. Insights into constructing a stable and efficient microbial consortium. *Chinese Journal of Chemical Engineering* 30, 112–120. https://doi.org/10.1016/j.cjche.2020.12.012

Yao, X., Guo, H., Zhang, K., Zhao, M., Ruan, J., & Chen, J. 2023. *Trichoderma* and its role in biological control of plant fungal and nematode disease. *Frontiers in Microbiology* 14, 1160551. https://doi.org/10.3389/fmicb.2023.1160551

Zahran, H.H. 2001. Rhizobia from wild legumes: diversity, taxonomy, ecology, nitrogen fixation and biotechnology. *Journal of Biotechnology* 91(2–3), 143–153. https://doi.org/10.1016/S0168-1656(01)00342-7

Zhang, Y., Wang, E., Feng, B., Xu, L., Xue, Y., & Chen, Y. 2025. Siderophore production capability of nitrogen-fixing bacterium (NFB) GXGL-4A regulates cucumber rhizosphere soil microecology. *Microorganisms* 13(2), 346. https://doi.org/10.3390/microorganisms13020346

Zimmermann, G. (2007). Review on safety of the entomopathogenic fungus *Metarhizium anisopliae*. *Biocontrol Science and Technology* 17(9), 879–920. https://doi.org/10.1080/09583150701593963

Index

Note: Page numbers in *italics* and **bold** refers to figures and tables respectively.

n

q

r

Printed and bound by CPI Group (UK) Ltd, Croydon, CR0 4YY
17/09/2026
14964299-0001